# YERKES OBSERVATORY, 1892-1950

# Yerkes Observatory, 1892-1950

## THE BIRTH, NEAR DEATH, AND RESURRECTION OF A SCIENTIFIC RESEARCH INSTITUTION

## Donald E. Osterbrock

THE UNIVERSITY OF CHICAGO PRESS / CHICAGO AND LONDON

The University of Chicago Press, Chicago 60637
The University of Chicago Press, Ltd., London
© 1997 by The University of Chicago
All rights reserved. Published 1997
Printed in the United States of America
24 23 22 21 20 19 18 17 16 15    2 3 4 5 6

ISBN-10: 0-226-63945-2 (cloth)
ISBN-13: 978-0-226-63946-8 (paper)
ISBN-13: 978-0-226-63944-4 (e-book)

Library of Congress Cataloging-in-Publication Data
Osterbrock, Donald E.
    Yerkes Observatory, 1892–1950 : the birth, near death, and
resurrection of a scientific research institution / Donald E.
Osterbrock.
       p.    cm.
    Includes bibliographical references and index.
    ISBN 0-226-63945-2 (cloth : alk. paper).

    1. Yerkes Observatory—History.   I. Title.
QB82.U62W556   1997
522′.19775′89—dc20                                    96-25450
                                                      CIP

⊚ The paper used in this publication meets the minimum requirements of the
American National Standard for Information Sciences—Permanence of Paper
for Printed Library Materials, ANSI Z39.48-1984.

# CONTENTS

Preface      vii

1    Birth, 1868–1897      1
2    Infancy, 1897–1904      25
3    Near Death, 1904–1932      47
4    The Savior, 1897–1931      77
5    The Boy President, 1929–1932      107
6    The Boy Director, 1932–1936      133
7    Resurrection on the Campus and
     at Yerkes, 1893–1937      159
8    Birth of McDonald Observatory, 1933–1939      187
9    An Extraordinarily Fine Group, 1936–1942      211
10    World War II, 1939–1945      245
11    Golden Years, 1945–1950      267
12    Epilogue: To the Centennial, 1950–1997      303

Abbreviations Used in Notes and Bibliography      325
Notes      329
Bibliography      363
Index      367

## PREFACE

**Y**erkes Observatory celebrates its centennial in 1997. When it was dedicated and went into operation in 1897, it was America's second big-science establishment, built around a large, expensive scientific instrument, the forty-inch refracting telescope—"the largest and best . . . in the world," in the words of its donor, Charles T. Yerkes. Like Lick Observatory's thirty-six-inch refractor a decade earlier, or a huge particle accelerator, an Antarctic research base, or the Hubble Space Telescope of today, it was intended to be used by a group of top-flight scientists, working more or less independently, to unravel the secrets of the universe we live in—"who we are, where we came from, where we are going"—on the cosmic scale.

Surely Yerkes Observatory deserves its first book-length history, and the time is ripe to write it. As the first professional observatory in America planned from the start to be devoted to astrophysics (the study of the physical nature of the stars, nebulae, galaxies, and planets), it is particularly relevant today. To bring its history down to the present, however, would be very difficult. A certain distance and perspective are necessary to see what is important, what should be emphasized, and what were the most meaningful discoveries. Rather than attempting to cover a hundred years of its history, I decided it would be preferable to confine this book to the time of the first three directors: George Ellery Hale, who built Yerkes Observatory, Edwin B. Frost, who let it decay, and Otto Struve, who revived it. And of course it did not miraculously spring into existence on October 21, 1897, the day James E. Keeler gave his invited address "The Importance of Astrophysical Research, and the Relation of Astrophysics to Other Physical Sciences" and Yerkes handed over the keys to the observatory to William Rainey Harper, first president of the University of Chicago. The story begins even before the founding of "Harper's University" (actually John D. Rockefeller's) in 1891, with Hale's first observatory, Kenwood, and the steps that led up to it. The period of this book ends with Struve's resignation and departure for California in 1950, but a brief epilogue completes his story and the stories of some of the most important staff members and ideas of his time.

I should make my background and prejudices clear at the start. I am an unabashed admirer of Yerkes Observatory and the University of Chicago; I was an undergraduate and graduate student there and am the fortunate holder of five Chicago degrees. Many of the scientists of the latter Struve years were my teachers, and though Frost died when I was only ten years old, I met his widow, Mary H. Frost, and lunched with her and with their daughter, Katherine B. Frost, on several occasions when they came to Williams Bay and visited their friends who had made their home into the boardinghouse where I lived for three years. My teachers included, among others, Enrico Fermi, Gregor Wentzel, and Thornton L. Page in Chicago and Struve, Subrahmanyan Chandrasekhar, Gerard P. Kuiper, William W. Morgan, and Bengt Strömgren at Yerkes. I admired them all and learned much from them.

Some readers may find a few of my judgments harsh, however, and think me overly critical. I do not mean to be, but I believe it would be false to present a rosy picture in which every decision was the right one, all motives were pure, and all great scientists were also perfect human beings. I cannot write like that; I believe an author's duty is to write the truth as he or she sees it, and I think readers will accept no less. When you read my criticisms of these great figures of the past, remember that as a research scientist and an observatory director I have walked in their footsteps, although at by far a lower standard of achievement than theirs. I can see myself all too clearly in some of their less noble actions, taken, they believed, from the highest of motives, for their observatory, their university, and their science. Henrik Ibsen's words, "To write—that is to hold judgment over one's self," very definitely apply to this book.

Many persons helped me in preparing and writing this history of Yerkes Observatory. Most of all I thank Helen Wright, who more than two decades ago, and many times afterward, encouraged me to go on with my efforts on the history of American astronomy in the big-telescope era. Her book *Explorer of the Universe* contains an excellent short history of the early days of Yerkes Observatory, and if she and I have differed in some of our interpretations of George Ellery Hale's later activities, I hope that in this book I have depicted him fairly as the founder of a great observatory.

Almost all of my source material came from letters, scientific papers, and newspaper articles from the time my book describes. I am especially grateful to Judith Lola Bausch, curator of the Yerkes Observatory Archives, for her unfailingly friendly, efficient help every time I came to Williams Bay to work there. I am also most grateful to Daniel W. Meyer, Richard L. Popp, and their coworkers in the Special Collections Department of the Regenstein Library at the University of Chicago; Ralph L. Elder, head of public service at the Center for American History at the University of Texas at Austin; Dorothy Schaumberg, curator of the Mary Lea Shane Archives of the Lick Observatory here at the University of California, Santa Cruz; William W. Roberts, University of California archivist at the Bancroft Library in Berkeley; Ronald Brashear, curator of History of Science and Technology at the Henry Huntington Library, San Marino, California; Clark Elliott, associate curator of the Harvard University Archives in the Pusey Library; and Bernard Schermetzler at the University of Wisconsin Archives, in the Memorial Library at Madison. Finally, I thank the librarians at the Storrs B. Barrett Memorial Library, Williams Bay, Wisconsin, the Lake Geneva Public Library, and the State Historical Society Library, Madison, Wisconsin, as well as the staff members and officials of the Williams Bay School Board. They all helped me greatly in locating information for this book.

Many of my fellow University of Chicago Ph.D.'s in astronomy granted me interviews (either in person or by telephone) or sent me letters and other documents to help in preparing this history: Philip C. Keenan, Paul Rudnick, William P. Bidelman, Anne B. Underhill, Nancy G. Roman, Arne Slettebak, Arthur D. Code, H. Lawrence Helfer, Hugh M. Johnson, and Robert F. Garrison, as well as the late Nicholas T. Bobrovnikoff, William W. Morgan, and Franklin Roach. Several former Yerkes Observatory staff employees—Theodosia Belland, Donna D. Elbert, Maude Laidlaw Kelso, Barbara Perkins, and Jeanette Ringstad Johnson—helped me greatly with similar interviews, letters, documents, and suggestions, as did Jesse L. Greenstein, a former Yerkes postdoctoral fellow and faculty member. Years ago my late friend Su-shu Huang, another Yerkes Ph.D., contributed several important insights into some of the persons and events of this history. D. Harold McNamara gave an interview on his exiriences working for Struve at Berkeley and McDonald Observatory, David W.

Dewhirst provided valuable material on Struve's stay at Cambridge, and Charles J. Peterson allowed me to use his fascinating unpublished manuscript on T. J. J. See as a source. Laurel A. Andrew sent me important information on her aunt, Lillian Ness, and Edwin Brant Frost III on his grandfather. Maxine Hunsinger Sullivan kindly provided transcripts from her Office of the University Registrar in Chicago, and successive Yerkes Observatory directors Lewis M. Hobbs, D. A. Harper, and Richard G. Kron welcomed me whenever I could come and granted me office space while I was working on this book. Richard D. Dreiser, Elizabeth Cox, John W. Briggs, the late Joseph Tapscott, and the late Subrahmanyan Chandrasekhar were all extremely helpful in providing many of the photographs used to illustrate this book. Robert A. McCutcheon kindly translated several letters in Russian to Struve from members of his family and from Bobrovnikoff for me. To all of them I am most grateful.

Jesse L. Greenstein, Lewis M. Hobbs, and the late Thornton L. Page each read one or more draft chapters, and Kevin Krisciunas did the same for the entire book; their comments and suggestions helped me greatly, but any remaining errors of fact or interpretation are my own.

Last, I am deeply grateful to my wife, Irene Hansen Osterbrock, a former assistant, computer, and secretary at Yerkes Observatory, who made me aware of many perspectives on its history I would not otherwise have realized. She also entered the entire first draft of this book on our word processor, proofed and corrected every chapter several times, discussed them all with me, and prepared the index for the book. Without her extremely effective help, as well as her forbearance while I absented myself in the observatory on countless family "holiday" and "vacation" visits to Williams Bay, this book could never have been written.

# CHAPTER ONE

## Birth, 1868–1897

Yerkes Observatory began on October 7, 1892, a fine autumn day in Chicago. That morning William Rainey Harper and George Ellery Hale called on Charles T. Yerkes at his office at 444 North Clark Street. When they walked out less than an hour later, they had his authorization to build "the largest and best . . . telescope in the world . . . and send the bill to me." Harper reported, "The whole enterprise will cost Mr. Yerkes certainly half a million dollars. He is red hot and does not hesitate on any particular. It is a pleasure to do business with such a man." Completed in 1897, the observatory did cost Yerkes more than $500,000, the equivalent of roughly $7 million in purchasing power today.

Yerkes, known in the Chicago newspapers (except his own) as "the Boodler," was a manipulator of streetcar and railroad franchises in some of the largest cities in the English-speaking world. He was the model for Theodore Dreiser's *The Financier* and *The Titan* and was the despair of "the great and good few who represent the great and good part of our city," as he contemptuously referred to them. However, only a few months before that fine October day, Yerkes had married a chorus girl whom Harper, a biblical scholar, described to Frederick T. Gates, a Baptist minister who was John D. Rockefeller's personal adviser on educational matters, as "the most gorgeously beautiful woman I have seen for years." Yerkes was temporarily in a euphoric mood, and he liked the idea of being the "owner" of the largest and best of anything. Harper and Hale, two of the most persistent and successful educational fund-raisers of all time, succeeded in playing on his aspirations and coming away with the money for the telescope. It had not been easy; a few months earlier when Harper

Figure 1   Charles T. Yerkes. Courtesy of Yerkes Observatory.

had been trying to get money for a different academic project from Yerkes, he had been unsuccessful: "We worked with him the hardest we knew. . . . He did everything but say yes. . . . I am encouraged & more, with the help of God. I begin the campaign. It is by no means certain; but we must succeed." Harper never admitted failure. He had gone back with Hale and his plan for the observatory, and they had succeeded. That October night Harper's aide telegraphed Gates, "Yerkes builds observatory with the largest telescope in the world." Harper soon moved on to other millionaires whom he persuaded to give money to finance other university buildings; Hale began planning Yerkes Observatory.[1]

Harper, then thirty-six years old, was the short, roly-poly, intense president of the brand new University of Chicago, which had held the first classes in its history just one week earlier. Hale, then twenty-four years old, was his short, slight, intense associate professor of astrophysics, probably the first faculty member in America, if not the world, to hold that title. The story of the birth of Yerkes Observatory is their story.

Hale was born in Chicago in 1868, three years before the Great Fire destroyed the whole downtown area. His parents, immigrants to the Midwest from New England, lived in Kenwood, a wealthy suburb that was unharmed by the conflagration. Hale's father, originally the Chicago agent of a paper company, had founded a prosperous elevator firm, and it soon became enormously successful, supplying elevators for the skyscrapers erected as Chicago sprang like a phoenix from the ashes of the fire. Young George was a weak, sickly boy, the apple of his parents' eye, their firstborn after two children who had died in infancy. They showered attention on him. From his earliest days he was interested in science, particularly the experimental and technical sides, and as his family moved up in the world he always had a laboratory and a shop, both well equipped, in each house they owned. Hale had a private live-in tutor for several years, went to a private day school, and also attended the Chicago Manual Training School, a private institution where boys who would work in trade and boys who would run companies rubbed shoulders. At fourteen he made his first telescope, built around a single cheap lens. It failed because of the inherent inability of such a lens to bring light of different colors to a single focus ("chromatic aberration," he would later learn to call it), magnified by the long focal length of a telescope. Hale sought out an expert, Sherburne W. Burnham, an outstanding visual double-star observer, then an amateur astronomer who lived on Vincennes Avenue, a more modest area not far from Kenwood. The kindly, middle-aged Burnham told young Hale where an excellent four-inch refracting telescope, made by the best American telescope makers, Alvan Clark and Sons, could be obtained secondhand. Hale begged his father to buy it for him—he needed it to observe a transit of the planet Venus across the sun's disk, which would take place on December 6, 1882. How could his doting father refuse? He bought the Clark refractor, and George observed the transit of Venus. His career was launched—as an astronomer and as a fund-raiser. The same scenario was to be acted out several more times before William E. Hale died in 1898. The need for an instrument to make a specific observation, the opportunity to "secure" it, and the deadline were all parts of it, as Yerkes, John D. Hooker, Andrew Carnegie, and the Rockefeller Foundation executives were all to learn from Hale.

Figure 2    George
Ellery Hale (1892).
Courtesy of American
Institute of Physics/
Niels Bohr Library.

Burnham became Hale's astronomical mentor and showed him
Dearborn Observatory with its 18-1/2-inch Clark refractor, then situ-
ated in Douglas Park, on the Lake Michigan shore on the South Side
of Chicago, not far from their homes. Hale, however, was less interested in the old visual astronomy, as practiced by Burnham, than in
astrophysics, as the "new astronomy" of spectroscopy and other applications of physics to understanding the nature of the sun and stars
was just beginning to be called. Hale learned about it from a book he
received as a Christmas present, and soon he was building spectroscopes, then getting his father to buy one for him. When he graduated
from the Allen Academy in the summer of 1886, Hale's parents took
him on a trip to Europe, during which he visited several observatories
and met the pioneer French astrophysicist Jules Janssen. Before they
came home, Hale's father bought an expensive, professional-quality
spectroscope in London.

In the fall Hale went east to enter Massachusetts Institute of Technology. He was interested in astronomy, spectroscopy, and research and tolerated his courses in physics, which seemed dull but necessary. In his first summer back home in Chicago, he fitted up a laboratory for solar research and spectroscopy in the attic of the family mansion. On his return to Cambridge for his sophomore year, Hale began working on Saturdays as a volunteer assistant at the Harvard College Observatory. It was the outstanding astronomical research observatory in the United States (except possibly for the Naval Observatory, a highly specialized institution), and its director, Edward C. Pickering, trained as a physicist, was a pioneer of astronomical photometry and spectroscopy. The Harvard College Observatory had no large telescope and like all American observatories was strongly organized around the director, who closely supervised the work of all his staff members. Hale learned many of the practical details of photographic and spectroscopic astronomical research from Pickering, and at the same time he broadened his horizons and deepened his love of research. In the East Hale met Henry Rowland, the famous physicist who was then working on the solar spectrum at his laboratory at Johns Hopkins University in Baltimore, Charles A. Young, the pioneer astrophysicist at Princeton, and John A. Brashear, the leading astronomical instrument maker of the day, whose shop was in Allegheny, Pennsylvania (today a suburb of Pittsburgh).

With their encouragement, Hale arranged to do his MIT senior thesis on a study of solar prominences. His father ordered a huge astronomical spectrograph from Brashear, which Pickering was glad to let Hale use at the Harvard College Observatory. Hale invented the spectroheliograph, a brilliant instrumental idea that made it possible for him to photograph solar prominences in the light of a single emission line for his thesis. He did not really have enough time to do much observing with the brand-new instrument, a frequent problem with his later research projects, but he completed a thesis and graduated from MIT in June 1890. He then married his childhood sweetheart, took her on a honeymoon trip to New York and Niagara Falls, and then "home" to Chicago, where he and his bride stopped briefly with his parents. After a few days the newlyweds left on a whirlwind trip of the West, culminating in a visit to Lick Observatory, which Hale regarded as the high point of the honeymoon.

Lick Observatory, the research station of the University of California, was on Mount Hamilton, near San Jose. Completed in 1888, its thirty-six-inch refractor was the gift of James Lick, an extremely wealthy, eccentric miser who in 1876 had provided the funds for a telescope "superior to and more powerful than any telescope yet made." A decade and a half later the Chicago newspapers were to exult that Yerkes would "lick the Lick" with his gift of the forty-inch refractor. In 1890 Hale's mentor Burnham, now a member of the Lick staff, showed him through the observatory and introduced him to its director, Edward S. Holden, and to the young pioneer American astrophysicist James E. Keeler. Hale observed spectroscopically that July night at the thirty-six-inch with Keeler and was enchanted by the experience. They were to become close friends and the leading American astrophysicists of their generation (Keeler was eleven years older than Hale) until Keeler's tragic early death in 1900.

Then it was back to Chicago, where the young couple moved in with Hale's parents. His father had a handsome brick observatory dome and attached spectroscopic laboratory matching their home built in the backyard. He ordered an expensive telescope "mounting" (everything except the lens), built to George's specifications by the Warner and Swasey firm (which had made the Lick telescope mounting), with a twelve-inch lens made especially for it by Brashear. This was the Kenwood Physical Observatory, better equipped and outfitted than most university observatories of its time. With his spectrograph, brought home from Harvard, Hale was soon doing excellent solar research.

Just then, in February 1891, the thirty-five-year-old Bible scholar Harper accepted the presidency of the new University of Chicago, which he was to build in Hyde Park, some two miles from Kenwood. There had been an earlier university with this same name, which failed financially and closed its doors in 1886. It had been a Baptist institution, and through a complicated series of events the fantastically wealthy Rockefeller, a Baptist, was persuaded to provide the money to build a new University of Chicago. He wanted it to be a four-year college, strongly oriented toward teaching, and he wanted Harper as its president. This child prodigy had finished high school at New Concord, Ohio, at age ten, then graduated from Muskingum College, where his favorite subject was Hebrew, at age fourteen. To let the rest of the world catch up he marked time for three years, clerking in his father's store and studying languages on the side. He entered Yale as a graduate student at seventeen, studied Sanskrit, Greek, and Chaucer, and earned his Ph.D. a month before his nineteenth birthday. Harper became an inspiring teacher and a great scholar, and in 1886 he returned to Yale as a professor. He was a

Figure 3   William Rainey Harper, the boy president of the new University of Chicago. Courtesy of the Department of Special Collections, University of Chicago Library.

leader in the Chautauqua movement, and in his spare time he edited two journals, one on Hebrew, the other on Old and New Testament studies.

Harper was interested in becoming the founding president of Rocke-feller's new university-to-be, but only on his own terms—that it be a secular (though religiously oriented), great university from the start, emphasizing graduate education and research. In the end Rockefeller wanted Harper so badly that he accepted his conditions, and the University of Chicago was born. Harper immediately began recruiting top-notch faculty members, using Rockefeller money for salaries and research support as bait. He persuaded a whole string of college and university presidents, including geologist Thomas C. Chamberlin from the University of Wisconsin, to come to Chicago as department heads.

For the Physics Department, from Clark University Harper brought Albert A. Michelson, who in 1907 would become America's first Nobel laureate in a scientific field.[2]

Compared with those of these renowned scientists, in 1891 Hale's scientific accomplishments were negligible, but Harper sought him out and offered him a faculty position. The condition was that Hale's Kenwood Observatory be moved to the campus and thrown open to the students and public. Hale and another "young man," expert in the classical astronomy Harper understood from his Yale days, would share the teaching and research, their salaries and the expenses of the observatory to be provided by an unnamed donor, whom the president clearly expected to be William E. Hale. Young George indignantly declined this offer; he wanted to win the position on his merits, he wrote, not buy it with his father's money. Then he and his wife went off on a tour of Europe, visiting observatories and meeting astronomers and astrophysicists. After their return to Chicago, Hale continued his solar research at Kenwood. But the idea would not go away. In 1892 Asaph Hall, discoverer of the two moons of Mars, wrote Harper to recommend Hale as "a young man . . . who is devoted to his branch of astronomy, and who has already shown good ability. It is said that Mr. Hale's father is a man of great wealth, and is very generous in the support of his son's investigations. It may be well for you to know Mr. Hale." Harper, always persistent, had another session with Hale. The terms he offered were basically the same, but this time Hale accepted. The university would have the use of his Kenwood Physical Observatory, but only for graduate students under his supervision. Hale would be named an associate professor and director of the observatory but would receive no salary for the first three years unless an endowment came through for him. The university would pay the operating expenses at Kenwood, but Hale's father agreed to give the entire observatory, telescope, spectroscope and all other instruments to the University of Chicago if Harper kept his son on the faculty and they succeeded in raising $250,000 or more for a larger observatory.

Very soon after accepting this position, Hale went on a summer vacation in the Adirondacks and then attended the summer meeting of the American Association for the Advancement of Science. There he learned of an unexpected opportunity of the kind he loved. Two forty-inch glass disks, intended for an achromatic telescope lens, were available for purchase. They had been ordered and made for a planned new southern California observatory, proposed to be built on Mount Wilson. Edward F. Spence, a trustee of the University of Southern California, had announced in 1887, as Lick Observatory neared completion, that the Southland must have an even larger telescope, and that he would contribute $50,000 to start a fund drive for it. Others joined him, and the pledges rolled in, based on the idea that the Spence Observatory would become a tourist attraction, luring hordes of free spenders to visit from the East. The trustees of the planned observatory commissioned Alvan Clark and Sons to make a forty-inch lens, and after many attempts a French glass firm produced the two disks, crown, and flint glass necessary for it. The Clarks began working on the glass, but in 1891 the land boom in Los Angeles, on which the whole project was based, burst. Spence and several of the other prospective donors went broke and could not redeem their pledges. They defaulted in early 1892, and Alvan Clark and Sons were left holding the partly ground disks and a bill for $16,000 due to the Paris glass firm.

Hale, as he heard the story, recognized it as an opportunity for the University of Chicago. He hurried back to Harper, who in turn realized this was just the project to appeal to Yerkes. Under Harper's tutelage, Hale drew up a letter explaining in very simple terms how much more powerful the forty-inch telescope Yerkes could build would be than any existing instrument, including the famous Lick Observatory thirty-six-inch refractor. Hale estimated the cost of the proposed new observatory as $300,000 and claimed he already had a pledge from an anonymous donor (his father) for $30,000 (his estimate of the value of Kenwood Observatory) on condition that the university raise the full amount.

Hale's letter caught Yerkes's attention. He was in a receptive mood, and when Harper and Hale called on him on October 7, they had little trouble closing the deal. Within a few days Alvan G. Clark arrived in Chicago and signed a contract to figure the lens. (His father had died as the Lick lens was completed, and he himself was to do the same just as the Yerkes lens was finished.) Warner and Swasey were to build the mounting, as they had for the Lick thirty-six-inch.

From the beginning, Hale planned Yerkes Observatory for astrophysics—a very new subject in 1892. Back into the dim past astronomy had been concerned with the positions of the stars in the sky and the motions of the planets. Classical astronomy, over the centuries, had become progressively more technical, dominated by mathematicians, visual observers, and expert measurers of positions on the sky. But in the nineteenth century Josef Fraunhofer discovered spectral

Figure 4   Kenwood Observatory (1891). From the *Sidereal Messenger.*

lines in the light from the sun. Spectroscopy advanced rapidly. Soon Gustav Kirchoff and Robert Bunsen were able to recognize the same familiar elements in the sun and stars that they had studied spectroscopically here on earth. The great pioneers of astrophysical spectroscopy were Hermann Vogel in Germany, William Huggins and J. Norman Lockyer in England, Angelo Secchi in Rome, and Pickering and Young in America, while Samuel P. Langley pioneered in astrophysical infrared measurements of the sun and moon. They were all members of the generation just before Hale's. He had read all about them and their discoveries, beginning with a book his parents gave him for Christmas when he was a boy. Hale was resolved to follow in these pioneers' footsteps, and he did. He started Yerkes Observatory on the path of astrophysics, in which it has played such an important role throughout its history.

Hale had begun doing astrophysical research at his Kenwood Observatory in 1891; a year later he hired his first assistant, Ferdinand Ellerman. Born in Centralia, Illinois, after high school Ellerman had come to Chicago to seek his fortune. In his first jobs there, he became skilled in photography and machine-tool work, an unusual combination that qualified him ideally for the observing assistant position. He was hardworking, quick to learn, friendly, and personable. Soon Ellerman was doing much of the observing, leaving Hale more time for travel and for the organizational activities for which he was so well suited. Ellerman, a year younger than Hale, idolized him and stayed with him through thick and thin for more than forty years, until Hale's death.[3]

A few years later the young Kenwood director was to hire his second assistant, George Willis Ritchey, a manual training teacher with exceptional skills in optics and telescope design. Born in a tiny village in eastern Ohio, Ritchey was the son and grandson of skilled woodworkers and experts in machinery, immigrants to America from Northern Ireland. He had worked with them both in Cincinnati, taught himself how to make excellent mirrors for reflecting telescopes, attended the little University of Cincinnati for two years, and moved to Chicago, where he met Hale in 1891. Before long Ritchey, five years older than Hale, was helping him in his shop and in his photographic darkroom in his spare time. In 1896 Ritchey went to work full time at Kenwood Observatory, paid by Hale's father just as Ellerman had previously been paid until he went on the University of Chicago payroll. Even before Yerkes Observatory was completed, Ritchey was grinding, figuring, and polishing mirrors for Hale and designing even larger telescopes for the future.[4]

However, Ritchey was still just a part-time volunteer in 1892, when, with the forty-inch refractor ordered, Hale plunged into organizing what became the World Congress of Astronomy and Astro-Physics, the first international astronomical meeting to be held in the United States. Chicago was proud of its position as the rapidly growing metropolis of the Midwest, and its leaders were determined to show it off in the World's Columbian Exposition, more commonly known as the World's Fair. Its stated purpose was to exhibit the material progress and richness of America, four centuries after Christopher Columbus had discovered it. The organizers did not begin planning the fair quite early enough, so it actually took place in the summer of 1893, the 401st anniversary of Columbus's landing on San Salvador.

The Columbian Exposition was a vast popular attraction, a combination of spectacular temporary buildings, landscaping and inlets, uplifting exhibits, restaurants, eating halls, carnival shows, sights, and rides, all erected on the South Side, in immediate proximity to the new University of Chicago. The more high-toned accompaniment of the fair was the series of world congresses, or meetings, held in conjunction with it. They were the brainchildren of Charles C. Bonney, a leading Chicago reformer who had been an educator in his younger days. In each of these congresses the leaders of government, science, religion, finance, and literature were supposed to discuss their fields, describe problems of their age, and outline the solutions to them, all in one week. Over one hundred congresses were held in the summer of 1893, many simultaneously. They all took place in the Art Institute on Michigan Avenue, the building just completed in May 1893, barely in time for the first week of congresses to begin.

Hale, whose father was an important Chicago businessman, well tied in to the civic boosters who dominated the World's Congress Auxiliary that organized these meetings, was appointed to its Committee on Scientific and Philosophical Congresses, as a member of the Special Committee on Astronomy. It was set up in October 1890, a few months after Hale's graduation from MIT, but before his twelve-inch telescope had been completed and installed at Kenwood. Hale was then twenty-two years old, but he became the main spark in organizing the World Congress on Astronomy and Astro-Physics. His first step was to have George W. Hough, the dignified elderly professor of astronomy at Northwestern University, appointed senior member of the Committee on Astronomy. He was the figurehead behind whom the tireless Hale operated efficiently as secretary. By 1892 the subject of the congress had been defined as mathematics and astronomy, the classical combination that appealed to Bonney and the other senior members of the World's Congress Auxiliary. Hale did not fight the concept, but he emphasized the astronomical, and in particular the astrophysical side, in his organizational efforts. He wrote hundreds of letters to astronomers and astrophysicists throughout the United States and Europe, urging them to come to Chicago to take part in the sessions and present scientific papers.

Finally the week of August 21, 1893, dawned, and all ten World Congresses on Science and Philosophy began. The opening event was a short speech by Hermann von Helmholtz, the great German physicist, then seventy-three years old. He delivered it to all the attendees of all the congresses, assembled in the vast Columbian Hall that made up most of the first floor of the Art Institute. Bonney replied with a long, uplifting address of welcome. Next the chairmen of all the congresses responded, including Felix Klein of Göttingen University on behalf of the mathematicians and Hough for the astronomers. Then the group split up into their various congresses. The mathematicians and astronomers stayed together only long enough to hear Klein's ad-

Figure 5   World Congress of Astronomy, Astro-Physics and Mathematics, held in the Art Institute, Chicago (1893). John A. Brashear (with long beard) and James E. Keeler are seated at far left in front row. T. J. J. See (wearing light suit and with hat on his knee) is seated just to right of center, Sherburne W. Burnham is third to right of him, and Henry Rowland all by himself at far right, all in this same row. George E. Hale and Edwin B. Frost are seated in second row just behind Brashear. Standing to right of center in second row from rear, George W. Hough and Alvan G. Clark are the two men with white beards. Courtesy of Yerkes Observatory.

dress "The Present State of Mathematics" and then separated for the rest of the week.

About thirty astronomers and astrophysicists were present, including Max Wolf from Heidelberg, Pietro Tacchini from Rome, Egon von Oppolzer from Vienna, and Eugen von Gothard from Hereny, Hungary. Two Americans recently returned from South American observatories were also at the meeting, John Thome from Argentina and William H. Pickering from Peru. Rowland, the great Johns Hopkins spectroscopist, attended some of the sessions of the Congress of Astronomy and Astro-Physics, spending the rest of his time at the Congress of Electricity, which was meeting in another room in the Art Institute. Among the other astronomers present were Keeler, the leading American astrophysicist of Hale's generation and new director of Allegheny Observatory, and Edwin B. Frost of Dartmouth University, who was soon to join the original Yerkes staff and ultimately to succeed Hale as its director. Telescope makers Alvan G. Clark, Worcester R. Warner (of the Warner and Swasey firm), and Brashear were also at the meeting. Several papers were presented each day, and on Wednesday the entire group went out to the fair, where the Yerkes forty-inch telescope mounting, already completed and assembled, was on display in the huge Manufacturers and Liberal Arts Building. The giant tube, fifty-five feet long and set on a thirty-foot-high pedestal, loomed over the assembled astronomers. Hale, Clark, and Warner gave short speeches at the base of the mounting. Clark's talk was titled "Great Telescopes of the Future"; in it he proudly stated that nearly all the most important discoveries in astronomy had been made with the largest telescopes in use at the time and predicted that the same would be true for the forty-inch when its lens was completed and installed. He traced the history of astronomy completely in terms of refracting telescopes, beginning with two fifteen-inch refractors his father had made in 1846, then regarded as "monster telescopes," moving up, step by step, a few inches in aperture each time, through the Lick thirty-six-inch to the forty-inch lens he was then figuring. Clark seemed to believe the process could go on forever, but Hale and Keeler knew this was not so and that huge reflecting telescopes, not refractors, would be necessary for the astrophysics they saw as the future of astronomy. In fact the forty-inch was the largest refracting telescope ever made and used successfully for research— the "last of the dinosaurs."

When the last paper had been given on Saturday, August 26, and the wiser but poorer astronomers left Chicago and its fair, Hale was established as an important figure in American astronomy. Barely twenty-five years old, he was the director-to-be of the world's largest telescope and the secretary who had successfully organized America's first world astronomy meeting.[5]

By that time Harper and the University of Chicago trustees had decided on the site where Yerkes Observatory would be erected. At first everyone had assumed it would be on the campus, and a location in Washington Park, a huge open area beginning a few blocks west of the university, was seriously considered. But professional astronomers strongly advised Harper against any site in the city, which was low, hazy, and worst of all smoky, the bane of telescopes in those days of soft-coal heating and industrial plants. Harper thought of his young director as "George" (in contrast to his father, "Mr. Hale") and had not completely trusted his advice, which had been the same, but he listened to the outside experts. He actively solicited information on possible locations for the new observatory. Invitations or proposals came in for twenty-seven possible sites, most of them in northern Illinois. One invitation came from the Pasadena Board of Trade, in far-away California, suggesting the Sierra Madre range, and even the specific peak, Mount Wilson. Harvard University had operated a temporary station there in 1889, when it had been considered the potential location for the planned Spence Observatory. Harper and Hale never seriously considered any site more than a hundred miles from Chicago, however, although a decade later Hale was to move on to Mount Wilson to found his third observatory. Among the nearby sites, Hale favored Lake Forest, north of Chicago, but no one stepped forward to offer the necessary land, so that "deal" (as Harper called it) did not become a reality.

John Johnston Jr., a wealthy Chicago real estate speculator, proposed a site in Wisconsin, on Lake Geneva, approximately halfway between Chicago and Madison. Johnston owned several large tracts around Williams Bay, on the north shore of the clear, deep lake, and he offered to donate fifty acres for the observatory. As university property it would be free of any state or local taxes. A further advantage, Johnston wrote Harper, was that any needed legislation could be had "easier & cheaper" in Wisconsin than in Illinois. Several wealthy Chicago families, including one University of Chicago trustee, George C. Walker, owned palatial summer homes on Lake Geneva, which was then just at the limit of leisurely commuting distance by train from the center of the city. Chamberlin, former University of Wisconsin president and former state geologist of Wisconsin, whom

Harper had lured to Chicago to head the Geology Department, knew the Lake Geneva area well and recommended it strongly to the president. He emphasized its high hills (to get the observatory above the haze level), the absence of factories, its relatively low population density (which meant little smoke), and above all that it was a "resort for the choicest people of Chicago." This was what made it attractive to Harper; he hoped to raise endowment funds to operate the observatory and pay the astronomers' salaries from the wealthy Chicagoans with summer homes around the lake.

The one remaining possible problem was the proximity of the proposed site to the lake. Some feared the water might cause local fog, clouds, or bad observing conditions. But Burnham, the double-star observer who was the best-known astronomer in Chicago, had spent one year on the staff of Washburn Observatory, on the University of Wisconsin campus on the shore of Lake Mendota in Madison. He gave his expert opinion that Lake Geneva would not cause any problems, and Harper and Martin Ryerson, president of the University of Chicago Board of Trustees, decided definitely on the Williams Bay site. Within a few years Ryerson and his friend and fellow trustee Charles L. Hutchinson were to build beautiful lakeshore estates within a few miles of the observatory. On a cold December day in 1893, Harper, Johnston, Walker, Burnham, and university architect Henry Ives Cobb came out from Chicago and, with Edward E. Ayer, another wealthy Lake Geneva resident, picked the exact spot that, in the words of the local newspaper, "without doubt will be the location of the world's greatest wonder."[6]

Hale was not at Williams Bay with them because he had gone to Germany. He had decided that, since Clark would need two years to complete the forty-inch lens, he himself might profitably spend some time in postgraduate study. He intended to live for a year in Berlin, studying German and attending advanced physics lectures. German was then as much the language of astrophysics as was English; Keeler and Frost, who had both done graduate work in Germany, could read it easily and converse passably with Helmholtz, Wolf, Klein, and other visiting German scientists. Hale said he wanted to learn German that year, but he never got around to doing so. He found he did not like the country or the food, and he spent most of his time in his apartment, working on conceptual plans for the Yerkes Observatory building and on "schemes" for Harper to use in persuading Yerkes to provide the funds for additional spectrographs and a small telescope. From the beginning, Hale designed Yerkes Observatory for astrophysical research. He included spectroscopic and physical laboratories in the plans for the building, as well as several photographic darkrooms. Also he planned optical and mechanical shops for full-time in-house instrument makers. Lick Observatory, designed scarcely a decade earlier by a gentleman of leisure with advice from classical astronomers, had not had even a single darkroom in its original plan and had suffered for it. Hale was not as successful in his schemes for getting additional funding from Yerkes; the financier had cooled off greatly, and though he met his commitment to build the observatory, Harper could never persuade him to endow it.

Hale also spent much of his time in Europe lining up support to start a research journal for astrophysics. The time was ripe. Many new American journals came into existence in all branches of science around the turn of the century. Until then, in astronomy each of the larger observatories had one or more publications of its own, for papers by its director and scientists to report work they had done. Thus Harvard College Observatory scientists published their results in the *Harvard Observatory Annals* and the *Harvard Observatory Circulars;* Lick Observatory astronomers in the *Lick Observatory Publications* and *Lick Observatory Contributions.* But smaller university observatories and privately owned observatories could not afford this method of publication; Hale faced this problem himself at his Kenwood Observatory.

The *Astronomical Journal,* founded by Benjamin A. Gould in Cambridge in 1849, had existed for many years. He was a well-trained classical astronomer who had graduated from Harvard in 1844 and then studied mathematics and astronomy under the great Carl F. Gauss in Göttingen. Gould was in charge of the longitude work of the U.S. Coast Survey and was director of Dudley Observatory in Albany, New York, for a few years, but after a blowup with its trustees he returned to Cambridge in 1859. All the while he kept the *Astronomical Journal* coming out, but in 1861, soon after the Civil War began, he suspended its publication to join in the "fight against treason" (from well behind the lines). He was one of a small group of scientists who founded the National Academy of Sciences in 1863, as an arm of the Union war effort. Soon after the war ended, however, he went to Argentina as first director of its National Observatory, and the *Astronomical Journal* remained dormant. He finally returned to the United States in 1886, and put the journal back into operation after its quarter-century hiatus. Gould did not consider astrophysics to be within the purview of the *Astronomical Journal,* however, and only rarely published astrophysical papers. Thus the pioneer American astrophysicists Young and Langley had no recourse but to publish nearly all their early papers in the *American Journal of Science, Reports of the U.S. Coast Survey, Popular Science,* or European journals such as *Comptes Rendus* or *Astronomische Nachrichten.* This was the situation that Hale, who published his first papers in 1890 and 1891 in the *Astronomische Nachrichten* and the *American Journal of Science,* wanted to change.[7]

There was one other astronomical magazine in America, the *Sidereal Messenger,* a monthly published in Northfield, Minnesota, by William W. Payne. He was professor of mathematics and astronomy at Carleton College, a former schoolteacher and school superintendent who had no research pretensions whatever but a driving urge to spread the gospel of astronomy. His *Sidereal Messenger* was a general astronomical magazine that published short papers by professional astronomers and advanced amateurs and many news items, editorials, and opinion pieces on astronomy written by Payne himself. It was not at all a satisfactory research journal, but many astronomers read it, and Payne would accept astrophysical papers if they were not too technical, so both Keeler and Hale had published in the *Sidereal Messenger.*

In the fall of 1891, soon after the dedication of his Kenwood Observatory, Hale had hoped to start a professional research journal, to be called the *Astro-Physical Journal.* He soon found that although plenty of prospective authors would gladly have let him publish their papers, the *Sidereal Messenger* was too solidly established. Hale could not raise enough money, or sign up enough charter subscribers, to get his proposed new journal off the ground. Payne, to head off a potential rival, proposed that they join forces, and Hale agreed. Thus *Astronomy and Astro-Physics* was born, published by Payne in Northfield as the successor to the *Sidereal Messenger,* but with two separate sections in each issue. One, on astronomy, was edited and largely written by Payne, from the amateur point of view; the other, on astrophysics, was edited by Hale and contained professional research papers. Keeler became an associate editor for astrophysics and filled in for Hale while he was gone to Europe.

By that time the number of paid subscriptions to *Astronomy and Astro-Physics* was dropping, as the amateurs found the astrophysical part of their magazine becoming too demanding. Payne, who depended on his editorial work for part of his income, began a new magazine, *Popular Astronomy,* aimed at schoolteachers and amateur astronomers, leaving the more technical papers for *Astronomy and Astro-Physics*. As soon as Hale arrived in Europe, he began conferring with leading astrophysicists and laboratory spectroscopists, telling them his dream of a completely professional astrophysical journal. He found strong support for his idea if he would leave out the articles for and by the amateurs. Hale adopted a wide definition of astrophysics and planned an editorial board large enough to include at least one representative from each of the astronomically significant European countries. As soon as he returned to Chicago in the summer of 1894, he took the idea to President Harper. The latter, a research scholar and editor, was committed to building up the University of Chicago Press as a journal publishing house. He authorized Hale to go ahead with the new journal; the Press would publish it if he could buy out Payne's interest in *Astronomy and Astro-Physics*. Payne agreed to sell his share, including the remaining 420 subscriptions, for $600. Hale had to raise the money himself because his father, although he was probably the largest contributor, would not provide the whole amount. Keeler obtained a $100 contribution and put up $50 of his own money; Hale talked Payne into reducing the price to $500, and in the end he had to accept $400. The deal was consummated, and the first issue was scheduled for January 1895. Hale and Keeler were the editors, and the associate editors (basically an advisory board whose members lent credibility to the journal) were five Americans, among them Rowland, Young, Pickering, and Michelson, together with one Englishman, one German, one Italian, one Frenchman, and one Swede. In addition, there were five younger American assistant editors, to help out with the work of the new journal. Hale and Keeler named it the *Astro-Physical Journal,* but the University of Chicago Press decided the hyphen did not belong there, and it was the *Astrophysical Journal* from its birth. Costs were a constant worry in the early years, but Hale's father was available to provide the funds to meet the deficit if it was not too large. The *Astrophysical Journal* quickly became the leading journal of its field, as it has remained to this day. It belonged to the University of Chicago Press, and the director of Yerkes Observatory was automatically its editor for more than half a century.[8]

Cobb, the university architect, converted Hale's general design for Yerkes Observatory into detailed building plans. Actual construction began at Williams Bay in the spring of 1895. Clark finished the forty-inch lens in September at his Cambridgeport, Massachusetts, optical shop, and Keeler, who acted as the independent expert, tested it on stars, using an awkward temporary mounting erected on an outdoor brick pier. He found the lens satisfactory, Hale forwarded his formal report to Yerkes, and the tycoon paid Clark for it but left it at his shop for safekeeping until the building was completed. Finally, a year and a half later, on May 19, 1897, Clark himself, with his head optician, Carl A. R. Lundin, brought the lens to Williams Bay in a special train. Packed in cotton and excelsior inside a series of strong wooden boxes, the lens came in the same car with them, accompanied by one of the University of Chicago trustees. At the station Hale and staff astronomer Edward E. Barnard met the train. The lens and its boxed cell (the metal ring that would hold it in the telescope), were unloaded into a horse-drawn wagon, and a teamster drove it up the hill to the observatory. The next day Clark and Lundin, assisted by Ritchey and two Warner and Swasey mechanics, unpacked the lens and cell and installed them in the telescope.

That night was cloudy, but the next evening Harper and a large delegation of University of Chicago trustees and officials came out to Lake Geneva in the private car of the president of the Chicago and North Western Railroad. Several steam yachts took them from the station to the landing below the observatory, and they walked up a path to it. As they entered the dome they marveled at the newly installed monster telescope. Harper was the first to view a star through it, and he instantly pronounced the lens, telescope mounting, and observatory tremendous successes. After everyone else had looked through the telescope too, the Chicago party returned to the station and took their special train back to the city. Then Hale, Barnard, and the other astronomers observed the rest of the night and confirmed that the giant refractor and its lens were indeed well made.[9]

The astronomers began regular nightly observing, but little more than a week later, disaster struck. Early in the morning, just a few hours after Barnard and Ellerman had closed up the dome at dawn and gone home to sleep, the giant wooden floor, which was raised and lowered by electric motors to keep the observer near the telescope eyepiece, broke loose from its cables and crashed to the ground, forty-five feet below. The floor was smashed to pieces. Had it happened on the evening of May 21, Harper, half the high officials of the University of Chicago, and the whole Astronomy Department, all of whom were on the floor, might have been killed. Investigation revealed that the cause of the accident was not sabotage, as Yerkes had feared, but that the cables had not been securely fastened to the floor. Once one of them had worked loose, the increased load on the others quickly pulled them out, and the floor dropped, tearing itself apart against the pedestal and base of the mounting. The telescope and lens, far above the floor, had not been harmed, but they could not be used without it. Warner and Swasey sent a team of workmen to build and install a new floor, but the telescope was out of operation for nearly two months. This time the cables were very securely fastened.[10]

Hale had begun organizing a dedication ceremony for Yerkes Observatory even before the lens was delivered. It finally took place in October, a beautiful month in southern Wisconsin. Hale and Harper had chosen it so that Yerkes would be back from his regular summer trip to Europe. About sixty astronomers and physicists were present for the week-long series of conferences, papers, lectures, and viewing Hale had scheduled. The forty-inch refractor and Hale's twelve-inch, brought out from Kenwood the previous December, were in their domes in the large buff-colored brick building on the spacious open height overlooking Lake Geneva, surrounded by forests in full autumn color. Most of the visiting astronomers were from midwestern or eastern observatories and universities, but spectroscopist Carl Runge had come all the way from Germany to be present. Thursday, October 21, was the high point of the week. Yerkes had arrived the previous night, and Hale had shown him through "his" observatory. Two special trains came out from Chicago on the great day, bringing nearly seven hundred trustees, faculty members, and friends of the university to Williams Bay. A fleet of carriages, omnibuses, and steamboats met them at the station and transported them to the observatory or the landing below it. As the guests stepped out onto the new wooden floor in the dome they must have been slightly apprehensive, even though they knew it was down on blocks in its lowest position. Harper had arranged a full academic procession, with all the official party clad in caps and gowns. After a prayer and a selection by a string quartet, Keeler delivered the main invited address, "The Importance of Astrophysics, and the Relation of Astrophysics to Other Physical Sciences." It was a masterly, forward-looking formal talk, describing what in fact became the main problems of astrophysics for the next half century. The local reporter was impressed but remarked that "the address was rather long which must be excused by the fact that it was a long subject and one on which the Professor was full to the brim."[11]

Next came another selection by the string quartet, followed by a speech by Yerkes, who was wildly applauded as he was introduced. He expressed his appreciation to Harper, the University of Chicago,

Figure 6   Yerkes Observatory dedication (1897). George E. Hale (wearing light hat) is at far right of second row from rear, Edward E. Barnard (wearing derby) is standing farthest left, George W. Ritchey (wearing dark hat and with mustache) is standing just in front of door in rear. James E. Keeler (wearing derby) and Edward C. Pickering (wearing light hat) are standing in front, to right of center, with Carl Runge in front and to right of Pickering. Courtesy of Yerkes Observatory.

and the builders of the telescope and then launched into a five-minute synopsis of the history of astronomy. Undoubtedly written by Hale, it started with the ancient Greeks and culminated in astrophysical spectroscopy. Next Yerkes turned the telescope, observatory, and grounds over to the University of Chicago. He was cheered to the rafters, not least because he had finished so quickly. Ryerson and Harper followed with speeches on behalf of the trustees and the administration, and after a final benediction everyone adjourned to a gargantuan luncheon, served in the halls of the observatory, courtesy of Yerkes himself. Then, after inspecting the telescopes, shops, and laboratories, they all took the special trains back to Chicago.

On Friday a dedication ceremony was held on the campus, with a tour of the Ryerson Laboratory, the physics building, hosted by Michelson, followed by a lunch at which Harper presided. Then Simon Newcomb, the sixty-three-year-old expert on planetary orbital theory, gave his address titled "Aspects of American Astronomy." He was, in the words of a newspaperman, "the most distinguished astronomer of the age," at least in the United States, and had been the chief astronomical adviser to the Lick Trust, although he had little personal experience in observing with large telescopes. His speech had much more oratorical polish than Keeler's the previous day, but very little vision of the future. Newcomb's history dealt almost entirely with celestial mechanics and positional astronomy. He did not even mention Young or Langley, the American pioneers of astrophysics, and he wrapped up spectroscopy in two brief sentences. But his impressive, white-haired figure and his oratorical powers made his speech "well in keeping with his reputation" and earned him long and vigorous applause.[12]

That evening Yerkes gave a sumptuous banquet for over two hundred guests, including all the visiting astronomers and physicists and many university officials, at his favorite Chicago restaurant, Kingsley's. A long series of toasts closed the evening and the week-long celebration of the official opening of Yerkes Observatory. Consciously planned as the world's first university astrophysical observatory, America's second big-science institution, with its forty-inch refractor, built to surpass Lick Observatory's thirty-six-inch, "superior to and more powerful than any telescope yet made" just a decade and a half earlier, was ready for action.

# CHAPTER TWO

## Infancy, 1897–1904

On Friday evening, October 22, 1897, the party ended in Chicago. Yerkes Observatory had survived the dedication; now it was time for the science to begin. Actually, George Ellery Hale and his assistants had been doing research at Kenwood Physical Observatory from the moment its telescope had arrived and had gradually transferred their activities to Williams Bay, Wisconsin, as the big new observatory was being built. Hale had dreamed since boyhood of detecting the corona—the faint extended envelope of the sun, visible only at total solar eclipses—by some brilliant new method that would reveal it daily. In the summer of 1895 he, George Willis Ritchey, and Ferdinand Ellerman had taken a bolometer (a device for measuring radiant energy) and a small reflecting telescope to the Yerkes Observatory site. They set up the telescope there, and Hale and Ellerman tried to measure the far infrared radiation of the corona with this system. Infrared radiation is much less susceptible to scattering by dust and molecules in the atmosphere, and therefore at these wavelengths the corona would not be lost in the bright glare of the sun itself. But in the infrared the corona proved too faint to be measured with this small telescope. Nevertheless, observing had begun at Yerkes Observatory.

The following summer Hale tried again, this time with a larger telescope, a twenty-four-inch reflector with a long focal length, mounted horizontally and fed by a heliostat. It was on the second floor of the partly completed observatory building. The roof directly over the telescope could be rolled away, and the optical parts were mounted on heavy, stable laboratory benches, built on pillars that ran through the building to the solid earth below. The detector was an improved, water-cooled model. Once again, however, the corona proved too faint. Hale did not detect it.[1]

By November 1896 the observatory building was close to completion. Hale had Ritchey and Ellerman begin to disassemble his Kenwood twelve-inch telescope and pack it for shipment to Williams Bay. Workmen removed the rotating dome for the same trip. In December Ritchey moved to a boardinghouse in the little village, to be on hand to install the twelve-inch and the drive clock for the forty-inch, which arrived just after Thanksgiving. Hale, with his wife and their little daughter Martha, moved into the director's house that the university had built for him a few hundred feet from the observatory. Ritchey and Ellerman mounted the twelve-inch, and it was ready for visual observations by the end of January 1897. As soon as the big telescope had been assembled and the forty-inch lens was installed in May, nightly observations began. The disaster to the floor put a stop to them before the end of the month. Little more than a month later, the construction crew completed the new floor, and observations began again at the forty-inch, three months before the dedication.[2]

Hale's own main research interest was the sun. Although he was very busy organizing the observatory, preparing for the dedication, and seeking additional financial support, he found time for observing the chromosphere, the outer envelope of the sun, with the forty-inch and his grating spectrograph brought from Kenwood. Just as soon as it was mounted on the big telescope, in September, he pointed it at the sun, and he very soon saw a host of new bright lines in the chromosphere, never previously observed. He soon recognized, in emission, the closely packed lines of a "fluting," or molecular band, of carbon. That was the first discovery, a highly astrophysical one, made with the big new refractor. Hale showed it to Henri Deslandres, the French astrophysicist who was visiting the observatory, who confirmed that it was indeed the carbon band. Hale gave a paper on his discovery at the dedication, less than a month later.[3]

He was less fortunate with his sally into stellar astrophysics. He had long planned to use the largest telescope in the world for problems where its great light-gathering power was especially needed. Hence, after the dedication Hale began a program of observing the spectra of the "fourth-type stars." This was the name given to an obscure family of very red stars whose spectra were not at all well understood. One reason was that not many of them were known, and all were very faint—not one as bright as fifth magnitude. Hence a large telescope was needed to study them. The forty-inch, a visual

Figure 7   Yerkes forty-inch refractor in its dome (1897). Courtesy of Yerkes Observatory.

refractor, worked best in the green, yellow, and red spectral regions, making it well suited for work on these stars. For this program Hale used the "universal" spectrograph of Yerkes Observatory, a prism instrument modeled closely after the spectrograph James E. Keeler had designed and used very successfully with the little thirteen-inch Allegheny observatory refractor.

To obtain spectrograms of the faint fourth-type stars, long photographic exposures were necessary, ranging from two to nine hours, with four or five hours typical. Hale depended heavily on his assistant, Ellerman, to take the spectrograms. They confirmed what was already known, that these stars are dominated by carbon molecular compounds, and somewhat extended this conclusion. But beyond that, little in the way of scientific conclusions emerged from years of observations on this program. The basic reason was that the problem was too complicated; these low-temperature carbon stars have thousands of bands and lines, whose relative strengths depend on temperature and the abundance of the elements, all of which vary from star to star. Only decades later did astronomers with still bigger telescopes, more powerful spectrographs, and much more complete laboratory data begin to understand the fourth-type stars, by then called R and N stars, or C stars.[4]

Ellerman, a jolly, talkative jack-of-all-trades, got along very well with the local farmers, summer resort operators, and small-town storekeepers in Williams Bay. In 1899 he was elected clerk of the local country school district, which extended all along Lake Geneva, Delavan Lake, and the neighboring countryside, the first of a long succession of "observatory people" to be involved in the area's educational affairs.[5]

Ritchey, who brought his family out from Chicago in late winter of 1897 to a house that he, like Ellerman, bought or rented, was the instrument maker and optician. He could make anything, and he could make anything work. In charge of the instrument design and shop was Frank L. O. Wadsworth, an assistant professor whose expertise was in engineering and design. He had been physicist Albert A. Michelson's assistant at Clark University and had come to the new University of Chicago with him as an assistant professor in 1892. Wadsworth had a talent for not getting along with people, and Michelson was glad to hand him over to Hale in 1896. Wadsworth designed several instruments for Yerkes Observatory, but had an inflated sense of his own importance. In 1898 he demanded a promotion to full professor and a large salary increase. When President William Rainey Harper countered with an offer of an associate professorship and a much smaller raise, Wadsworth angrily rejected it and accused the president of dishonesty. Hale, who had no immediate need of more designs for spectrographs and bolometers, had perfunctorily recommended Wadsworth's "proposition" but now made no move to save him. Harper and pious old Dean Thomas W. Goodspeed, his right-hand man, were glad to let Wadsworth go.[6] Hale, by now confident of Ritchey's design skills, shifted him from his father's payroll to the university's and made him superintendent of instrument construction.

Although from the beginning Yerkes Observatory was planned and designed for astrophysics, its two senior faculty members were representatives of the old-time astronomy. One was Sherburne W. Burnham, the great visual double-star observer who had been Hale's childhood mentor. Burnham, who was nearly fifty-nine years old when

Figure 8
Sherburne W.
Burnham.
Courtesy of
Mary Lea
Shane Ar-
chives of
the Lick
Observatory.

Yerkes Observatory was dedicated, had no scientific training whatever. Completely self-taught, he had a full-time job as a court reporter
in Chicago, where he had observed with his own telescope in his
backyard for many years. His keen eyesight, dedicated entirely to one
research project, measuring the separations of close binary stars, had
brought him recognition from the professionals. In 1876 he had spent
two months with his telescope at Mount Hamilton, testing it as a site
for the Lick Trust. He pronounced the location excellent, and they
built their observatory there. Then Burnham had taken a professional
job at the University of Wisconsin's Washburn Observatory on a one-
year trial, 1881–82. Convinced by the end of the period that the director, Edward S. Holden, was exploiting his discoveries, Burnham
had gone back to his position in the Chicago court.

In 1888, when Lick Observatory went into operation, the opportunity of observing with the "largest, most powerful telescope in the
world" on the mountain he knew was so well suited for his work

lured him back. He joined the Lick staff even though his nemesis, Holden, was now its director. Burnham lasted just four years at Lick, increasingly at odds with his superior. Then in 1892 he resigned with a public blast against Holden and returned once again to the Northern District of Illinois court, convinced he would never observe again. But just then Harper and Hale pulled off the deal with Charles T. Yerkes. Burnham joined the committee Hale had put together to organize the World Congress of Astronomy and Astrophysics, lending it tremendous credence among classical astronomers. After the congress and the World's Fair had ended, Burnham was appointed professor of practical astronomy as an unsalaried volunteer. He accepted on the understanding that he would begin observing when the forty-inch was completed. At the dedication in 1897 Burnham was very much in evidence, insisting to the reporters that Williams Bay would be as good a location for an observatory as Mount Hamilton. His real opinion, published in 1900, was that "there is probably no place in the world, where an observatory has been established, which can compare favorably with Mount Hamilton." On the other hand, he reported that with the forty-inch Yerkes telescope "much of the time the most difficult of these pairs could not be observed under the conditions present."[7] Williams Bay was a good midwestern site, but a California mountaintop was much better, both for atmospheric steadiness ("good seeing") and for number of clear nights.

By the time of the dedication Burnham had already begun observing with the forty-inch, coming out from Chicago on the train each Saturday afternoon, working two nights, sleeping on a settee in his office whenever he could, cooking his own meals in the building, and returning to the city and his courtroom on the Monday train. Burnham continued observing double stars at Yerkes for more than fifteen years, changing his routine only in coming to observe on Wednesday and Thursday nights after he had retired from his regular job. He is generally considered the leading visual double-star observer of his time.[8]

The other senior old-time astronomer on the Yerkes staff was Edward E. Barnard, another expert visual observer as well as an outstanding astronomical photographer. He was ten years older than Hale but twenty years younger than Burnham. Barnard had grown up in poverty in Nashville, Tennessee, in the defeated South after the Civil War. He had worked from early childhood to support his widowed mother and had practically no formal education. Working as a photographer's assistant, he had become intensely interested in as-

Figure 9    Edward F.
Barnard (1893).
Courtesy of Mary
Lea Shane Archives of
the Lick Observatory.

tronomy and had discovered several comets, making himself locally famous. His eyes were extremely keen, and his dedication made him an unrivaled observer. Friends arranged for him to receive a scholarship as a special student at Vanderbilt University, where he taught astronomy and, a married twenty-six-year-old, sat in freshman classes in English and history. Holden hired Barnard on the initial Lick staff, along with Burnham, and the young Southerner began his brilliant career in wide-field photography of nebulae and the Milky Way. Like Burnham, and under his tutelage, he soon came to dislike, distrust, and eventually hate the Lick director. After Burnham left Mount Hamilton in 1892, Barnard felt isolated and threatened. He wanted desperately to get away from Holden, but he dreaded the thought of leaving the big telescope in the land of the clear skies and good seeing. Burnham, who stayed in close contact with Barnard, advised Hale that they could get him on the Yerkes staff whenever they were ready for him.[9] The wily Burnham also leaked the news that Barnard was restive to Arthur Edwards, a Methodist minister and editor in Chicago, who passed the news on to President Harper at the university. Harper's aim was to have one outstanding expert in each field of knowledge on his faculty. He hardly knew what astrophysics meant, and Hale seemed young and unformed to him. Barnard, on the other hand, had just made himself famous by discovering the fifth satellite of Jupiter, a feat that Harper, well schooled in classical astronomy, could understand very well. He decided that he wanted Barnard at the University of Chicago. Hale, who no doubt had other priorities, was far too diplomatic to contradict the president. He recommended that appointment, and Harper offered Barnard the job.

After stewing about the offer for months, the neurotic Barnard finally told Harper and Hale that he would decide to come, but he had not yet done so. He begged Harper to keep it secret, but the president "inadvertently" announced the good news at a convocation and printed his name in the catalog for the coming year as a professor of astronomy. Then Barnard had to resign his Lick position. He moved to Chicago in the fall of 1895, almost two years before the forty-inch was ready for operation, just as he had jumped the gun and arrived in California a year before Lick Observatory was ready. In Chicago he marked time, observing with Hale's Kenwood refractor and giving popular lectures on astronomy at the university, throughout the city, and on tours of the East.[10] Barnard was eager to move to Williams Bay and believed that the university would build a house for him there. Harper stalled on this, so the impatient Barnard bought a lot next to the observatory and had the house built himself. He and his wife moved in just before Christmas 1896, and he was the first observer (after Hale) to use the forty-inch refractor. He and Ellerman were the observers who were walking home in the morning of May 19, after observing, when they heard the rising floor, which they had been standing on less than an hour earlier, crash to the ground. At the dedication, Hale cited Burnham and Barnard as the experienced observers who had tested the forty-inch refractor on close double stars and found it performed perfectly.[11]

Barnard continued observing with the forty-inch refractor, both visually and photographically, throughout his quarter-century career at Yerkes Observatory. He was famous for his dedication, on more than one occasion climbing up the outside of the dome on dark winter nights to chip away ice deposited by a storm that had just passed, so that he could begin observing before the sun rose.[12]

Figure 10   Edwin B.
Frost as a young
man. Courtesy of
Yerkes Observatory.

Hale had wanted to hire James E. Keeler, the pioneer American astrophysicist who had given the main invited lecture at the dedication, to the observatory's staff in 1897, as its spectroscopist. Harper saw the needs of the many other departments of the university as more pressing than astronomy's and insisted he could not fund Keeler's salary from the university's budget. Hale succeeded in persuading Catherine W. Bruce, an elderly, wealthy recluse, to contribute $15,000 to pay Keeler for the first five years, and on this basis Harper was willing to hire him. Keeler, however, was offered the directorship of Lick Observatory just at this time, in succession to Holden, who had finally been forced out, partly as a consequence of his feuds with Burnham and Barnard. Hale then offered the position to the German astrophysicist Carl Runge, but he did not want to leave Europe. Next Hale offered it to Edwin B. Frost, who accepted. Hale then managed to convince Bruce, the source of the funds, that "next to Professor Keeler . . . Professor Frost [was] better qualified than anyone else we could secure for the place." She accepted this argument, and Frost became the second astrophysicist on the initial Yerkes staff.[13]

Two years older than Hale, Frost was a graduate of Dartmouth, where his father had been a professor and then dean of the Medical School. Charles A. Young was a family friend, and after finishing his bachelor's degree, Frost went to Princeton and studied astronomy with him there, more or less as an unofficial graduate student. Next he went to Europe for two years, visiting several observatories and then spending a year at Potsdam, the capital of German astrophysics. He worked with Hermann C. Vogel, the Potsdam observatory's director, and Julius Scheiner, a leading spectroscopist, and after returning to Dartmouth as an assistant professor in 1892, he translated Scheiner's treatise on astronomical spectroscopy into English. It was for many years the main book on the subject. Frost had been at the World Congress in 1893 and the dedication in 1897; he agreed to take the Yerkes position in 1898 on condition that he be allowed to go back to Dartmouth each winter to teach the astronomy course, as he did until 1902.[14]

Frost's main task, assigned by Hale, was to measure the radial velocities of stars spectroscopically with the forty-inch. He began this program using the universal spectrograph, but the initial results were not very good. One reason was that the spectrograph was used alternately for this program and for Hale's program on carbon stars, necessitating frequent changes in the adjustments of the prisms, lenses, and cameras, all fatal to precise measurements, which demand a stable system. Just as Frost started the program, William Wallace Campbell at Lick Observatory published a classic paper reporting the first results of the radial-velocity program he had just begun there. He had designed and built a spectrograph optimized for the one function of measuring radial velocities and nothing else. With his engineering background, single-minded drive, determination to succeed, and dictatorial personality, Campbell was himself optimized to carry out this program, and Mount Hamilton was a superior site for doing it. For the rest of his career Frost was vainly trying to bring his radial-velocity program up to the level of the Lick program, but Campbell, who became the director there in 1901, always kept ahead of him.

In 1899 Walter S. Adams came to Yerkes as one of its first graduate students. The son of American missionaries in Syria, he had been educated in New England, to which his parents had returned when he was eight. He had studied astronomy under Frost at Dartmouth, graduated in 1898, spent a year on the campus of the University of Chicago studying mathematics and celestial mechanics, and then moved on to Williams Bay. Adams worked with Frost on the radial-velocity program. They both recognized the need for a spectrograph specially designed for this work, and in 1899 Hale persuaded Bruce to provide the necessary funds to have one built. It cost her $2,300; she had earlier donated $10,000 to enable Barnard to have a ten-inch wide-field photographic telescope built for his work. Yerkes Observatory depended heavily on her money in its early days, but she died in 1900, leaving the Bruce spectrograph and the Bruce photographic telescope as her legacy. Frost and Adams began using the new spectrograph in 1901 and got better results, but still not comparable with those from Lick.[15]

Another brilliant young astronomer—Frank Schlesinger—came to Yerkes a few years later, supported by the first grant Hale succeeded in obtaining from the new Carnegie Institution of Washington. Born in New York of German immigrant parents, he had done his under-graduate work at City College of New York, then worked full time several years while doing graduate studies at Columbia. Eventually he got a fellowship that allowed him to study full time. He came to Yerkes as a summer research assistant in 1898 and wanted to stay, but Hale could not raise the money to keep him there, so Schlesinger went back to Columbia. He became an expert in measuring stellar positions accurately on photographic plates taken at the telescope. Schlesinger wanted to measure the distances, or "parallaxes," of stars in this way, from their minute changes in position as photographed from the earth as it moves in its orbit around the sun. The forty-inch refractor, with its long focal length, was ideal for this work. Finally, when Hale got the Carnegie grant in 1902, Schlesinger was able to come to Yerkes and begin this very promising program.[16]

Hale himself was less enthusiastic about the forty-inch refractor than Burnham, Barnard, and Schlesinger. He knew that a large reflecting telescope would be much better suited for his astrophysical research than the giant refractor, and he had published a paper to this effect early in 1897, even before the forty-inch was dedicated. In it he detailed all the advantages of a reflector for astrophysics: that a parabolic mirror, unlike a lens, brings light of all colors to the same focus, that is, it is free of chromatic aberration; that a large mirror absorbs and thus loses less light than a lens of the same diameter, particularly in the violet and ultraviolet spectral regions; that it is possible to use a mirror with a much shorter focal length, thus keeping the tube of the telescope short and therefore lighter and less expensive; and that the cost of a completed parabolic mirror was considerably lower. He especially mentioned the great advantage of a reflecting telescope with a coudé mounting, in which the light from a star is brought down the polar axis, which only turns but does not move, to a large fixed spectrograph, mounted in a laboratory-like room below the ground. With this arrangement it was possible to use very long focal-length cameras in the spectrograph, spreading out (or "dispersing") the light of a star so its individual spectral lines could be analyzed in detail. No similarly long focal-length camera could be mounted on a telescope that moves, like the forty-inch, for it would be badly affected by "flexure," bending as the telescope moved during the long exposure and ruining the definition of any truly high-dispersion spectrograph. Always the diplomat, Hale also pointed out the advantages of a long-focus refractor for the kind of astronomy Burnham and Barnard did and concluded that an observatory needed both kinds of telescopes.[17]

In 1892, after Hale had accepted his appointment to the University of Chicago faculty but before he and Harper called on Yerkes, the young director had planned a new university observatory built around a thirty-inch reflector.[18] That plan went out the window when "the tycoon" came through with his pledge to finance building the forty-inch refractor, but Hale did not forget it. He helped Ritchey (who was to have made the thirty-inch mirror) order a twenty-four-inch glass disk, which the optician turned into an excellent telescope mirror in his own shop at his home over the next few years.[19] Raising his sights, Hale persuaded his father to order and pay for a sixty-inch disk in 1895. (The only supplier of large glass disks was in France.) It belonged to Hale, and in 1897, at the dedication, the visiting scientists saw Ritchey rough grinding it into a parabolic mirror in the Yerkes optical shop. That was for a telescope far in the future, but soon after the forty-inch went into operation, Ritchey began building the twenty-four-inch mirror he had made, and now sold to the university, into a reflecting telescope. Wadsworth made the original design for it, which Ritchey modified after the engineer's angry departure. When completed, the new twenty-four-inch was mounted in one of the smaller domes of Yerkes Observatory. This reflector had a fast optical ratio, f/4, making it ideal for photographing faint nebulae. After he had completed it in 1901, Ritchey took a spectacular series

Figure 11   Yerkes Observatory staff and visitors (1898). In group of five at top, George E. Hale is farthest right, then Edwin B. Frost, then Edward E. Barnard. Seated in front of them are Ferdinand Ellerman (left) and Frank Schlesinger (right). Seated on wall to right of them are John A. Parkhurst (left) and George W. Ritchey (middle). Courtesy of Yerkes Observatory.

of photographs of nebulae with it, demonstrating to skeptical astronomers just how useful reflecting telescopes could be.[20]

As he was completing the twenty-four-inch reflector, Ritchey was also designing a sixty-inch reflecting telescope, based on his own ideas from the beginning, as well as Hale's. According to the original design, it was to be erected in a new, separate dome at Yerkes Observatory. That was for the future though, after a donor had been found to provide money to build the telescope. Hale was the first director to grasp the principle that a large "objective," like the forty-inch lens or the sixty-inch mirror, provided a tangible, understandable graphic demonstration that a telescope would be built, much more convincing to a rich but scientifically illiterate financier than a hundred plans and testimonials.

Hale abhorred anarchy and was a master at organization. Just as his own father had consolidated the elevator business in Chicago and Yerkes consolidated street railway, underground, and traction company franchises in many of the large cities of the world (including London and Philadelphia as well as Chicago), Hale burned to consolidate astronomy. He hated to see individual astronomers working on their own little projects, often at cross-purposes, rather than on a common front under wise, overall direction—preferably his own. The World Congress of Astronomy and Astrophysics of 1893, the *Astrophysical Journal,* founded in 1895, and the sessions for papers at the dedication in 1897 had all been steps in this consolidation. The Yerkes meeting had been a great success.[21] Many of those present thought there should be a second "conference" or meeting the following year. Hale publicized the idea in the *Astrophysical Journal* and kept it alive through his voluminous correspondence.

Edward C. Pickering, another great organizer, invited the astronomers and astrophysicists to the "Second Annual Conference of Astronomers and Astrophysicists" at Harvard in August 1898, just before the meeting of the American Association for the Advancement of Science (AAAS), which many of them would attend. About ninety-five scientists turned up for the Harvard astronomy meeting. In addition to the papers, at one of the sessions (at which Hale was presiding) the question of forming "an astronomical and astrophysical society" was discussed. There was unanimous agreement that the annual meetings should be continued, either ad hoc or under the auspices of a permanent society. Hale referred the question to a committee, consisting of Pickering, Simon Newcomb, George C. Comstock, Edward W. Morley, and himself. Pickering and Newcomb were the two outstanding leaders of American astronomy. Comstock, director of the University of Wisconsin's Washburn Observatory, was a positional astronomer with legal training, who could be counted on to draft a constitution. Morley was Michelson's collaborator in the famous ether-drift experiment, and Hale, the youngest of the group, was the dynamic pusher behind the scenes. The committee reported back the following day, suggesting that a good deal of their work had been done in advance. They recommended forming a permanent society and presented the first draft of a constitution. A few days later sixty-one charter members signed up in anticipation, though the society had not yet even been named.[22]

Naming the society was in fact the most controversial part of organizing it. Hale insisted that astrophysics should be included in the title, to make the physicists welcome. He proposed the "Astronomical and Astrophysical Society of America." Newcomb and other astronomers of the old school were not happy about this idea and preferred "American Astronomical Society." But Hale held out, and in the end his version was adopted.[23] (Putting "American" first would have caused confusion with the already existing AAAS.) The name was ratified at the "Third Conference of Astronomers and Astrophysicists," which was also labeled the first meeting of the new AASA, held at Yerkes Observatory September 6–8, 1899. About fifty astronomers and physicists were present, and thirty-one papers were read. By then, a grand total of 113 charter members had signed up, including all the members of the Yerkes faculty except Burnham, who never joined the society, probably because of its astrophysical implications.[24] Over the years, astrophysics gradually became accepted by everyone as "real" astronomy. Even the astrophysicists were willing to change the awkward title, and although Newcomb's original favorite, the "American Astronomical Society," lost on a vote in 1909, it was adopted nearly unanimously in 1914.[25]

The year after the founding of the new society, Yerkes sent its first expedition to a solar eclipse. It was on May 28, 1900, in the southeastern United States, and nearly every major observatory had a camp somewhere along the narrow track of totality. The Yerkes group, led by Hale, also included Barnard, Frost, Ritchey, and Ellerman. They set up their instruments at Wadesboro, North Carolina. Hale and Ellerman were trying to measure the infrared radiation of the corona with a bolometer, using a fixed, horizontal long-focus reflecting telescope fed by a two-mirror coelostat. However, a series of accidents beginning with damage to the bolometer en route to Wadesboro and ending with someone's accidentally kicking one of the coelostat mirrors just as the eclipse began, ruined this attempt. Nevertheless Frost obtained a number of good spectrograms of the corona and the chromosphere, and Barnard and Ritchey, using another fixed, horizontal telescope, this one with a six-inch, sixty-one-foot focal-length lens, corrected for photographic light, took several excellent photographs of the corona.[26]

These coronal images, which showed more fine detail than even the Lick expedition's, taken with its forty-foot focal-length lens, led Ritchey to propose building a monster two-hundred-foot focal-length horizontal reflecting telescope, to be permanently mounted at Yerkes Observatory and used for lunar and planetary photography, double stars, and globular clusters. It would be built around a twenty-inch

mirror so that it would be achromatic. Hale wanted to use such a telescope with physicist Ernest F. Nichols of Cornell, who had developed
a sensitive radiometer, to measure the heat radiation of stars. Most
important, it would be perfect for use with a large, heavy fixed spectrograph to obtain high-dispersion spectra of stars, still Hale's chief
aim. He succeeded in getting a grant of $500 from the AAAS, and he
poured Ritchey's time and the resources of the observatory into building it, illustrating how much less expensive a reflecting telescope of
this type is than an equatorially mounted refractor. The telescope,
eventually made with a twenty-four-inch aperture and 165-foot focal
length, was completed and installed late in 1902, in a long wooden
shed on the observatory grounds. Adams, who had gone abroad for
an academic year as a graduate student in Germany, had returned
to Yerkes in the summer of 1901 and was soon back at work with
all the observatory's spectrographs. Assisted by Ellerman, he began
testing the horizontal telescope, but in December 1902 a short cir-
cuit in the wiring set the wooden building on fire. It was completely
destroyed, along with the optics and grating of the telescope and
spectrograph.[27]

By then Hale had definitely had enough of trying to do astrophysics
in Wisconsin. He felt the severe winters were bad for his daughter's
health, and he knew that the clear skies and steady atmosphere
("good seeing," in astronomical terms) of California were much better for observing.[28] His father died in 1898. It was a terrible blow to
Hale and for a time made him question the reason for existence. Then
in 1899 his mother died. They were the last anchors holding him in
Wisconsin. His father had considered Williams Bay almost a suburb
of Chicago, but not so for California. As his father lay dying, Hale
wrote a legalistically phrased letter to Harper. In it he reminded him
that the sixty-inch mirror, which Hale stated was "now approaching
completion," belonged to his father, who had paid for the disk and
had provided Ritchey's salary while he worked on it. His father, Hale
said, now offered it to the University of Chicago, on condition it build
a suitable building and supply a staff to operate and maintain it. He
estimated the value of the sixty-inch mirror as $10,000 and the
amount the university would have to pay for the building, mounting,
and endowment at $60,000 to $75,000.[29]

Hale knew there was no chance that Harper would allot this
amount of money to Yerkes Observatory. There were too many other
hungry mouths to feed at the University of Chicago. The mirror may
have been approaching completion, but it was still very far from it.

Hale had written the letter to prove he had given the president the chance to keep the sixty-inch as part of the University of Chicago. Harper could not act, and though Hale remained on the faculty and worked through the president, he now felt free to take the sixty-inch elsewhere if he could bring about a better "scheme" (one of Hale's favorite words) to do so. He had undoubtedly decided, soon after his father died, to move it to California and erect it there. In January 1899 Hale asked N. B. Ream, a Chicago businessman and friend of his father, to contribute $130,000 to build the sixty-inch telescope and observatory in southern California and provide the funds to operate it. Hale's plan was that one staff member at a time would go to the proposed observatory, work there for a few months, and then bring his data, chiefly photographic, back to Chicago or Yerkes to work it up while another observer was at the telescope.[30] Ream did not come through with the donation, and next Hale prepared a prospectus for Harper to use to convince Yerkes to provide the necessary money for the southern California station. The "titan," however, was still recalcitrant, and nothing came of this approach either.[31] Keeler, by now director of Lick Observatory, invited Hale to bring the sixty-inch to Mount Hamilton and locate his remote station there. Hale, however, held out for southern California, where he knew the observing weather was better.[32]

In early 1902 Hale prepared a long memorandum for Harper to use in seeking $100,000 to $200,000 from Isabel Blackstone, a recently widowed (and very wealthy) Chicagoan, to complete the sixty-inch, erect it in southern California, and endow its operational costs.[33] She declined, but by now the fabulously rich Andrew Carnegie had set up his Carnegie Institution of Washington, with an initial capital of $10 million to support research. Hale quickly sent its executive committee an expanded version of the same proposal, complete with some striking photographs of nebulae taken with the twenty-four-inch reflector and less striking ones of the same objects taken with the longer focal-length but slower forty-inch refractor. They graphically illustrated his point, the need for a large, fast reflector at a clear mountain site in California.[34]

The Carnegie Institution formed an advisory committee on astronomy, with Pickering as its chairman and Newcomb, Samuel P. Langley, Lewis Boss, and Hale as members. In the end each of them got a grant to support his own research, but Hale got the biggest one. He submitted a proposal asking for funds to support three smaller projects at Yerkes, one of them Schlesinger's parallax work and another

Adams's high-dispersion stellar spectroscopy with the horizontal telescope (this was before the fire), and also the much larger project of
mounting the sixty-inch "at some more favorable point in the northern hemisphere, and ultimately in the southern hemisphere."[35] Hale
threw all his energies into persuading the advisory committee to recommend the sixty-inch, and in the end he was so successful that it endorsed the project even though he theoretically "withdrew" the proposal. He did so to avoid any charge of conflict of interest, but only
after he knew the other members would support it.[36] The committee
organized a rapid site survey of southern California, carried out by
William J. Hussey of Lick Observatory, under Campbell's direction
but with reams of advice from Hale. In the end it recommended
Mount Wilson, near Pasadena, undoubtedly the location he had favored from the first. It had been the planned site for the failed University of Southern California observatory for which the forty-inch
lens had originally been ordered, and Harvard had had a temporary
station there just after the solar eclipse of January 1, 1889, when Hale
had been working under Pickering's tutelage in Cambridge.[37]

While the site survey was under way in 1903, Hale had Ritchey
build a new horizontal telescope and spectrograph to replace the ones
destroyed in the fire. He had first tried to get the funds for this smaller
telescope from Isabel Blackstone, after she had declined to finance the
sixty-inch project, but again her answer was a polite, regretful no.[38]
Next Hale approached Helen Snow, another wealthy Chicagoan,
whose nephew, George S. Isham, was an astronomy buff. He had accompanied the Yerkes party to the Wadesboro eclipse in 1900. This
time Hale did succeed, and the horizontal telescope was built as the
Snow telescope, in memory of the donor's father. It had a twenty-
four-inch mirror and a focal length of sixty feet. This time it was
mounted in a long wooden house, on stilts above the ground to minimize thermal distortions and air currents. Adams tested it and had
obtained good spectrograms of the sun with it by the end of October 1903.[39]

By this time Hale had made up his mind to leave Yerkes and start a
new observatory on Mount Wilson somehow or other.[40] He was tired
of Williams Bay, tired of Chicago, tired of the cold, and tired of President Harper, who constantly thwarted his plans by allotting money
to other departments and other professors' projects instead of his
own. The grants for the smaller projects had come through from the
Carnegie Institution, and Hale had been able to hire three additional
young staff members, including Schlesinger, without Harper's help.
Undoubtedly Hale felt that the president was cramping his style, and

Figure 12    George E. Hale in his office at Mount Wilson (ca. 1904). Courtesy of the Observatories of the Carnegie Institution of Washington.

he was eager to get away from him.[41] In mid-December, taking his family with him, he suddenly left Chicago and moved to Pasadena. There, on Christmas Eve, he exulted in the sun, the birds singing, the flowers in bloom, the orange trees loaded with fruit, and his happy, healthy children playing outside in their shirtsleeves. He could not help contrasting it with Williams Bay, where the temperature had been −20°F the previous week. He was convinced that "the place to continue my solar work is on the summit of Mt. Wilson," which he could see, eight miles away, from his bedroom window. Hale was determined to bring the Snow telescope out in the spring and to get the observational results that would convince the Carnegie Institution officials to grant him the money for what he called "the solar observatory." Actually its main instrument was to be the sixty-inch reflector. At the least, he felt almost certain they would provide the funds for him to stay in California for several years, operating the Snow telescope as an "expedition" from Yerkes Observatory.[42]

Everything worked out pretty much as Hale wished. Helen Snow insisted she had given the Snow telescope to be used at Williams Bay, not in California, so Hale had to leave it temporarily at Yerkes, but he had the old six-inch lens sent out instead. He was sure he could persuade her to change her mind the next time he got back to Chicago—as in fact he did.[43] Yerkes, probably out of pique at Hale's departure, stopped his payments of $2,000 a year for two assistants' salaries, the only operating expenses he had been providing. Hale managed to find enough to pay Adams, who was one of them, but he let the other assistant, who had "done almost nothing of importance," go. Ellerman came to California in March to set up the solar telescope and start observing with it, and when he went to Williams Bay to bring his family back with him, Adams replaced him on Mount Wilson. Hale decided he needed a shop in Pasadena and sent for Ritchey to set it up in May. Barnard came in the summer, bringing the Bruce photographic telescope to photograph the Milky Way from the clear mountain site. Hale had stationery printed whose letterhead proclaimed he was at the "University of Chicago, Founded by John D. Rockefeller, Yerkes Observatory, Mt. Wilson Station, Office of the Director, Mt. Wilson, California." Hale was heavily committed to the Carnegie Institution and had run up bills for $30,000 to build the observatory, on the expectation that they would make a grant to him which would more than cover it. If not, he would have to pay it from his own personal income, which was considerable since his parents' deaths.[44]

Never one to sit idle, at the same time he was building the observatory on Mount Wilson Hale was organizing an international Conference on Solar Research, to be held at the St. Louis World's Fair in September 1904. Both the fair and the International Congress of Science, of which the Conference on Solar Research was a part, were modeled after those in Chicago a decade earlier. Among the foreign scientists present at the Conference on Solar Research were French mathematician Henri Poincaré and Dutch statistical astronomer Jacobus Kapteyn, along with some legitimate astrophysicists. Hale gave a paper titled "Co-operation in Solar Research," and the group decided to form the International Union for Cooperation in Solar Research. Hale was elected its president on the spot, Poincaré its vice president. Frequently called the International Solar Union, it continued to meet at three-year intervals, each time in a different country, until World War I. After the war it was the nucleus from which the International Astronomical Union of today was formed.[45]

In October Hale received encouraging news. Charles D. Walcott, secretary of the Carnegie Institution, told him that more money than expected would be available and encouraged him to apply for it. Hale presented his "big scheme" for $65,000 a year for five years to build an independent Mount Wilson Solar Observatory around the sixty-inch reflector. After an agonizing wait, just before Christmas he learned that the Carnegie Institution's executive board had approved it and he would get the money.[46]

He sent his resignation from the University of Chicago faculty to Harper, but the president, hoping to change Hale's mind, did not pass it on to the Board of Trustees. Harper, only forty-eight years old, was dying of cancer. He tried to persuade Hale to stay, and at one point he apparently suggested or stated that the Yerkes director owed the University of Chicago his and Ellerman's salary for part of the time they had been on Mount Wilson. Hale at once took his correspondence with Harper to his own attorney, who produced a written opinion that he had acted honorably and had no obligation, legal or moral, to make restitution. Harper now accepted the inevitable, presented Hale's resignation to the trustees, and sent him a handsome valedictory statement. Nine months later the president was dead.[47]

As soon as he received the Carnegie Institution grant, Hale switched Ritchey, Ellerman, and Adams to its payroll. They all were glad to become members of the Mount Wilson Solar Observatory staff. The first team was leaving Yerkes to found a new research institution, which would soon have the largest telescope in the world. There was no doubt it would be enormously successful. The second team was left behind at the seven-year-old observatory, which had lost its dynamic director. Its future was more in doubt.

CHAPTER THREE

# Near Death, 1904–1932

E dwin B. Frost became the astronomer in charge at Yerkes Observatory when George Ellery Hale left hurriedly for Pasadena, just before Christmas 1903. Hale remained director, however, and in theory was expected to return. Frost kept him fully informed on important events at the observatory, including George Willis Ritchey's attack of appendicitis, his dangerous operation, and his slow recovery and the visit of Carl A. R. Lundin, successor to Alvan G. Clark, to modify and improve the support cell of the forty-inch lens. In March, when Hale received an interim grant from the Carnegie Institution for development work on quartz mirrors, he assigned Ritchey to carry it out (as soon as he was well), just as he sent for Ferdinand Ellerman and Walter S. Adams when he needed them on Mount Wilson.[1] When Charles T. Yerkes stopped his $2,000 a year payments for assistants, Hale advised William Rainey Harper on how to continue Adams's salary, and he strongly urged the president to provide a salary for Sherburne W. Burnham (who had finally retired from his court reporting job) and to grant Ritchey a promotion. Harper was not sympathetic and asked Hale when he planned to come "home" from the "seductive . . . climate of California."[2]

Hale did come back to Chicago and Williams Bay briefly in the spring. In conferences with Harper on the campus and Frost at the observatory, he made it clear that he intended to return to California and remain there for some time, operating the remote station on Mount Wilson. He advised Harper to appoint Frost acting director at Yerkes, and on departing for Pasadena after a stay of only a few weeks, he told the spectroscopist that he was in charge. Frost, however, initially received no grant of authority from Harper (it was confirmed in June), and Hale continued to bombard him with letters reassigning staff members and asking for lenses, mirrors, gratings, and tools.[3] Harper, who was operated on for "appendicitis" in May (actually cancer, though he himself did not then know the diagnosis), moved to Hale's vacant house in pleasant southern Wisconsin to recuperate in July and stayed until December 1904. He used Hale's office in the observatory building but did not interfere in the observatory's operation.[4]

When Hale was certain he would get his grant from the Carnegie Institution for the "big scheme," enabling him to break away from the University of Chicago and found the separate Mount Wilson Solar Observatory, he warned Frost to nail down support from Harper before the word got out. But Frost was much less effective than Hale in such negotiations. When the grant did come through, the Carnegie Institution withdrew its support from the projects Hale had started at Yerkes. Harper, angry that his young star was deserting him, threatened to cut the Yerkes budget still further. Frost was caught in the middle. Four of the Yerkes staff members, one of them Frank Schlesinger, did not receive their salary payments for January 1905. Harper's position was that they were the Carnegie Institution's problem, not his. Hale and Frost shifted funds around (including $500 of Hale's own money) and managed to cover them temporarily, but everyone left at Williams Bay was depressed.[5] Schlesinger, a married man with children, needed to be sure he would be paid every month. Though he wanted to stay at Yerkes, he jumped at a job offer from the smaller Allegheny Observatory and did his pathbreaking parallax work there. Frost was bitter toward the Carnegie trustees, who had turned down his request for the funds needed to keep Schlesinger at Yerkes.[6]

Hale produced an accounting showing that the University of Chicago had spent only $337.20 for Ritchey's time working on the sixty-inch mirror. It still belonged to Hale, and he sent Ritchey with a check for this amount, asking him to pick up the mirror and bring it to Pasadena with other tools and supplies he was buying in the East. To Frost the amount seemed far too small, but Hale insisted he had figures to prove it. In the end Frost gave in and certified the figure as correct, and the university turned the mirror over to Hale.[7]

As soon as the Snow telescope arrived in California, Hale had it set up and Adams began using it. Hale had originally said he wanted to borrow it and would return it to Yerkes Observatory, but as soon as it was installed on Mount Wilson he raised the idea of buying it, because it would be so much more useful there and would produce so much better results. Helen E. Snow, the donor, was willing, if Frost was agreeable and Yerkes Observatory got a fair price for it. George S. Isham, her nephew and Hale's friend, handled the negotiations. He suggested $10,000 as a fair price, but Hale came up with a figure of $5,700 and argued so vociferously for it that in the end Isham, Frost, and Martin Ryerson, president of the University of Chicago Board of Trustees, all gave up and agreed to his price. Hale used all the ruthless tactics of polite intimidation and tenacity he had learned from his businessman father to force the price down to his own starting level.[8]

Frost, a gentler soul, found it harder to argue for his observatory against his younger friend, who had given him his job there. Hale, a wealthy man, knew that his own future was assured whatever happened. Frost had to worry about his own salary as well as those of his staff members, and he recognized that the personal gifts Hale had left him, including much of his own library, might be seen by others as clouding his judgment.[9]

In January 1905, just after Hale's declaration of independence, Edward E. Barnard had his ten-inch wide-field Bruce photographic telescope shipped to Mount Wilson. He had gotten $10,000 to build it from Catherine W. Bruce shortly before her death, and after a prolonged investigation of several rival optical firms, had had John A. Brashear construct it for him. Mount Wilson, with its clear sky and transparent atmosphere, was a better site for it than Williams Bay. Frost had been unable to raise $500 to finance Barnard's expedition, but Hale easily got $1,000 in Pasadena for this purpose. Barnard stayed at Mount Wilson for nearly a year and obtained two-thirds of the photographs there that were published posthumously in his *Atlas of the Milky Way* many years later. Frost and others at Yerkes worried that he would not return, and Barnard probably would have stayed if Hale had offered him a job. But this time Hale was determined to have a real astrophysical observatory.[10]

In January 1905, with Hale settled in California and Harper back in Chicago, the geologist Thomas C. Chamberlin spent the winter quarter at Williams Bay, occupying the director's office (which he liked to call his "Telluric Annex"). He was working on his "planetesimal hypothesis" of the origin of the solar system and wanted to escape from distractions of the campus (chiefly students!). Chamberlin considered Yerkes Observatory "an atmosphere of tranquility" and joked to Frost that he was "invading, like the Vandals of old, the temple of Urania."[11]

Harper, after considerable resistance, finally accepted Hale's resignation in March 1905 and moved Frost up from acting director to director, effective July 1, 1905. Although the president had periods of apparent remission, the course of his disease was unrelenting. He died in January 1906, leaving a great university as his monument.[12]

Frost was to retain the directorship for twenty-seven years. During that entire time he brought no new scientific ideas to the observatory. His policy seemed to be to keep Yerkes Observatory as nearly as possible the same as it had been when Hale left and he himself took over. In 1904 the University of Chicago published (a year late) its Decennial Publications, a series of volumes, one for each major department, in which the professors summarized their recent research findings at length. The Yerkes volume contained papers by Hale, John A. Parkhurst, and Ellerman on the spectra of carbon stars, by Frost and Adams on their radial velocity work, by Ritchey on his astrophotography with the forty-inch refractor and the twenty-four-inch reflector, and by Barnard on his wide-field photography of the Milky Way.[13] It was the second volume of the *Publications of the Yerkes Observatory;* the first had been Burnham's double-star catalog, published four years earlier.[14] These were the programs Frost continued until he reached retirement age.

Frost himself remained in charge of the radial-velocity program, assisted by a succession of younger staff members such as Philip Fox and Oliver J. Lee, all far less gifted than Adams. The Yerkes program specialized in what were then called early-type stars, of spectral types O, B, and A, which turned out to be important astrophysically because they are hot and luminous. They are observable even at great distances and are relatively young, as the Yerkes radial-velocity data eventually helped to prove. There are also many spectroscopic binaries among them, too close to be seen or photographed separately and hence appearing as a single star. However, these pairs of stars are actually in bound orbits about one another, revealed spectroscopically by the periodic variations in the measured velocities that go with the motion. Frost and his collaborators, who included many of the graduate students who passed through the observatory, discovered many such spectroscopic binaries. But at Lick Observatory, William Wallace Campbell was carrying out a similar program with its thirty-six-inch refractor. Although his telescope was smaller, Campbell's spectrograph was considerably better for this type of work. Mount Hamilton, with its long periods of clear skies and fine seeing, was a greatly superior site. Campbell's drive, engineering approach, and ruthlessness in pushing himself and his assistants made the Lick radial-velocity program much more successful than Yerkes Observatory's. Campbell's radial-velocity factory clearly outclassed Frost's cottage-industry program.[15] Within a few years Adams had a radial-velocity program started with the sixty-inch reflector, completed in 1908, on Mount Wilson. This bigger telescope on a better site could produce even better spectrograms faster, so although many other frontier astrophysical problems competed for telescope time at Mount Wilson, its radial-velocity program also outclassed the one at Yerkes.[16] Yet Frost apparently never considered breaking out into newer, more speculative lines of investigation with the spectrograph on the forty-inch refractor.

Burnham continued his double-star measurements long after he had retired from his court job in 1902. That gave him time to compile his four-volume *General Catalogue of Double Stars*,[17] published in 1906. He kept on observing until he retired from his professorship in 1914, at age seventy-six. He had gone on the payroll in 1902 but had finally become too feeble to make the weekly trips from Chicago to Williams Bay and put in his two nights observing. The eagle-eyed observer had written an old friend, "I can't write worth a cent nowadays. . . . I haven't been able to go anywhere, and so have cut out the Y. O. As I am fired on 1st July, I shall probably never go there again." Nevertheless, he lived on another seven years in Chicago and received an honorary doctor's degree from Northwestern University before he died in 1921.[18]

Soon after Burnham retired, Frost hired George Van Biesbroeck to replace him as a visual double-star observer on the Yerkes faculty. Van Biesbroeck, a Belgian, had started with an engineering education and had worked several years as a government civil engineer in Brussels. But then he had shifted to astronomy and joined the staff of the Royal Observatory in 1908. In 1915, after the Germans invaded Belgium, he had managed to get out and became a visiting faculty member at Yerkes. In 1917 Frost had him appointed to the regular faculty. Double stars were his first love, but he also worked on positional measurements of asteroids and comets.

Barnard's main effort went into his long-exposure photographic program on the Milky Way. He discovered many dark markings on it, which he first believed were voids—long tubular regions empty of stars, pointing at the sun's position in the Galaxy. Gradually he came to realize that they represented obscured regions, nearby dense dust clouds that cut off the light of stars behind them. It was a very impor-

Figure 13  Edward E. Barnard (1917). Courtesy of Mary Lea Shane Archives of the Lick Observatory.

tant step in understanding that there is interstellar matter in the Galaxy, strongly concentrated to its central plane. Barnard also did a great deal of visual observing with the forty-inch refractor, searching for undiscovered faint satellites of the planets and measuring the positions of the known ones. Toward the end of his life he suffered from diabetes and kidney disease; he died in 1923 at age sixty-five. Always impatient, Barnard had built his own house on a lot he bought near Yerkes Observatory so that he could hurry to the telescope as soon as the skies began to clear. In his will he left the house to the observatory after his wife's death, and many famous astronomers have lived in it since his time.[19]

After Barnard's death, Frost wanted to hire a new staff member to carry on his work. The nation was prosperous, and the University of Chicago endowment was doing very well. Dean Henry G. Gale, head of the Physical Sciences Division, advised Frost that he might think in terms of adding one or two new faculty members at Yerkes. He urged the director to seek a senior astrophysicist, whom he could appoint as a full professor. Frost, however, could think only in terms of continuing the immediate past.[20] What he had in mind, in a vague way, was to hire a person who would use the hundreds of Milky Way photographs Barnard had left to make a statistical study of the star distribution in the Galaxy. He knew of no one who was working in this field, but he decided that Seth B. Nicholson, of the Mount Wilson Observatory staff, might be a suitable candidate. Nicholson was thirty-two years old, handsome, intelligent, and a good speaker. His training at the University of California had been almost entirely in celestial mechanics; as a graduate student he had discovered a faint satellite of Jupiter and made determining its orbit his thesis, and he was now working on solar observational research at Mount Wilson. Frost thought that Nicholson could easily switch fields. True to the directors' "no raiding" code of those days, Frost consulted Frederick H. Seares at Mount Wilson before approaching Nicholson. Seares bluntly advised Frost that he doubted Nicholson would wish to learn a wholly new branch of astronomy or to leave Mount Wilson for Yerkes Observatory. That ended that idea, as far as Frost was concerned.[21]

The Yerkes director also consulted Joel Stebbins at the University of Wisconsin, Armin O. Leuschner at Berkeley, and Henry Norris Russell at Princeton. Only Russell was an astrophysicist. Hale was having severe mental problems and had hidden himself away in England, trying to avoid astronomy and astronomers.[22] Stebbins recommended one candidate and Leuschner six, but none of them except Nicholson (one of Leuschner's suggestions) had accomplished much in research. Stebbins and Leuschner did not really regard Yerkes Observatory as a first-class scientific institution but saw it as a haven for their friends or former students who needed jobs.[23]

Russell gave Frost the best advice. He could not recommend any individual, but he told Frost he should not try to get a person who would carry on Barnard's work. What Yerkes needed was not a great observer who would take more wide-field photographs, as Barnard had, with no scientific plan behind them, but rather someone with research ideas for problems to be solved. If the great collection of photographic plates Barnard had left could be used, so much the better, but the research drive was essential. Unfortunately Russell had no suggestions for where to find this paragon. He hinted that such a "first-rate man," if he existed, would probably not want to take the Yerkes job.[24]

Frost ignored Russell's advice (the younger theorist was generally regarded as a talkative busybody by "solid" directors) and went right on looking for a wide-field photographer. He did not find one, but the photographer found him. This was Frank E. Ross, an amusing, irreverent astronomer for all seasons who turned out to be an excellent researcher. Ross had gotten his Ph.D. in mathematics at the University of California in 1901, although most of his work had been in astronomy. An expert in celestial mechanics, he had been Simon Newcomb's chief assistant at the Nautical Almanac Office. Ross continued working there after Newcomb's death but also became director of the International Latitude Observatory at nearby Gaithersburg, Maryland. There he perfected the photographic zenith tube he had invented, an instrument to measure the latitude of a station very precisely and thus to chart, from observations at several stations, the minute motions of the axis of rotation of the earth. Hence he had two essentially full-time paid jobs simultaneously. Ross's strategy was to have his brother work at one or the other of them for him from time to time, to keep the results flowing. As a third source of income, the two of them briefly operated a tobacco shop in Washington, but this proved too much and Ross sold it. Eventually, however, through the machinations of a rival, Ross was forced out of the Nautical Almanac Office just as the International Latitude Program declared its problem solved and closed its observatories.

In 1915 Ross, a born survivor, came up with a new job at the Research Laboratory of the Eastman Kodak Company in Rochester, New York. He published many papers on the physics of the photographic process, especially on distortions of images, the resultant errors in measured star positions, and the best methods to minimize them. Ross also became a lens designer, and during World War I he invented a four-element wide-field lens for aerial photography. It was a great improvement over any wide-field lens then used in astronomy, and after the war ended Eastman Kodak allowed him to publish his design, which became known as the Ross lens. He wanted to get back into astronomy and had been angling for a job at Lick Observatory for years.[25]

Evidently Ross learned that Frost was seeking a successor to Barnard, probably indirectly through Leuschner, his former teacher at Berkeley. Ross began writing to the Yerkes director about new types of photographic plates and an especially fast camera he had developed and recommended for very faint, very wide-field subjects such as the Milky Way and the Zodiacal Light. He asked Frost for specific data on the forty-inch refractor, for a study he said he was making. In these letters Ross subtly emphasized his expertise in astronomical photography and hinted at his eagerness to get back into research.[26] Frost, who had nearly given up hope of finding the right replacement for Barnard, finally realized that Ross might be a possibility. The Yerkes director feared Ross was making so much money at the industrial laboratory that he would not consider a university position, but he decided to ask. Ross telegraphed back ten words (the minimum-rate message): "Your offer attracts me. Write details. Have family small children." He followed it with a confirming letter in which he claimed he was considering offers from Lick and Yale but doubted that his health could stand the rigors of the California climate! Frost was overjoyed. He explained to Ross that he wanted him to work up Barnard's plates, and he dashed off a delighted letter to the dean in Chicago saying he thought he had found the right man.[27]

Ross agreed to carry on Barnard's work but made it clear to Frost that he had ideas of his own. He outlined his plans for taking photographs of Mars in various colors, including the infrared. To Frost's suggestion that he begin a nebular spectroscopy program at Yerkes, Ross replied that "we are handicapped about three miles by Mts. Wilson and Hamilton . . . in the matter of equipment and air, [but] I am perfectly willing to devote my best thought to the problem and see what can actually be done." He was always realistic scientifically, but he was most unbusinesslike. Ross told Frost he would accept an associate professorship at a salary of $4,000 a year. The director wanted to work fast and put the appointment through before Campbell at Lick and Schlesinger at Yale could raise their supposed offers. Frost therefore solicited only one outside letter of reference, from William S. Eichelberger at the Naval Observatory. He knew Ross's skills in orbital calculations very well but knew nothing about his photographic qualifications. Eichelberger praised Ross, and Frost recommended the appointment. By now the University of Chicago was not so flush. The administration held back and suggested that Frost start a fund-raising drive among the wealthy Lake Geneva residents to endow Ross's salary. Otherwise they could not guarantee his appointment for more than one year. Ross, who stopped off to visit Yerkes Observatory on a trip to Los Angeles, made it clear that he would not accept such a short-term post. Then the administration weakened, and the fifty-year-old Ross got his permanent tenure position.[28]

Figure 14   Frank E. Ross at the forty-inch Yerkes refractor (1925). Courtesy of Mary Lea Shane Archives of the Lick Observatory.

The new faculty member arrived at Yerkes with his family on October 1, 1924, and moved into Barnard's house, now the property of the observatory. Ross plunged into a program of wide-field photography, making a systematic survey of the sky, repeating the same fields Barnard had taken years earlier. Comparing the plates taken at the two different epochs with a "blink" microscope (which enabled him to see one exposure and the other in rapid succession, so that any changes leaped to his attention), he discovered many new "high proper-motion stars." They are mostly very nearby stars, whose angular motions in the sky are therefore unusually large, together with a certain fraction of "high-velocity stars" with large space motions. Both groups are interesting from an astrophysical point of view, and these Ross stars served as the grist for many detailed studies over the years.[29]

When he accepted the job at Yerkes, Ross wrote to his good friend William H. Wright at Lick that Frost was "certainly . . . a charming man." Ross went on to say that he was sure that he would enjoy Williams Bay "as a permanent abode, supplemented by trips to more favorable climes for observational data." He made it plain that he hoped to go to California often to observe, and to stay as long as he could. That is just what he did.[30] All the Yerkes astronomers were in theory allowed to be away from Williams Bay for three months each year, although most of them took only a month vacation. Ross was gone at least three months every year and stayed away longer whenever he could get permission to do so. He did a great deal of planetary photography, especially of Venus, Mars, and Jupiter, in various colors, all on visits to Lick and Mount Wilson Observatories, using their big telescopes rather than the forty-inch refractor. The skies were certainly clearer in California, and the seeing was better, but Ross's travels did not help build a cohesive group at Yerkes. After little more than a year on its staff, he had confirmed his original impressions. He was working hard and getting results. But he was determined to spend his three months each year "collecting plates which on account of the poor atmosphere cannot be obtained here." Yerkes Observatory, he concluded, was "a pleasant and easygoing place where we sometimes see the stars."[31]

Ross was constantly trying to get a half-time job at Lick or Mount Wilson. In 1928 his efforts were at last successful. Hale crowned his own career by persuading the Rockefeller Foundation trustees to provide $6 million to build his fourth "largest telescope in the world," the two-hundred-inch reflector that eventually bore his name. Almost immediately Hale directed John A. Anderson, who headed the project, to hire Ross, the best astronomical optical designer in America, to work on it. Frost was agreeable to letting Hale have him half-time, and from then on Ross spent half of each year in Pasadena. By this time he had a five-inch Ross lens and camera of his own design with which he could take photographs of the Milky Way that were even wider field, and better, than Barnard's. The University of Chicago Press published the first part of his *Atlas of the Northern Milky Way* in 1934, and he dated it from Yerkes Observatory, but all twenty plates in it were ones he had taken at Mount Wilson and Lowell Observatories.[32]

Both Van Biesbroeck and Ross turned out to be very good at the research they did as the successors of Burnham and Barnard, but under Frost's faculty replacement policies there was no chance for Yerkes

Observatory to move into new fields. Frost himself suffered an especially grievous tragedy for an astronomer. He had a family history of vision problems and was strongly nearsighted from an early age. Early symptoms of a retina defect in his right eye began when he was forty. Then, on December 15, 1915, while Frost was observing with the forty-inch, he suffered a detached retina in his right eye. His vision in that eye gradually decreased and after about a year was entirely gone. He stopped observing and tried to rest his left eye as much as he could, but in 1918 a cataract appeared in it. This gradually impaired his sight, and in 1921 he suffered a hemorrhage in his left eye. When it cleared he was almost completely blind. He estimated his remaining vision in that eye as 3 or 4 percent. From then on students, secretaries, and colleagues read letters and scientific journal articles to Frost. He had a fine memory, and his blindness helped him develop it to an even higher level. But it was not the same as being able to see, and he became more and more out of touch with what was going on in astronomy and science outside Yerkes Observatory. Frost had not gone to many astronomical meetings after he became director; now he attended even fewer. He delighted in speaking to school groups and writing admirable little essays on nature for the *Chicago Tribune.* He was a wonderful, warm, sympathetic human being, but he was not doing his job as the head of what the University of Chicago wanted to be one of the outstanding astronomical research institutions in the United States. Nevertheless Frost continued as director for eleven more years, until he reached retirement age in 1932. He had brought few new ideas to Yerkes Observatory before 1921, but after that he brought none.[33]

In this Frost was not completely unlike the president of the University of Chicago, Harry P. Judson. Seven years older than Harper, Judson had started at Chicago as a professor of political science and dean of the colleges when the university opened in 1892. In 1894 he had stepped up to head of the Department of Political Science and dean of the faculties. Judson was thus Harper's right-hand man and had taken over as acting president during the latter's final illness. Soon after Harper's death in 1906, the trustees appointed Judson as the second president in the university's history. He was then fifty-seven years old, and he continued in office until he retired in 1923 at age seventy-four. Judson's successor was Ernest D. Burton, a theologian. He had also been on the university faculty since the opening, as a professor of New Testament, and had comforted Harper on his deathbed. Burton was sixty-seven when he became president but died in harness just

Figure 15   Harry P.
Judson, second pres-
ident of the Univer-
sity of Chicago.
Courtesy of the De-
partment of Special
Collections, Uni-
versity of Chicago
Library.

two years later. Next came Max Mason, a mathematician and theoretical physicist who had spent almost all of his career as a professor
at the University of Wisconsin. Mason was forty-eight when he became president at Chicago in 1925, but he lasted only three years. His
personal life was "very disturbed" during this period, with his wife
dying of cancer and rumors permeating the campus of his liaison with
the attractive spouse of a professor on his faculty. Mason's term as
president was "an exciting adventure," which ended with his resigning in 1928 to accept a top position in the Rockefeller Foundation.
He became its president in 1930 and then left in 1936 to become vice
president of the California Institute of Technology. There he was effectively the Rockefeller Foundation agent who presided over the
completion of the Palomar two-hundred-inch telescope, which Hale
had begun.[34]

As presidents, both Judson and Burton had followed laissez-faire policies, allowing each department to go its own way with little direction from above. Yerkes Observatory, seventy-five miles from the campus, was especially easy to overlook. Frost allowed it to drift slowly downhill as an effective force in astronomical research. Mason was too preoccupied and disorganized to attempt to seize control, and the drift continued. Frost's main contact with the higher levels of the administration was through Frederic Woodward, who became vice president in 1926. Then fifty-two, he was a law professor who had taught at Dickinson College, Northwestern, and Stanford (where he was dean of the law school) before coming to Chicago in 1916. Woodward was a suave, diplomatic believer in the status quo. He did not make waves at Yerkes Observatory either, as it drifted on.

As the University of Chicago grew, its administration grew with it, and from 1922 on Frost reported through a dean rather than directly to the president. The dean was Henry G. Gale, a physicist. Eight years younger than Frost, Gale was born in Aurora, Illinois, almost a suburb of Chicago. He had entered the University of Chicago in its first class in 1892, receiving his bachelor's degree there in 1896 and his Ph.D. in 1899. Gale immediately became an assistant in the Physics Department, was promoted to instructor in 1902, and spent his entire life at the University of Chicago, going up through the ranks to full professor in 1916. He worked as a junior faculty assistant to the two great research scientists who were America's first Nobel Prize winners in physics, Albert A. Michelson and Robert A. Millikan. Gale resented them both, believing they had exploited him and prevented him from getting the credit (and salary raises) he thought he deserved.[35] His specialties were optics and spectroscopy, but he did very little research on his own. Chiefly Gale devoted himself to teaching, administration, and revising the immensely popular high-school and college physics texts known to a generation of students as "Millikan and Gale." A varsity football player and captain of the Maroons in his senior year, after graduation Gale earned pocket money by refereeing college football games in the Chicago area. During World War I he rose to the rank of lieutenant colonel in the army, improved the tables used in long-range artillery firing, and served in France, where he was cited by General Pershing and received the French Legion of Honor.[36]

Gale and Adams had been fellow graduate students on the campus in 1898–99, and they kept up their friendship until Gale's death in 1942. At Mount Wilson, soon after the arrival of the Snow telescope,

Hale and Adams had begun obtaining high-dispersion spectrograms of the sun and of the bright stars Arcturus ($\alpha$ Bootis) and Betelgeuse ($\alpha$ Orionis). The exposure times for the stars were as long as fourteen hours, requiring several nights for completion. Adams, who lived on Mount Wilson, did most of the observing. They found that the spectrum of Arcturus was not the same as the sun's but was very similar, line by line, to the spectrum of a sunspot. Furthermore, the spectrum of Betelgeuse deviated from that of the sun even more than Arcturus's did. Hale had always planned to have a spectroscopic laboratory on Mount Wilson, to aid in interpreting the spectra of stars. No doubt at Adams's suggestion, he brought Gale out from Chicago in the summer of 1906 to set up the laboratory on the mountain and do the experimental work. In a series of excellent papers, the three of them conclusively proved that the main effects were due to temperature, that sunspots are cooler than the undisturbed "surface" of the sun, and that the spectral sequence from the sun through Arcturus to Betelgeuse is a sequence of decreasing temperature.[37] This was an important new result for the sunspots and a complete confirmation of earlier results obtained, for instance by James E. Keeler, with lower-dispersion spectrographs.

Gale went back to California in the spring of 1909 to do more experimental astrophysical work of this same type. Fortunately for him, Hale had by then moved the spectroscopic laboratory to the observatory's office building in Pasadena, for in late April Gale was knocked unconscious by a 15,000 volt jolt from the secondary of a transformer he brushed against. Luckily the current was not powerful enough to kill him, but he was badly burned on the hip by a rheostat he fell on. He had been working alone and remained unconscious for ten or fifteen minutes, but apparently he convulsively kicked away the rheostat before it charred his leg. Hale happened to come into the laboratory, saw him, and pulled him free. Gale was taken to Pasadena Hospital, where he underwent several skin graft operations, and was finally released after a month-long stay. In later years he liked to say he had one hundred square inches of Adams, Ellerman, and three other donors from Mount Wilson Solar Observatory on his hip and leg. He recovered completely and worked in the Pasadena laboratory again the next summer, and the following year in Chicago, measuring and analyzing spectroscopic plates he had brought back with him.[38]

Thus when Gale was appointed dean of the Graduate School of Science at Chicago in 1922, he knew far more about astronomy and astrophysics than most scientists. But he hated Hale, or at least disliked him intensely. Twice, once in 1906 and again in 1912, Gale had wanted a job at Mount Wilson Solar Observatory, and each time Hale had put him off and given the position to another. The first time he had hired Arthur S. King to a permanent post in the Pasadena spectroscopy laboratory; the second time he had brought Carl Störmer from Sweden as a visiting scientist rather than Gale. The Chicago physicist felt Hale had humiliated him. Gale vowed never to ask the Mount Wilson director for anything again, but he remained a close friend of Adams, who did his best to smooth over the situation. Eventually Gale brought himself to speak to Hale, politely if superficially, but he retained his dislike of the great man for the rest of his life. As dean he discounted any outside advice he received from Hale. On the other hand, Gale trusted Adams completely and frequently sought his counsel on Yerkes Observatory and University of Chicago appointments he was considering.[39]

Frost had inherited the editorship of the *Astrophysical Journal* when Hale resigned his University of Chicago faculty position. The new director had done much of the editorial work whenever Hale was off on his travels, so the transition was not a difficult one. The *Journal* continued to appear in ten monthly issues each year, collected into two semiannual volumes. All the Yerkes papers and most of the Mount Wilson papers appeared in it. Papers that Hale, or later his Mount Wilson associate editor, first Charles St. John, then Seares, sent to the *Journal* were published without question. On most other papers Frost, perhaps aided by a consultation with a Yerkes staff member, made the decision himself. The number of submissions was small enough, and the range of subjects limited enough, that he did not have to send many papers out to referees. Many laboratory spectroscopy papers came in, and it was natural for Frost to send them to Gale to referee and decide on himself. The dean (who also became chairman of the Physics Department in 1925) was the liaison with the University of Chicago Press, publishers of the journal, located only a block from his office. Gale's role became increasingly important after Frost lost his sight and came to Chicago less frequently for conferences at the Press.

Harper, Judson, Burton, Woodward, and Gale all encouraged Frost to try to get funds for the observatory from the wealthy Chicagoans who owned summer homes on Lake Geneva. Occasionally he was successful in obtaining small grants, but the really big donors eluded him. Two University of Chicago trustees, Ryerson and Charles L. Hutchinson, had large estates on the shore, but they were already heavily committed to other departments on the campus and did not want to divide their efforts. Another lakeshore property owner, Edward E. Ayer, was interested in astronomy but also gave his money elsewhere.[40]

But Frost was successful, on one occasion, with William Wrigley Jr., owner of Green Gables, probably the largest estate on Lake Geneva. Frost got the money from Wrigley for an eclipse expedition. In 1918 Frost and Barnard had gone to Green River, Wyoming, to observe the total solar eclipse of June 8. They had located this site, on the narrow path of totality, the previous summer. Frost planned a comprehensive program of photographic, photometric, and spectroscopic observations of the eclipse and tried to use it as leverage to obtain funds for a very long-focus twelve-inch lens to photograph the corona and to use at Yerkes after the eclipse. He did not get the money for the lens, so they took the lens from the twelve-inch refractor instead and made the trip. On the day of the eclipse "passing clouds prevented complete success but some results of value were probably obtained. Enough could be seen through the clouds to make a superb spectacle," as Frost telegraphed home. (He still could see with his left eye at that time.) In fact they were nearly completely unsuccessful because of the clouds. Frost had to beg a photograph of the corona for a talk he was to give later that summer from a Chicago amateur who had gone to the eclipse as a tourist. The amateur had been clouded out too but was able to send Frost a picture of the corona taken by a commercial photographer in Rock Springs, using his Graflex camera. This local man had succeeded where most of the astronomers from distant observatories, camped along the eclipse track in Wyoming, had failed.[41]

In spite of this experience, when Frost learned that the track of the total solar eclipse of September 10, 1923, would pass over Catalina Island, off the California cost near Los Angeles, he resolved to send an eclipse expedition there. It would be one of the longest eclipses of the century visible from the United States, and in addition Wrigley, whom Frost had met, owned Catalina Island. The director appealed to the millionaire for financial help, and Wrigley agreed to provide it. Then Frost got down to specifics, writing that the cost of the 1918 expedition had been $5,000, that he thought he could conduct a successful expedition to Catalina for $8,000, but that $10,000 would be much better. Wrigley, whose high-pressure advertising tactics and the very nature of his chewing-gum business made him somewhat suspect in the world of bankers and industrialists like Hutchinson, Ryerson, and Ayer, was willing to help, but only within limits. "I do not understand that I am to do the entire financing of making a picture of the sun and eclipses. What I did say was that I would finance it up to $5,000.00 and that's as far as I will go. As I understand it, if it should happen to be cloudy at the time this eclipse is going on, the money is all spent for nothing. . . . Not being an astronomer, I, of course, do not feel that the world is going to lose very much whether the photo is made or not, but I am willing to give you a chance to make it up to the sum mentioned," he wrote Frost. The director wrote President Judson, in Chicago, that he would not ask Wrigley for more. He hoped in time to educate the tycoon on the scientific value of observing eclipses. In spite of repeated efforts, however, Frost did not manage to teach Wrigley anything, or even to see him again. The rich donor avoided any further meeting with the Yerkes director, writing, "I am afraid that a business man has a hard time realizing the importance of Scientific investigation as far as the Sun is concerned. If it does not interfere with his business he is not interested."[42] Frost's tactics were completely ineffectual compared with those of Hale, who would have flooded Wrigley with letters telling him of the wonderful opportunity this eclipse presented and how the results the Yerkes party might achieve would open new vistas to humanity. He would have sent Wrigley striking photographs of past eclipses and letters of endorsement from famous scientists all over the world. If the businessman had ignored his letters, Hale would have gone to Chicago, met him in his club, and delivered a brilliant monologue on solar eclipses from the time of the Egyptians and Greeks to the present opportunity-laden moment. Instead Frost wrote two or three letters mostly devoted to budgets and past failures.

With the $5,000 in hand, Frost prepared another comprehensive program of measurements at the eclipse. Through Wrigley's sponsorship, he controlled access to Catalina Island, and other astronomers who wanted to locate there had to clear their plans with him. Stebbins, at the University of Illinois, planned a much more modest program. He had successfully observed the 1918 eclipse from Rock Springs, measuring the total brightness of the solar corona with a photoelectric cell, his specialty. At Catalina he planned to repeat this measurement, to see if the corona's brightness varied with time. "It may be that we would undertake something additional to this but I should want to do a good job of that which we are especially prepared for," he wrote, expressing the lifelong philosophy that made him such a successful research scientist. Moreover, he was philosoph-

Figure 16   Yerkes Observatory (1922). Courtesy of Yerkes Observatory.

ical. "It looks as if we shall have a grand time, and the chances for a clear sky are apparently as good at Catalina as anywhere else," was his final opinion.[43]

In spite of his blindness, Frost headed the expedition himself. Barnard was dead by now, but two other Yerkes faculty members and a host of former students and friends, volunteers from institutions as close as Northwestern and as distant as Kyoto, rendevoused at Catalina. Over twenty-five names were mentioned in the press release that Frost wrote before the eclipse, and a group picture, taken on the great day, shows many more.[44] But Wrigley's veiled foreboding turned out to be all too true. The skies were completely cloudy, and not a single scientific result, or even observation, was obtained. At least they could be happy in the thought that they were in the same boat as everyone else; all the eclipse parties in the narrow strip of totality along the southern California coast and adjoining Baja California were clouded out. They might all have just as well stayed at home, except for the "grand time" they had on their travels.

That same summer Frost was caught in an embarrassing revelation of his scientific isolation. Albert Einstein's general theory of relativity was very much in the news, because of the Lick Observatory's confirmation of its prediction of the gravitational deflection of light. British astronomers had first confirmed it at the solar eclipse of 1919, but their evidence was somewhat unconvincing; Campbell and Robert J. Trumpler had just published extremely solid results from the much more favorable 1922 eclipse. Relativity, and Einstein himself, had a radical aura; many conservative physicists and astronomers wanted him to be wrong. A certain amount of anti-Semitism was also involved.[45] Thomas Jefferson Jackson See, a failed classical astronomical theorist, was Einstein's most vociferous critic. See claimed that Einstein had not been the first to make this prediction but had stolen it from an article by Georg Soldner, an "Aryan" German, published in 1804. (In fact Soldner had made the same classical derivation as Isaac Newton and obtained the same result, different by a factor of two from the general relativity result that had been verified by the observation. But Soldner had made an error in arithmetic, and so the published result in his paper was the same as Einstein's.) See had sent his criticism of Einstein to Frost, who had someone read him Soldner's paper. He did not check the numbers himself but sent a "personal" letter to another friend, stating that Soldner had found the result a century before Einstein. Unfortunately for Frost, the friend to whom he sent this missive was the editor of *Popular Astronomy* magazine. He immediately published the news, attributing it to Frost. Hence the Yerkes Observatory director found himself publicly quoted in opposition to Campbell, Hale, and all the other high priests of American astronomy, who rightly considered See an all-too-vocal crackpot. All Frost could do was plead that his letter, sent in confidence, should not have been published without his consent and that he did not trust See either. It was a lame excuse.[46]

In 1927, when America was enjoying prosperity under President Calvin Coolidge and the University of Chicago endowment was again in fine shape, Frost came forward with a plan that epitomized his administration. It was to convert the twelve-inch Kenwood refractor, Hale's original telescope that his father had turned over to the university when the Yerkes gift came through, into a twin telescope. When Hale had first ordered the twelve-inch, he had specified a visually corrected lens, achromatized for yellow and red light. Almost immediately after putting the telescope into operation in 1891, he had decided that what he really needed was a photographically corrected lens achromatized for blue light, and his father had ordered that for him too. The two lenses were interchangeable, so either one could be used in the telescope. Hale had planned some day to have the telescope converted into a twin instrument, with two tubes mounted parallel to one another, so that both lenses could be used simultaneously. Now, over a quarter of a century later, Frost proposed to carry out this plan. He listed a series of reasons, but they all referred to old-fashioned observing programs that had been going on unchanged for years. The hard fact was that the twelve-inch was a small telescope that would not produce any exciting new results, no matter what instruments were on it. The forty-inch refractor was Yerkes Observatory's main telescope, and the observers who used it desperately needed a new, more efficient spectrograph than the Bruce, installed more than twenty years before. Frost was living in the past, trying to carry out plans Hale had left for him. Gale weakly endorsed the request for $12,000 to $15,000 "if there is no urgent need for the money for other purposes." Woodward was lukewarm at best, but Frost pressed valiantly on. He wrote directly to President Mason, emphasizing the need for "rapid action" to put into simultaneous use the two twelve-inch lenses that, as he said himself, the observatory had owned for thirty years. Frost had no other project to propose, the administration had the money and felt that the Astronomy Department should get something, and so this program from the past was approved.[47] It dragged on forever, absorbing all the money the university put into it as well as much of Frost's time and energy and the administration's patience.[48]

In 1928 Frost toyed briefly with the thought of building a sixty-inch or seventy-two-inch reflector at Yerkes Observatory, using a quartz disk for the primary mirror to minimize thermal distortions of its parabolic figure.[49] This idea combined two of Hale's plans, the original sixty-inch for Yerkes, and the quartz mirrors then under consideration for the Palomar two-hundred-inch. Either a sixty-inch or a seventy-two-inch reflector at Yerkes, in a new building, would have been much more expensive than the twelve-inch remounting project. Probably most astronomers of the time would have advised against building a large, expensive telescope at a midwestern site rather than in the West or Southwest. Certainly Van Biesbroeck, Ross, and at least one other Yerkes faculty member, Otto Struve, favored a western site by then. Gale had concluded years before that any telescope larger than forty inches would have to be built in a better climate than Williams Bay's.[50] Frost did not follow up on the new telescope idea, and by then Gale and other long-term administrators at Chicago were simply waiting for him to retire.

Few American astronomers criticized one another publicly, but within their community Frost's deficiencies were well known. Writing frankly to a close friend, Joseph H. Moore, effectively in charge of the Lick Observatory radial-velocity program by 1928, described the Yerkes results in this field as "not as good as they should have been."

Struve, who was doing most of this work, was "an excellent man," Moore wrote, but he had had to take it over "without any experience or the benefit of much advice or assistance from others" (meaning Frost). And, in a letter to his former fellow graduate student, Ralph H. Curtiss, who was seeking advice as the new director of the University of Michigan Observatory, Stebbins was quite critical of Frost's lack of leadership and his failures in dealing with an unproductive staff member, Oliver J. Lee.[51]

In 1930, two years before Frost's expected retirement, his right arm, Storrs B. Barrett, reached retirement age. He had worked with Hale at Kenwood and Yerkes, mostly as an observer and as part-time secretary of the observatory. In this role he handled much of its business correspondence, relations with suppliers, and all the myriad practical details necessary to keep a distant outpost of a university running. Barrett was also the librarian. He continued under Frost, who depended on him increasingly after losing his sight. Barrett read many letters to him and took care of all the less important ones for him.[52]

Frost would have been lost without the faithful Barrett, except that he thought he had the perfect candidate to replace him. This was Clifford C. Crump, then nearly forty years old. Born in Indiana, Crump had done his undergraduate work at Earlham College and then his graduate work at the University of Michigan, receiving a Ph.D. in astronomy in 1916. In 1917 he joined the faculty of Ohio Wesleyan University as an assistant professor of mathematics and astronomy. He quickly became the protégé of his predecessor, the very old, very rich Hiram M. Perkins. His fortune was based on his speculations in hog futures during the Civil War, multiplied many times by subsequent investments in farms and land in southern Ohio. Perkins had already given one telescope, a 9-1/2-inch refractor, to Ohio Wesleyan, but he wished to leave a larger one also, as a memorial to himself and his wife. He was thinking of a twenty-inch or twenty-four-inch refractor, on campus, but under the smooth, caring tutelage of Crump, what emerged was a sixty-nine-inch reflector, the third largest telescope in the world when it was completed, installed on a low hill a few miles from Delaware, Ohio. Perkins turned over the first shovel of earth for Perkins Observatory at the age of ninety, just before he died in 1924. Crump was very poorly equipped to plan an observatory, but he was artistic, and so the building and its furnishings made a very attractive memorial to the aged donor. Before the telescope was completed, however, in the summer of 1928, Crump was very suddenly sent off on a year's sabbatical tour of the Far East and then summarily dismissed from his faculty position. He was single, sensitive, poetic, and handsome; this was suspicious enough in the straitlaced Methodist college town of 1928, but that summer he was accused of having a homosexual relationship with either a student or a young boy. That was grounds for instant dismissal; Crump did not fight the charge then or explain it later. He was gone from Ohio Wesleyan University.[53]

Whatever his faults may have been, Crump certainly had a charming manner and the ability to gain the confidence of older, powerful men. First he had been the protégé of William J. Hussey, director of the University of Michigan Observatory, then of Perkins, then of Frost. Crump had first come to Yerkes for a few months in 1915–16, to work on measuring velocities of spectroscopic binaries, the subject of his thesis. He kept in touch with Frost and, after getting the Ohio Wesleyan faculty position, consulted the Yerkes director frequently about his plans for Perkins Observatory. Frost was the only important figure in American astronomy who took Crump seriously. In 1916 (when he was briefly on the faculty of Carleton College), the young man spent a summer at Yerkes; he returned frequently in subsequent years. In 1918 he accompanied Frost to the solar eclipse in Wyoming, acting as Frost's "eyes" in focusing and adjusting the spectrograph. Soon he was staying in Frost's house on his visits, first for a few days, then for entire summers, as a close, trusted family friend. He again accompanied Frost to the southern California eclipse in 1923 and by now was known to his elderly adviser as "Crumpo."[54]

Thus when Crump left Perkins Observatory in 1928 on his sudden sabbatical leave, Frost was glad to write warm letters of introduction for him to observatory directors and to old Dartmouth classmates. Frost referred to Crump as "my friend, . . . director of the Ohio Wesleyan Observatory . . . [who] has often been a member of our family on his visits at the Yerkes Observatory." He did not mention that Crump was in fact on terminal leave, or the reason for it.[55]

Crump landed a job as professor of astronomy at the University of Minnesota for the 1929–30 academic year, undoubtedly on Frost's recommendation. Probably word of Crump's real reason for leaving Ohio Wesleyan caught up with him there in April 1930, for late in that month Frost suddenly telephoned Vice President Woodward to recommend his protégé for the job as Barrett's successor. Frost wanted Crump appointed as associate professor, secretary, and librarian. According to the director, Crump was willing to give up

$1,000 a year (his salary at Yerkes would be $4,000 annually) in order to have the opportunity to do research. In fact Crump had published only a single scientific paper since his Ph.D. thesis, in spite of all the summers he had spent at Yerkes with no teaching duties. Woodward was extremely skeptical. No doubt Gale had heard from Adams, an ascetic puritan, all about Crump's "misdeeds" at Ohio Wesleyan. Woodward told Frost that he doubted Crump would accept the position, because the appointment would be made without tenure, for a three-year term, the way all beginning associate professors were then appointed at Chicago. Crump's reappointment (due in 1933, one year after Frost's retirement date) would be "largely in the hands of your successor," Woodward emphasized. Everyone on the campus who knew Crump liked his personality, but no one except Frost was enthusiastic about adding him to the staff. One astronomer "not connected with the University ha[d] expressed considerable doubt as to [Crump's] ability." (This was probably Adams.) Woodward asked for an explanation of the full circumstances under which Crump had left Ohio Wesleyan University, an indication that the rumors had reached the Midway. Frost simply stonewalled, ignored the question, insisting that he had looked over the field and was certain that Crump was the best man for the job. Woodward had emphasized that his "earnest desire" was to do "the very best possible thing for Yerkes Observatory"; Frost replied that Crump was the best possible person for both the university and the observatory. Barrett also wrote "independently" to Woodward, at the director's suggestion. His letter outlined the qualifications for the job, as he saw them, and then said they exactly fitted Crump, "a high type of Christian gentleman." Then Barrett added, "I could not believe anything unworthy of him without losing my last vestige of faith in the whole human tribe." No doubt he meant it sincerely. Woodward had no choice but to make the appointment or to disavow Frost and Barrett. Crump began work as an associate professor on September 1, 1930.[56]

Even before Crump's appointment, physicist Arthur H. Compton had passed on to the president of the University of Chicago Harvard director Harlow Shapley's highly negative appraisal of Yerkes Observatory as a research institution and his warning not to trust Frost to select its staff members of the future.[57] After it had been made, Ross wrote a friend that "things are going from bad to worse here, especially in the matter of . . . filling vacancies, so much so that a new Director will have a hopeless mess on his hands." To Adams he referred to the rumors about Crump as a "foul-smelling scandal which the Director deliberately closes his mind to." From Mount Wilson, Frost's former student replied that he feared "there is little hope for radical improvement [at Yerkes] until there is a new director and a general house-cleaning."[58]

Graduate-student training provided a somewhat more positive picture during Frost's long tenure as director. From the beginning Harper's vision of the University of Chicago had been of an institution oriented to training graduate students and to research. There had been a few students in Hale's time, including William H. Wright, Adams, and Schlesinger for brief periods. Two other, later famous astronomers who applied for fellowships as graduate students at Yerkes but ended up elsewhere were Heber D. Curtis (who went instead to the University of Virginia) and Stebbins (who went to Lick Observatory).[59] On one occasion Harper, who liked to see results, wrote Hale to ask how long it was going to be "before we get some doctors" from Yerkes Observatory.[60] The first one did not come until 1912, eight years after Frost had succeeded to the directorship. He was Curvin H. Gingrich, who became a longtime teacher of astronomy at Carleton College. The next was Lee, who received his Ph.D. in 1913 and immediately joined the Yerkes staff. In thirteen years there, he accomplished very little. Generally there were two or three graduate students at the observatory, their numbers swelled in the summer quarter by a few older men and women who had undergraduate or master's degrees and held teaching jobs at colleges not far from Chicago. They came back for years, enjoying a summer vacation on Lake Geneva as they worked slowly toward the Ph.D. degree, which would mean a raise in pay and status at their home institutions. Most of the graduates were far from inspiring, or inspired, at Yerkes. The courses consisted entirely in reading up on a subject, discussing it with the professor in charge of it, and doing research with him. Each student spent a few quarters on the campus in Chicago, taking the required celestial mechanics, mathematics, and physics courses. Not many of the graduates of Frost's era made much of an impression in their later careers, but a few excellent research workers came out of the program. Yet a survey of graduate schools, published in 1925, reported that Chicago was generally considered second only to the University of California, with its Lick Observatory, as a training center for astronomers. According to it the University of California astronomy program received seventeen first-choice votes from a panel of senior astronomers, Chicago eight votes, and Michigan and Princeton three votes each.[61]

Figure 17 Yerkes Observatory group (1916). Storrs B. Barrett standing at far left in rear row, John A. Parkhurst fourth from left, Clifford C. Crump sixth, Oliver J. Lee eighth, Edwin Hubble tenth, Edward E. Barnard eleventh, and Edwin B. Frost twelfth, all in rear row; Mary R. Calvert second from left in front row. Courtesy of Yerkes Observatory.

The five Yerkes Ph.D.'s of Frost's era who had the greatest impact on astronomical research were Edwin Hubble (1917), Otto Struve (1923), Nicholas T. Bobrovnikoff (1925), William W. Morgan (1931), and Philip C. Keenan (1932). Hubble, perhaps the greatest observational astronomer of this century, who had been an undergraduate at the University of Chicago (class of 1910) and then a Rhodes scholar, returned to Yerkes as a graduate student in 1914. He did his thesis, almost entirely on his own, on "Photographic Investigations of Faint Nebulae." He was not quite sure they were galaxies, but he became almost convinced that they were. His thesis was not very good, but it was full of creative new ideas. Hubble finished his dissertation in a great hurry in 1917, to volunteer as an officer in World War I, and joined the Mount Wilson staff as soon as he was discharged in 1919. Many of the ideas of his subsequent research on the expanding universe of galaxies were foreshadowed by his Yerkes thesis.[62]

Struve[63] and Bobrovnikoff were both Russians who fled their homeland after the First World War. Bobrovnikoff, born in 1896 near Kharkov, Russia, started his university career as a student at the Institute of Mining Engineers in St. Petersburg. He became an infantry officer in the Russian army in 1917, and afterward an officer in the White army fighting the Reds in the Bolshevik revolution. Bobrovnikoff escaped to Cyprus after the Whites were defeated in 1920, then managed to make his way to Prague University, where he completed a B.S. degree. He succeeded in getting to the United States in 1924. At Yerkes he did a pioneering Ph.D. thesis on the physics of comets, then continued this work as a postdoctoral scholar at Lick and at Berkeley. It was an unusual subject for a research astronomer of his generation, and he made many important advances in it.[64]

Morgan came to Yerkes in 1926, after three years as an undergraduate at Washington and Lee University, because Frost was desperate for a research assistant to take the routine solar spectroheliograms he considered important. Morgan's physics professor, Benjamin A. Wooten, one of the elderly summer visitors at Yerkes, recommended him for the post. Morgan completed his B.S. at Chicago simply by taking graduate research courses at Yerkes. In 1928 he married Helen Barrett, daughter of the longtime observatory secretary.[65] Then Morgan did a very creative thesis, under Struve's supervision, on stars with lines of rare-earth elements in their spectra. By this time Morgan was committed to stellar spectroscopy as his life's work. He did all the observing for his thesis with the forty-inch refractor and its old radial-velocity spectrograph. In spite of his minimal formal training in physics, Morgan had made himself an expert in recognizing subtle patterns, similarities, and differences in spectra—what he later called "the thing itself." He classified and arranged these stars, almost purely in observational spectroscopic terms, and described their spectral peculiarities. It was a good thesis. Morgan completed it and took his final examination in December 1931, when the Great Depression was at its worst. The University of Chicago was operating under deficit conditions, and the administration was applying severe pressure to reduce spending. Yet Frost was able to keep Morgan on the payroll as an assistant, at $1,500 a year.[66]

Keenan had a more conventional astronomical education. Born in Bellvue, Pennsylvania, a suburb of Pittsburgh, he had grown up and attended high school in Ojai, California. Then he entered the University of Arizona, where he took many advanced courses in physics and mathematics, along with seminars and reading courses in astronomy. Keenan completed his B.S. at Tucson in 1929, then continued with a master's thesis under Edwin F. Carpenter, working on lunar photographic photometry with Steward Observatory's thirty-six-inch reflector, which was then on the campus in Tucson.

Keenan received his M.S. in 1930 and entered Yerkes Observatory that same summer. (He had spent the previous summer there as an assistant.) He took more graduate courses and held the same solar assistantship Morgan had had.[67] Life was not easy for a graduate student in the Great Depression. Keenan, as Morgan had before him, slept in a large room in the attic of Yerkes Observatory with the other students. It was called "the battleship" because of its small, widely separated round porthole windows, but the roof leaked and the room became a cold, dark tomb on winter weekends, when the electric power (except to the telescope) and heat were turned off in the building. Water dripped in during summer thunderstorms, and snow drifted in during the winter. The only bathtub in the building was in the middle of the men's room in the basement, adjoining the instrument, optical, and woodworking shops. There were no showers. The students took their meals at a nearby boardinghouse, where the more affluent visitors to the observatory stayed.[68]

During Frost's directorship, he, the other staff members, and their families were strongly involved in the community life of Williams Bay. Frost himself published many poems and little essays on nature in the columns of the *Chicago Tribune,* widely read in southern Wisconsin at that time. For several years in the 1920s he contributed a regular weather column to the *Lake Geneva News*. Barrett, who had been a member of the local school board from 1901 through 1903, was elected its clerk from 1909 through 1919. He played a key role in bringing about the consolidation of the schools of several other nearby towns (the Wisconsin term for townships) with the Williams Bay school and keeping the resulting consolidated school there. All the observatory people signed the petition he circulated to get the ball rolling, and no doubt they voted as a bloc in favor of better education for their children. Van Biesbroeck was elected clerk of the school board in 1928 and was reelected term after term through 1948, long after all his children had graduated from college and left home.

When Williams Bay was incorporated as a village in 1919 (before that it had been a populated settlement, governed as part of the town of Walworth), Barrett was elected president of its first village board (the near equivalent of mayor), and he was reelected for a second term in 1921. He was defeated for a third term by one vote in 1923. Frost was one of the elected trustees for that first five-member board, and Lee was its clerk, so the observatory faculty members formed the majority from 1919 to 1921. Frost did not run for a second term in 1921, but Lee was reelected, then defeated in 1923. John A. Parkhurst was elected a county supervisor (the equivalent of a member of a county board) in 1921, and Barrett was elected to the same office in 1927, then reelected in 1929. Until well after his retirement, Frost was also a longtime appointed member of the Walworth County Park Commission, and for many years a member of the board of directors of the Walworth County YMCA. A friendly, outgoing personality who projected old-fashioned charm and education, his memory is revered to the this day in Williams Bay by the children and grandchildren of the villagers of his time.[69]

In October 1931, less than a year before he was to retire, Frost suffered a sudden illness, with fever, nausea, and severe abdominal pain, probably related to his gallbladder. He was rushed to Billings Hospital at the University of Chicago, X-rayed, and diagnosed, and an operation was prescribed. But he was not judged strong enough to survive it and instead was treated with the medicine of the time. His condition began to improve, and in the end his physicians decided the operation was not necessary. Frost remained in the hospital for six weeks, with his wife in an adjoining room. On November 20, their thirty-fifth anniversary, radio station WBBM in Chicago broadcast a wedding march in honor of the famous patient, and in early December he was discharged from the hospital.[70] Ross reported that Frost was very much weakened, but before Christmas the old director was well enough to walk several miles along the Lake Geneva shore.[71] Nevertheless, his successor would take over the observatory in six months. The University of Chicago administration was still trying to decide who that successor would be.

# CHAPTER FOUR

# The Savior, 1897–1931

Otto Struve, who was to become the most successful director Yerkes Observatory ever had, bringing it back from its nadir and taking it to its peak, was born in Kharkov, Ukraine, in 1897. He represented the fourth generation of an outstanding astronomical family.[1] His great-grandfather, Wilhelm, born in 1793 in Altona (near Hamburg, Germany), became professor and director of Dorpat (now Tartu) Observatory in Estonia, then part of the Russian empire. In 1839 he founded Pulkovo Observatory, the imperial (or national) observatory near St. Petersburg. His son, Otto Wilhelm Struve, born in Dorpat in 1819, came with him and succeeded him as director of Pulkovo in 1862. Otto was a noted visual double-star observer who came to the United States in 1879 to order the Pulkovo thirty-inch refractor, made by Alvan Clark and Sons before the Lick and Yerkes telescopes. This Otto's older son, Hermann Struve, accompanied him to America. Hermann later became director of Königsberg Observatory in East Prussia from 1895 to 1904, then moved to Berlin Babelsberg Observatory as its director until his death in 1920. Otto's younger son, Ludwig Struve, became director of Kharkov Observatory in 1897, the same year his son, our Otto, was born.[2] Otto's generation was the first in the family to be educated in Russian rather than German; he, his brother, and his sister spoke both languages from early childhood.[3]

Otto Struve was taught at home until he was twelve; then he entered a *Gymnasium* (university preparatory high school) in Kharkov. In 1914, as World War I began, Struve graduated from the *Gymnasium* and entered Kharkov University, studying astronomy and mathematics. Two years later, at age nineteen, he joined the army and was sent to artillery officers' school in Petrograd, the capital. From there he went to a regiment in the Caucasus, fighting the Turks, who were allied with the Germans. The Treaty of Brest-Litovsk ended the war for the Russians, and Struve returned to Kharkov University. Nicholas T. Bobrovnikoff, one year older than Struve and later a student after him at Yerkes, was in some of the same classes, including one taught by Struve's father, Ludwig.[4] All the astronomy Struve and Bobrovnikoff learned at Kharkov was the older, positional kind. There were no lectures on astrophysics, and Struve later wrote that he knew very little about spectroscopy and had no experience in it. The Russian provisional government (which had taken over from the czar) was overthrown by the Bolsheviks, and the country was plunged into civil war. Struve joined the White army as an officer, was wounded in action, and suffered several serious illnesses. His father, ousted from his position as director by the Reds, died of a heart attack in 1920, and his brother, also serving in the White army, died of tuberculosis complicated by malnutrition.

Struve himself, however, was a survivor. As the White army collapsed in March 1920, he managed to get on one of the ships that evacuated some of the defeated soldiers and officers to the Crimea. Later that year Struve, along with the remnants of the White army, sailed down the Black Sea and the Dardanelles to Gallipoli, in Turkey, where he was safe but penniless. He managed to stay alive for a year and a half on food from relief agencies, together with what he could earn working as a woodcutter. His name and his family saved him. His aunt Eva had managed to escape from Russia to Germany and approached Paul Guthnick, his uncle Hermann's successor as director of Berlin Babelsberg Observatory. She told Guthnick of her nephew's plight and asked for his help. Germany was itself a defeated nation, and Guthnick knew there was no chance for Struve to get a job in astronomy there. But the Berlin director wrote Edwin B. Frost to seek his help. Frost was a great humanitarian who had earlier, while Ludwig Struve was still alive, offered to try to find him a position in the United States. Guthnick asked Frost if he could do so now for Otto, and the Yerkes director sprang into action. He had an open position for an assistant in stellar spectroscopy at Yerkes, paying $75 a month, enough to live on, and he would hold it for young Struve. Frost immediately informed University of Chicago President Harry P. Judson and began pulling strings through his contacts in Chicago, in the American Central Committee for Russian Relief, and in the State Department. He assured Judson that Struve would work out as an assistant in stellar spectroscopy, writing, "I am perfectly willing to take him on his lineage." At Frost's urging Charles R. Crane, the Chicago plumbing magnate, provided a local contribution of $200 which, with matching funds from the American Central Committee, was enough to get Struve from Constantinople to Williams Bay. He received his visa from the American consulate in late August 1921, thanks to Frost's continuous pressure on the State Department, and embarked for New York on the S.S. *Hog Island* in early September. It was a long, slow passage, over a month en route, but it was cheap.

Struve disembarked in New York on October 7. Alexander Kaznakoff, the American Central Committee agent there, met him and put him on the train for Chicago and Williams Bay the next day. Two days later Struve was at the observatory, clad in a nondescript outfit he had picked up as economically as possible along the way, but making an immediate "very pleasant impression" on Frost, who spoke German with him. The director reported that he was "sure that we shall be glad to have [Struve] as a member of our staff and of our social colony."[5] His famous name ensured his welcome. Even before Struve arrived, George Van Biesbroeck, the Yerkes double-star observer, speculated to Robert G. Aitken, his counterpart at Lick Observatory, that the young man might follow in the tradition of his great-grandfather and grandfather as a double-star man himself. However, he wrote, Frost wanted Struve for spectroscopic work, and that would probably be what he would do.[6] And so he did.

Struve immediately began his studies and observing on his assistantship. He followed the "program of instruction and research at Yerkes Observatory" that Frost had drawn up in 1919.[7] There were no other graduate students there with Struve until June 1922. Only three turned up then, just for the summer, and of these only Christian T. Elvey, then a young student at the University of Kansas who later came to Yerkes for his Ph.D., made a research career in astronomy. There were no lectures; Struve and the other students learned by reading and doing research and by discussing their work with their professors. There were six general topics: solar physics, stellar spectroscopy, celestial photography, stellar photometry, astrometry, and physical laboratory. The last was completely optional, described as work with a large concave grating "investigating spectra of certain elements." No faculty member had done any work in this field since Hale left for Pasadena, and no student did either. Frost was locked in

Figure 18   George
Van Biesbroeck at
forty-inch refractor
(1928). Courtesy of
Yerkes Observatory.

the past, Edward E. Barnard did not take students and was in his last
illness, and John A. Parkhurst, the visual and photographic photometry "expert" had been recruited locally by Hale and left behind.[8]
Only Van Biesbroeck was a stimulating teacher—of asteroid research
and positional astronomy. But Storrs B. Barrett taught Struve to use
the forty-inch refractor and its Bruce spectrograph to obtain spectrograms and to measure them for radial velocity. For this he used a
"measuring machine," a microscope with a moving stage, driven by a
small hand crank through a delicate screw. Struve's fantastic enthusiasm, drive, and desire to succeed enabled him to master these procedures, and the numerical work of reducing the raw measurements to
calculate the radial velocity of the star, very quickly. Frank R. Sullivan, a local man, was the night assistant who had begun with Hale
and worked with Struve, Barrett, and other observers at the telescope.

Five months after Struve's arrival at Yerkes he discovered that one of the stars he was observing, $\gamma$ Ursae Minoris, had a very rapidly varying radial velocity. Between February and August 1922 he obtained two hundred spectrograms of it, measured them, found the period of variation—two hours and thirty-six minutes—and analyzed the radial-velocity curve. As a spectroscopic binary, its calculated dimensions turned out to be impossibly small; Struve opined that it was in reality an intrinsically varying, pulsating star, as astronomers, led by Harlow Shapley, were just realizing that Cepheid variables were.

Struve gave his first paper on this star at the American Astronomical Society (AAS) meeting at Yerkes Observatory in September 1922, twenty-five years after its inaugural ceremonies.[9] Frost reminisced to the group about the founding and read aloud a letter from Hale congratulating the observatory and the society. Nearly seventy members were present, including the renowned Princeton astrophysicist Henry Norris Russell, just elected a vice president of the AAS, and Joel Stebbins, the photoelectric expert from the University of Wisconsin at nearby Madison.[10] No doubt they, and most of the other members present, carefully inspected the tall young astronomer with the famous name, standing rigidly as he delivered his paper in heavily accented English.

By his second year, Struve was taking more spectrograms than Barrett or anyone else at Yerkes; he was also measuring more. Keeping one eye at the microscope of the measuring machine and using the other to read the numerical setting from its scribed wheel, he strained his eye muscles, probably weak to begin with. As a result his two eyes often looked in slightly different directions, giving him a particularly forbidding expression. This condition, especially noticeable when he was tired, stayed with him all his life. When Frost, Barrett, Parkhurst, Oliver J. Lee, and several Yerkes summer students and volunteers headed to Catalina Island for the 1923 solar eclipse, Van Biesbroeck and Struve remained behind and worked.[11] The young Russian specialized in short-period spectroscopic binaries, the subject of his thesis. He submitted it along with a summary of his activities, which showed he had met all the requirements of the program. In addition to all his radial-velocity observations and measurements, he had made regular observations of asteroids with the twenty-four-inch reflector George Willis Ritchey had built, had made parallax observations, and had used the spectroheliograph on the forty-inch refractor. Struve had also made weather observations at Williams Bay and had tried his hand with nearly all the smaller telescopes at Yerkes. His previous work at the University of Kharkov had included courses in mathematics, physics, and astronomy, use of a meridian circle, and a "thesis" that was simply the accurate determination of the latitude and longitude of the site of the solar-eclipse station where he had been, with his father, in 1914. This previous work provided the basis for Frost's statement that Struve had a degree from Kharkov "about equivalent to a Doctor's degree," a claim the young Russian never made himself, then or later.[12] However, at the time it saved him from having to spend a quarter or two on campus in Chicago, taking formal graduate courses. More realistically, Frost also provided a statement that Struve was fluent in German, the language of his home, and French and had read numerous papers to him in both languages, on account of his "defective vision."[13] This excused Struve from having to pass reading examinations in these two languages, then a requirement for a Chicago Ph.D. in physical sciences.

Thus in early December 1923 Struve completed his thesis, titled "A Study of Spectroscopic Binaries of Short Period." It included several papers on individual stars of this class and a summarizing article on their general properties.[14] On December 8 he was examined at Yerkes by a committee consisting of Frost, Barrett, Parkhurst, Lee, and William D. MacMillan, a celestial mechanics theorist from the campus. Struve passed with flying colors and was granted his degree magna cum laude.[15]

Frost had long ago decided he would to keep Struve on the faculty. The director had managed to wangle a promotion for him to associate in stellar spectroscopy on July 1, 1923, which brought a salary of $1,350 a year, slightly higher than the other graduate students were making as assistants. After he got his Ph.D., Struve became an instructor, at $1,800 annually, effective January 1, 1924. He kept on working just as hard as before, and again in 1923–24 he took more spectrograms than anyone else at Yerkes.[16] He continued to obtain more radial-velocity spectrograms, measure and reduce more of them, discover more new spectroscopic binaries, and publish more papers than anyone else at Yerkes.[17]

In the midst of all this activity, Struve managed to bring his mother to America and found time to get married. It took long negotiations to get a visa for Elizabeth Struve, but she finally was able to come to Williams Bay in January 1925. In her early years there she worked part time at the observatory as a computer, doing the numerical work of reducing the measurements of radial-velocity spectrograms.[18]

Figure 19  Yerkes Observatory group (1925). Standing in rear, from left, Otto Struve, Christian T. Elvey, and Oliver J. Lee. In row standing in front of them, Elizabeth Struve (Otto Struve's mother) second, Alice H. Farnsworth, Edwin B. Frost, and Mary R. Calvert fourth, fifth, and sixth. Standing at right are Florence Lee (secretary), Frank E. Ross, and George Van Biesbroeck. Nicholas T. Bobrovnikoff is fourth from left in row seated in front. Courtesy of Yerkes Observatory.

Struve had met Mary Lanning at the observatory. Born in a small town in Michigan, she had attended Olivet College in Chicago. She considered herself a musician but worked part time as a secretary and had become acquainted with Yerkes Observatory when she visited one of the church summer camps near it along the lakeshore. Frost hired her from time to time to type his personal scientific correspondence, and she filled in for Florence E. Lee, the observatory secretary, during the latter's summer vacations.[19] Lanning was slightly older than Struve; she had been previously married and probably divorced. (She used her maiden name, which would have been unusual for a widow at that time.) Struve wooed her, singing romantic Russian songs softly to her and taking her skating on frozen Lake Geneva in the winter.[20] They were married by a probate judge in Muskegon, Michigan, the terminus of a lake ferry line from Wisconsin, not far from her birthplace of Wolverine, on May 21, 1925, the same day they obtained their marriage license there. Apparently none of Struve's friends from the observatory, or his mother, or any of Lanning's relatives were present at the ceremony.[21]

Two years later, on September 26, 1927, Struve became a naturalized American citizen at the Walworth County courthouse in Elkhorn, Wisconsin, near Yerkes Observatory.[22] He had taken out his first papers soon after his arrival and had waited little more than the minimum period to take the final step and swear allegiance to the United States. After witnessing the defeat of the White army in the Crimea in 1920, Struve probably had little hope of ever returning to a reborn czarist Russia, but by now he was certain that he was not going back. Now he could understand, write, and speak English perfectly, with only a slight remaining accent, which he retained for the rest of his life.[23] Marriage and becoming a citizen did not hold up Struve's research. He continued working very hard, publishing papers on spectroscopic binaries and completing a long joint paper, with Frost and Barrett as the senior authors, on the radial velocities of over three hundred B stars. It represented a major result of the Yerkes Observatory radial-velocity program and included measurements of nearly 2,500 spectrograms, taken from 1901 to 1925. Frost had begun the program, and he and Barrett had done most of the observational work in the earlier years, but Struve, with his persistent energy, brought it to a conclusion. It was a massive paper, consisting largely of data.[24]

All of Struve's research, and especially this radial-velocity paper, familiarized him with the spectra of the B stars (then often referred to as "helium stars"), the hot, high-luminosity stars that show relatively weak, broad absorption lines of helium and hydrogen in their spectra. He could see that many of these B stars also showed the narrower absorption lines of Ca II (calcium ions), first recognized as different and worthy of interest in 1904 by the German astronomer Johannes Hartmann in the spectroscopic binary δ Orionis, a B star. Although all the broader, diffuse lines of the star varied periodically in velocity as a result of the orbital motion of the pair of stars about one another, the calcium lines remained fixed and unchanging. Hartmann called them "stationary" lines. In 1909 Frost had announced his discovery of such narrow, stationary calcium lines in twenty-five more B stars. Some were spectroscopic binaries, others were single stars, but in each the calcium lines were fixed and completely different in appearance from the stellar absorption lines. The calcium lines clearly were formed not in the stars themselves, but somewhere in space between them and the observer on Earth. In more recent years John S. Plaskett, at the Dominion Astrophysical Observatory in Victoria, Canada, had observed them in many more B stars, with its new seventy-two-inch reflector. He concluded, in a paper published in 1924, that the calcium lines are formed in "vast clouds" of calcium outside the stars.

Struve realized that the large number of spectrograms of B stars, taken over the years at Yerkes, provided ideal material for investigating these clouds. He could measure the velocities and estimate the intensities of the stationary calcium lines in the spectra of many more stars than Plaskett had.

It was a new, exciting field. Struve jumped into it quickly. He gave his first paper on it at the AAS meeting at Carleton College, in Northfield, Minnesota, in September 1925.[25] Only thirty-eight members attended the meeting, but in addition to Struve, Van Biesbroeck, Lee, and Elvey went from Yerkes and Stebbins and C. Morse Huffer from nearby Madison. The president of the Society, William Wallace Campbell, did not make the trip east from Berkeley, where two years before he had been installed as president of the University of California. In his place elderly George C. Comstock, recently retired from the University of Wisconsin faculty, was in the chair.[26] Struve presented his paper on the calcium clouds and a second one on the changes that had occurred in the hydrogen emission line Hβ in the B star κ Draconis over the years. When the emission line became especially bright, a narrower absorption line, superimposed on it, also appeared. Like his calcium paper, this one also was based on his study of existing spectrograms in the Yerkes collection, dating back to 1902 in this case, together with more recent ones he had taken himself.[27] In it Struve told how he had been able to map, in rough outline on the sky, six different clouds, each with its own radial velocity and approximate strength of the calcium absorption line, related to the cloud's thickness and density.[28]

At just about this same time Struve reviewed, for the *Astrophysical Journal,* the new book by Cecilia Payne, *Stellar Atmospheres: A Contribution to the Study of High Temperature Ionization in the Reversing Layers of Stars.* Published in 1925, it was her Radcliffe Ph.D. thesis, done at Harvard under Shapley's supervision, with considerable guidance from Russell, the great Princeton astrophysicist.[29] As an undergraduate at Cambridge, England, she had studied under Arthur S. Eddington, learning much more physics than the typical astronomy student of her time. In her thesis she applied it to analyzing and interpreting the spectra of the hot O, B, and A stars, the spectral sequence, and the observed differences between hot "giant" and "dwarf" stars. She used the newest developments of Meghnad N. Saha from the early quantum theory but tied them much more directly to the observational data than he and other theorists had done. She showed how the measured strengths of spectral lines could be used to derive the physical conditions in the stars' atmospheres. It was a revelation to Struve. This "newest branch of astrophysics" would clearly take over more and more from the older "astronomical spectroscopy [which] consisted almost exclusively of the study of the positions and displacements of lines." Now the strengths of spectral lines, and why they had those strengths, were what was important.[30] Struve embraced the new astrophysics eagerly. Payne's book had converted him. From 1926 on he was an astrophysicist, although his own almost complete lack of training in physics made it very difficult for him to derive new theoretical results. But he tried desperately to understand them, to test them, and to apply them for the rest of his life. He sought collaborators who did know physics and worked closely with them. But with his prodigious energy and drive, he never stopped doing the "old spectroscopy" too, always continuing to grind out radial-velocity measurements and to derive orbits from them. Often his spectrograms taken for these programs yielded new results that he and his collaborators could interpret astrophysically too.

All this activity did not go unnoticed. In 1926 Lick Observatory's Aitken wanted to add a young assistant astronomer to its staff, "to devote himself largely to research in astrophysical lines." He naturally thought of Struve, the demon observer with the famous name, who was producing data, publishing papers and presenting them at meetings, and branching out into astrophysics. True to the directors' code of that time, before approaching Struve, Aitken wrote Frost to inform him of the contemplated raid on his staff and at the same time ask for his recommendation. The Yerkes director replied that Struve "certainly ha[d] inherited the ability and energy of his illustrious ancestors." Frost regarded Struve as "one of the ablest of the young men who have ever worked here" and said that his departure would "greatly cripple" Yerkes Observatory. He made it clear that the University of Chicago would match any offer California might make to him.[31]

The very next day, Frost fired off a letter to Dean Henry G. Gale in Chicago, enclosing a copy of Aitken's letter and of his own reply. Struve's loss would be "fatal" to the Yerkes program, he wrote. That Lick wanted him proved how good he was. Frost's own opinion was "that there is not any young man in the world who has published half as much good spectrographic work in the last two years as has Struve." The Yerkes director recommended promoting him from instructor to assistant professor and raising his salary from $1,800 to $2,400 a year. Gale concurred with the raise but balked at the immediate promotion. Struve's salary went up to $2,400 that summer, but he had to wait until July 1, 1927, for his assistant professorship.[32] Aitken, realizing that Frost was serious about keeping his rising young star, never directly offered Struve the job. In the end Donald H. Menzel, Russell's bright young Ph.D. from Princeton, got it, as he was later to get two other jobs that Struve had first declined.[33]

When Struve began receiving his new, higher salary in the summer of 1926, however, he was already in California. He was not there to stay, but was on an observing trip. In December 1925, when Walter S. Adams had visited briefly in Chicago, Struve had gone to see him and ask for his permission to visit Mount Wilson that summer, to learn about the work going on and to use the telescopes. Adams had agreed and, along with Frost, had sponsored Struve's application to the International Education Board, a branch of Rockefeller philanthropy, for a fellowship grant to support his travel. Struve wanted to extend his study of the calcium clouds, using existing spectrograms in the Mount Wilson collection and obtaining new ones with its sixty-inch reflector of stars too faint for the forty-inch.[34] Struve was in correspondence with Paul W. Merrill, another hardworking spectroscopist like himself and a member of the Mount Wilson staff, on problems of the "detached" lines. Merrill considered that word better than "stationary," because it did not imply that the stars were all spectro-scopic binaries. It caught on quickly. He was interested in this prob-lem himself, but he stood aside and let Struve come to Mount Wilson to work on it.[35]

The International Education Board came through with the grant, and in June Struve and Mary traveled to California for the first time. In Pasadena he mined the Mount Wilson Observatory plate files for spectrograms of B stars and made quantitative estimates of the strength of the detached Ca II K line on as many of them as he could. (It is in a clear region of the spectrum at wavelength $\lambda 3933$; the other Ca II line, H, at $\lambda 3968$ is often blended with a hydrogen line and is therefore not as suitable.) Struve also obtained spectrograms of many fainter B stars with the sixty-inch and used them in the same way. Be-tween observing "runs" of a few nights each month at the telescope, he traveled to Lick Observatory, on Mount Hamilton near San Jose, and to the Dominion Astrophysical Observatory, at Victoria, British Columbia, to examine and study their spectrograms, estimate the in-tensities of the K lines in the same way, and add all these data to his collection. He also found a little time to travel around southern Cali-fornia with Mary on weekends. They particularly enjoyed San Diego and Coronado Island but were repelled by Tijuana, just across the border in Mexico. On the way back to Williams Bay by train at the end of September, Struve allowed time for a three-day stop in Ari-zona, one day so he and his wife could see the Grand Canyon, the other two for his working visit to Lowell Observatory in Flagstaff. Refreshed by his summer of work in California, he wrote to let Frost know when he would be back to Yerkes to start using the forty-inch refractor again. Struve wanted to be sure the director would assign him "a lot of time" with it.[36]

Home at Yerkes again, Struve did get many nights at the telescope. In the daytime he worked unceasingly on analyzing his calcium-line data. In preliminary form, he discussed it in frequent letters to Mer-rill, since no one at Yerkes could help Struve with it. Merrill did sup-ply many useful suggestions. Struve wanted to study how the strength of the detached calcium lines increased with distance. He used the ap-parent magnitudes of the stars as distance indicators, their faintness indicating their relative distances. Merrill pointed out that the B stars do not all have the same intrinsic luminosity, as they should for this method of relative distance determinations to work. The "earlier-type" B0 stars were known to be more luminous than the "later-type" B3 stars and therefore appeared brighter than B3 stars at the same distance. Merrill gave Struve the correct "photometric distance" formula to use, which takes into account both the intrinsic luminosity and the apparent brightness. Every student in an elementary astronomy course hears of it today, but Struve had never encountered it in his Yerkes training or, if he had, did not know about the differing luminosities (in the astronomers' words, absolute magnitudes) of different types of B stars. Merrill told him where to find the best values for them, based on Mount Wilson Observatory results, of course.[37]

Struve's big paper "Interstellar Calcium," published in 1927 as a contribution from Mount Wilson Observatory, was an important one. It contained far more data than any of the previous papers in this field. In a general way, it showed that the strengths of the detached calcium lines increase with distance (although he thought they leveled off or actually decreased at "large" distances, not realizing that this was actually due to the combination of a saturation effect in the lines themselves and the fact that the distant stars with strong calcium lines were likely to be too faint for him to observe). Struve further noted strong regional effects in the strengths of the detached lines, indicating the presence of actual discrete clouds rather than a general, homogeneous medium, and that there was a wide range of strengths of the lines in stars at similar distances but in different directions, confirming this patchy distribution. He also saw the general, but rough, correlation between strength of the detached lines and the "color excess" or reddening of the stars, indicating that the calcium lines formed in the same regions in which small, interstellar "dust" particles absorb and scatter starlight. Struve thought there was conflicting evidence against this picture, however, because portions of the Milky Way could be seen through "calcium clouds" as revealed by the detached lines. He did not realize how little calcium is required to produce these lines, even in relatively small, low-density clouds that contain only a small amount of dust.[38]

This paper was like many others by Struve. It showed his ability to recognize an important new field, get into it fast, collect a lot of data from existing plates, do the necessary remaining work (including making more observations) in a hurry, make rough but adequate quantitative measurements, draw quick conclusions that were mostly right but often incomplete, and get them into print as soon as possible. His papers stimulated others to work on the subject, and he himself returned to it frequently, each time adding to the data and using them to refine, correct, and extend his conclusions. His method was almost the exact opposite of the motto of the great mathematician-astronomer

Figure 20    Yerkes Observatory (1925). Courtesy of Yerkes Observatory.

Carl F. Gauss, "pauca sed matura" (few [papers] but definitive [ones]). Struve's method could almost be summarized as "many papers, all slapdash but packed with new results," although "many papers, all subject to later correction and extension" would be a better way to put it.

Before he submitted his paper for publication, Struve showed it to Frost for his approval, as every observatory director expected his staff members to do then. He also sent copies of it to Adams, Aitken, and Plaskett, since he was using results he had obtained at their institutions. Adams and Aitken were very positive in their responses, finding the paper "interest[ing]" and "a good piece of work." Joseph H. Moore, the spectroscopist at Lick whose work was closest to Struve's, considered it "a splendid article." [39] Plaskett, however, was another story. An older man (then sixty-one), he was the leading astronomer in Canada and the recognized international expert on observations of the detached calcium lines. Conservative and cautious, he was affronted that Struve had worked so fast, examined so many of the spectrograms that Plaskett had given him permission to use, and gotten so many data out of them. He "wondered" if it had been possible for Struve "to do justice to the material" in so short a time. Plaskett, his son Harry H. Plaskett (also an astronomer), and his collaborator Joseph A. Pearce felt that Struve was "basing rather sweeping conclusions on quite insufficient evidence" and that his observational data were "both scanty and uncertain." Plaskett detailed his criticisms, which basically were that Struve had too few spectrograms, had measured them sloppily, and had drawn unwarranted conclusions. He suggested that Struve add a disclaimer to his paper stating that as he did not have enough data, his results were "questionable." The Canadian director hoped that Struve would take his criticisms, "which will probably be more severe than any others on account of my experience and knowledge of the B stars and of stationary calcium, in the kindly spirit in which they are intended" and then ended his letter, "with all best wishes for the New Year."[40] Plaskett wrote this letter just a few days before Christmas; after the holiday, in a slightly merrier mood, he reiterated that the remarks he had made had been intended "in the friendliest spirit," for, he added, he had "the highest admiration of [Struve's] energy and skill in tackling and bringing to a successful conclusion the measures of the old [spectrograms of B stars] at Yerkes."[41]

Struve had no intention of questioning his own conclusions in a paper on which he had worked so hard. He gave Plaskett's objections "careful consideration" and then in a polite but firm letter analyzed and rejected them. He carefully showed Plaskett how the new data confirmed and strengthened his previous results. This could not be a matter of chance; it was a scientific result. Yes, the probable errors were large, but the averages over them were valid, and that is what he had based his conclusions on. Struve showed Plaskett's letter and his own reply to Frost and sent copies of them to Adams. Both accepted his reasoning and advised him to go ahead and publish his paper. Adams wrote Plaskett a soothing letter, agreeing with a few of his minor criticisms but saying that Struve's measurements and methods of analyzing the data were good and the conclusions he had drawn from them valid. Struve revised his paper, taking account of some of Plaskett's criticisms as well as the suggestions sent to him by Adams and by Frederick H. Seares, the Mount Wilson astronomical editor. Plaskett, mollified, approved publication, and Struve did go ahead with it.[42] When it appeared, "Interstellar Calcium" was an important paper, widely used and quoted.[43]

Just after Christmas, when Plaskett was writing out his kindly criticism for Struve, the young spectroscopist was at the American Astronomical Society winter meeting, held in Philadelphia that year. Some eighty-odd members were present, a relatively large number because of the high concentration of astronomers along the eastern seaboard. The Society met each year between Christmas and New Year, sometimes overlapping the latter date, so that the members could attend the meeting without missing any of the classes that nearly all of them taught. Only Yerkes, Mount Wilson, Lick, Harvard, and the Naval

Observatories had research astronomers who did not have to teach undergraduate students. Struve gave an oral version of his interstellar calcium paper at the Philadelphia meeting.[44] While there he met, among others, Cecilia Payne, whose book he had reviewed. Always eager for more data, he asked her about spectrograms of B stars at Harvard College Observatory, which might show detached calcium lines. She had been working on the B-star spectra herself and told him that some of the Harvard objective-prism plates might be suitable for his purpose. That summer Struve wrote Harlow Shapley, the Harvard College Observatory director, to ask his permission to come to Cambridge and use this material.[45] Then in his early forties, Shapley was one of the most important men in American astronomy. Born in Missouri, he had worked as a newspaperman before he became an astronomy student, earning his undergraduate and master's degrees at the state university and his Ph.D. under Russell at Princeton in 1913. On the Mount Wilson staff Shapley had done brilliant work on variable stars and globular clusters until he went to Harvard in 1920 as acting director, then became director in 1921. There he was the charismatic leader of a large staff, who worked under his close supervision on photographic plates, taken mostly with small telescopes at Harvard and at its Southern Hemisphere station.[46] They included direct images, as well as plates taken with a large prism in front of the lens, or "objective" of the telescope, which showed the spectra of all the stars (that were bright enough) in the field. Shapley was a great believer in cooperation, especially if he was in full charge and Harvard College Observatory received full credit for the resulting research, and he invited Struve to come and use the objective-prism plates.[47]

First the young Yerkes spectroscopist attended the AAS meeting at the nearby University of Wisconsin in early September. That was the other regular time for meetings in those years, just after Labor Day, before colleges and universities (except the University of California) had started classes. Madison was farther from the eastern astronomers than Philadelphia, and only fifty members were present for this meeting. A large contingent came from Yerkes Observatory, and Struve gave a paper on the spectrum of Nova Aquilae, a "new" star that had flared up to brightness only a few days before the meeting.[48] He always had fresh research to discuss, and he never attended a meeting without presenting a paper.

Then Struve, accompanied by his wife, set off for his first visit to Harvard College Observatory. He was to return to it many times and ultimately to be offered its directorship, which he declined. But in the fall of 1927 Struve was working on the interstellar calcium lines. Frost gave him permission to spend two months at Harvard on this work, but he had to pay his own traveling and living expenses.[49] Again Struve inspected all the plates he could and estimated the intensities of the detached calcium lines in the B stars that showed them, for which he did not already have data. As soon as he returned to Yerkes, he worked up the new results into another paper, further strengthening his conclusions that the calcium lines arise in clouds in interstellar space, so that on the average their strengths increase with distance, but that there is a wide range of strengths of the lines in stars at comparable distances because of the irregular cloud distribution or "regional differences."[50] He sent the paper, a long one, to Shapley for his approval. The Harvard director circulated the paper to a few of his staff members and relayed their comments to Struve. They were mostly positive from Payne, who had discussed Struve's work frequently with him while he was at Harvard, but somewhat negative from Harry Plaskett, now at Harvard. Shapley's own main concern was that in addition to publishing the paper in the *Astrophysical Journal*, as Struve wished to do, he should prepare a short summary of it for the *Harvard Bulletin*, their own publication. This would be "appropriate" (meaning it would remind readers that the paper was based mostly on data Struve had secured at Harvard), and it would give him "the opportunity to present results in a concrete manner for the busy readers" (like Shapley himself). Struve was glad to oblige and drafted the short paper within a few days, sending it to Shapley with his rebuttal of Harry Plaskett's criticisms.[51] Mary Struve had been ill for much of the two months they spent in Cambridge; this was the first recorded sign of what later was to become a recurring motif of her life. She may have suffered a miscarriage at this time, but the very guarded language that educated people then used makes it impossible to do more than speculate.[52] Her illness was probably the reason Struve did not attend the AAS meeting at the end of December, at Yale that year, and give a paper as he had intended to do.

Nevertheless, Struve was eager to go farther away and to stay longer. In his latest publications on the detached lines, he had discussed the results in terms of a recent theoretical paper by Eddington. The great English astrophysicist, then in his early forties, had already made very important contributions to the theory of stellar structure, stellar atmospheres, stellar dynamics, and general relativity.[53] In 1926 he had published his long paper "Diffuse Matter in Interstellar

Space," a very general discussion of the subject. In it he showed that interstellar gas, because of its extremely low density, would be ionized appreciably by the very dilute (faint) ultraviolet radiation of distant hot stars. The calcium lines, technically Ca II H and K, were merely a convenient marker of this ionization, not a sign that calcium was an abundant element in interstellar space. It was not, but calcium was one of the very few elements that, in its ionized form, has absorption lines in the observable part of the spectrum that arise from its "ground" (or lowest) energy level. Again because of the very low density, all atoms and ions in interstellar space spend almost all of their time in their ground level, so that only absorption lines arising from it can be observed. The detached lines really showed the existence of interstellar clouds, in which the trace element calcium, for complicated atomic-physics reasons, happened to be observable. Likewise Eddington showed conclusively that the reddening, or color excesses, of distant stars indicated the presence of fine dust particles ("meteoritic" particles, he called them) in these same clouds. A true theorist, he admitted that he would have preferred it if only gas—so much easier to treat mathematically—were present, but the observations showed that the dust was there too.[54] Struve's data were fleshing out this picture, making it more specific, more detailed, and therefore more complicated. In particular, Eddington realized that the interstellar gas was strongly concentrated to the galactic plane and treated it in the simplest approximation as a uniform layer. Struve's observations showed it was much more patchy than this; it had a complicated cloud structure. In this aspect his data agreed more closely with J. S. Plaskett's almost purely observational picture of "calcium" clouds. The Canadian director, however, believed that each B star had its own cloud, which gave rise to the observed detached lines in its spectrum. Struve's measurements, which showed that the strengths of the lines increased, on average, with distance, disproved this simple model (and aroused the Plasketts' criticisms).

Struve wanted to go to the source and study with Eddington at Cambridge University. He applied for a Guggenheim Fellowship, which would pay for his travel to Europe and his living expenses there for a year. Besides his own director, Frost, who would recommend him as a matter of course, Struve asked Adams and Shapley, who now knew him and his work well, to write to the Guggenheim Foundation on his behalf. No American astronomer could conceivably do better than to have statements of support from the directors of the three leading observatories in the country (though the Lick director might have contested that statement). In his application to the Guggenheim trustees, Struve summarized the observational results he had obtained and his goal of using them to determine the physical conditions in interstellar space. Eddington had laid down the foundation for this study in his classical work and was ready to have Struve come to Cambridge and work on it. Furthermore, the Yerkes spectroscopist wrote, it was generally believed that although the younger generation of American astronomers were doing good observational work in astrophysics, there were almost no theorists among them. He was right; probably the only two capable in theory, and using it to interpret their observational work effectively, were Payne and Menzel, both protégés of Russell. Struve wanted to remedy this, he said, by studying under Eddington, for the future benefit of Yerkes Observatory and the University of Chicago.[55]

Adams and Shapley supported his application nobly. Both wrote that Struve's character and personality were of the highest quality and that he could do excellent research. He had outstanding drive and the ability to choose problems that were important and that he could solve. Shapley called Struve "one of the best men of his years" in American astronomy, "and possibly the best astrophysicist in the middle west." Adams, always a bit more restrained, stated that his only criticism of the young Yerkes observer was that his conclusions were "occasionally . . . a little too radical," echoing in subdued form J. S. Plaskett's complaints to Struve himself. The Mount Wilson director, however, opined that this fault was "due in part to [Struve's] comparative youth, and in part to the fact that he has been obliged to work quite individually at the Yerkes Observatory, without associate[s] capable of criticizing fully his results." Presumably, therefore, these "faults" would disappear as Struve matured and got away to Cambridge. Both questioned whether Struve, with his extremely modest background in physics, would be able to learn much from the notably "reserved," uncommunicative theorist Eddington, but both advised that he should get the fellowship anyhow, because he would surely accomplish a lot with it.[56]

Struve did get the Guggenheim Fellowship, and Adams and Shapley were both right. He did a lot of research, but he did not add appreciably to the theoretical understanding of interstellar matter or greatly improve his own knowledge of quantum mechanics. Struve and Mary sailed from New York on the *Aquatania* in early August.[57] They stayed in Cambridge for eight months. Struve rediscussed all his interstellar line data and, in a long paper he wrote in Cambridge, showed how the detached calcium line strength in an early B-type star's spectrum could be used to estimate, statistically, its distance.[58] Struve attended several of the regular monthly meetings of the Royal Astronomical Society (RAS) in London, gave a paper at one of them, and joined the Society. He was twice invited to the dinners held after the regular meetings by the RAS Club, the small, self-perpetuating in-group of officers, directors, and senior members of the Society.[59] While he was in Cambridge, he also got into correspondence for the first time with the young Dutch theorist Jan H. Oort, who was very interested in using the interstellar calcium lines as distance indicators.[60] In the spring Struve went over to Leiden for a short visit, met Oort in person along with his older, more famous colleagues Ejnar Hertzsprung and Willem de Sitter, and gave a talk on his own research. The Struves returned to America and to Williams Bay in May 1929. Again there are vague hints that they were hoping for a child, but they certainly did not have one. Mary was described as ill and slowly recovering afterward.[61]

At Harvard, and while in England, Struve had begun collaborating with Soviet Russian astronomers. Boris P. Gerasimovich, eight years older than Struve but four years younger than Shapley, spent the years 1926 to 1929 at Harvard College Observatory. He had been a student of Struve's father at Kharkov. Gerasimovich was better trained in mathematics than in physics but had converted himself into an astrophysicist. He had given the few lectures on spectroscopy that Struve had attended in Russia. Gerasimovich was renowned for fiercely criticizing almost every paper he read; he believed he was helping the author by doing so. At Harvard he became quite interested in Struve's work on the interstellar absorption lines. They began a theoretical paper while Struve was still there and completed it by correspondence just before he sailed for England. Gerasimovich provided many mathematical equations for this contribution, but the one new and significant point it added to Eddington's classic paper was that the velocities measured from the calcium detached absorption lines followed the expected galactic rotation formula recently derived by Oort, for an average distance half as great as the average distance to the stars. This indicated that these lines originated along the paths to the stars and that they therefore were truly interstellar absorption lines.[62]

A few years later, Gerasimovich was to return to the Soviet Union and become first the head of the astrophysics section at Pulkovo Observatory and later its director. But he was denounced, arrested, and shot in Joseph Stalin's great antiforeigner purge of 1936–37. In 1936 Struve was certain only that Gerasimovich had been arrested and disappeared, but he believed the rumors of his friend's execution, which were confirmed after World War II. Struve distrusted and hated the Communist regime; Gerasimovich's fate was, after his own father's and brother's deaths, probably the most significant specific case that fed this hatred.[63]

Grigory A. Shajn, five years older than Struve, was a son of the working class who had fought as an enlisted man in the czar's army against the Germans in World War I. After the Treaty of Brest-Litovsk, he had studied astronomy as a scholarship student in eastern Russia and had worked in Siberia but had not actively taken part in the Revolution. In 1921 Shajn was appointed to the staff at Pulkovo Observatory, and in 1924 he was put in charge of the new forty-inch reflector of the Simeis Observatory in the Crimea, which had been imported from England. With it, Shajn began an active program of radial-velocity spectroscopy. Struve and Shajn exchanged letters and found that they were both interested in measuring the rotational velocities of stars. Rotation broadens all the lines in a star's spectrum, because one edge, rotating away from the observer, has a positive radial velocity, the other edge, rotating toward the observer, has a negative radial velocity, and the parts of the star's atmosphere along the projection of its axis are all moving at right angles to the line of sight and hence have zero radial velocity. Struve and Shajn had realized independently that they could calculate the expected form, or "profile," of an absorption line "broadened" in this way by rotation. The larger the rotational velocity, the broader the line would be, and its "width" would measure its rotational velocity. Shajn checked this reasoning observationally by taking a spectrum of the entire image of the planet Jupiter, known from its surface markings to rotate with an equatorial velocity of approximately 25 km s$^{-1}$. The observed profile agreed well with their calculations. In their first paper, Shajn and Struve applied this method to the stars in spectroscopic binaries and showed that some of them have very large rotational velocities. They wrote their paper on these stars completely by correspondence, mostly while Struve was still in America, before he came to England. They completed it while he was in Cambridge, and he presented it orally at the meeting at which he became a member of the Royal Astronomical Society.[64] Struve always admired Shajn, who was famous for not criticizing other astronomers but instead trying to make the best of whatever they had to offer. He was widely regarded as one of the few heroes of Soviet astronomy during the Stalin years of terror. He gave Gerasimovich's widow a job as librarian at Simeis Observatory after she was released at the end of her term of eight years in a labor camp, her sentence for being the wife of a "traitor." Shajn had never denounced anyone but had quietly done his work as a scientist and an observatory director, and he had survived.[65]

Other Russian astronomers, especially the older ones who had known Struve's father, communicated with United States government officials through him, since the Soviet Union had not been recognized and had no ambassador in Washington. They could write him in Russian; he translated their letters into English and turned them over to Frost for transmission to the appropriate office or bureau.[66]

In addition to his work on stellar rotation (which he and Shajn realized applied to single stars just as well as to the individual components of binaries), Struve also began his research on the Stark effect in stars while he was in Cambridge. This is the broadening of spectral lines, especially of hydrogen (H I), and to a lesser extent of helium (He I), that occurs in stars whether they are rotating or not. It can therefore be most conveniently studied in nonrotating stars, identified by their extremely narrow lines of carbon, nitrogen, oxygen, magnesium, and other elements. In hot stars, especially O, B, and A stars, many atoms are ionized, and the free positive ions produce small-scale electric fields. Such fields slightly change the energy levels of atoms, especially H I, and to a lesser extent He I. The closer an ion is to an atom, the larger the electric field acting on the atom, and the larger the resultant shifts of the energy levels. These shifts split up the spectral line emitted by a single atom into several "components"; the spectrum observed from the billions of atoms in a gas or in a stellar atmosphere is a complicated average over all these components for all field strengths and thus broadens the line. This is called the "Stark effect" for the German physicist who predicted it and detected it in the laboratory. Before 1928 it had been suggested several times that this Stark effect was the cause of the observed broadening of H I and He I lines in stars, but the observational evidence seemed contradictory.

Just before Struve came to Cambridge, Eddington had reviewed this problem from a theoretical standpoint and concluded that the Stark effect must be the main broadening mechanism in nonrotating hot stars. Struve had probably been considering the problem since his conversion to astrophysics by Payne's *Stellar Atmospheres*. Now, stimulated by Eddington, he leaped into it. He had a lot of observational material on O, B, and A stars from the Yerkes radial-velocity spectrograms and was intimately familiar with the spectra of these stars. At Cambridge he quickly assembled his data and convincingly demonstrated that the observed details of the broadened lines did indeed fit the predictions of the Stark effect fairly well. None of his tests depended on detailed theoretical calculations; instead they were based on the ways the broadening of the H I lines and the He I lines were correlated with one another, the way one He I line's broadening was correlated with another, the appearance of "forbidden" He I lines when other "permitted" lines were greatly broadened, indicating especially high densities and hence large electric fields, and so forth. It was a very convincing paper, which he rushed into print to establish his own priority in the field.[67]

As soon as Struve got back to Yerkes Observatory that summer, he followed up with several more papers, confirming, extending, and applying his first results on the Stark effect. He had hit his stride as an amazingly productive observational astrophysicist.[68] Struve also finished a long paper reporting the results of the Yerkes Observatory program on the radial velocities of five hundred A stars. As in the earlier paper on the B stars, he was listed as the third author, after Frost and Barrett. They had done much of the early observing, but he had done much of the latter part of it, and most of the measurements, then pushed the paper through to publication.[69] He had demonstrated that he could continue the old work at the same time that he started the new.

But Struve had had enough of Yerkes Observatory. The big radial-velocity programs for which Frost had first hired him as an assistant were finished. It did not make sense to start new ones. The forty-inch telescope was antiquated and its spectrograph even more so. Yerkes could not match the much bigger telescopes and better spectrographs at Mount Wilson; Struve wanted to go there permanently. He wrote Adams and asked him for a job on the Mount Wilson staff or at "the new observatory." By this he meant the planned two-hundred-inch telescope to be built with the funds George Ellery Hale had secured the previous year from the Rockefeller Foundation. In the event, it was not to be completed and go into operation until 1948, after World War II, but in 1929 no one knew that. Struve told Adams that he was not dissatisfied with his present rank (assistant professor, promised a promotion to associate professor in 1930) and salary ($3,500), but that he was hungry for the research opportunities the big telescopes at Mount Wilson would give him.

Adams replied that he would like nothing better than to offer Struve a position. The Mount Wilson director knew that "with the opportunities which are available here [Struve] would make most important scientific contributions." But it was impossible to add any more astronomers to the Mount Wilson staff. The Carnegie Institution had a fixed budget, and new appointments were rare. Adams wrote that he would discuss with Hale the possibility of hiring additional staff members for the two-hundred-inch, but if so it would undoubtedly be years in the future. His answer amounted to a polite no.[70]

Everything Adams said was true. The Great Depression had already begun, and the Carnegie Institution was strapped for funds. Few new appointments were made at Mount Wilson Observatory in the next decade. But undoubtedly there were two other factors in his decision that he was too diplomatic to mention. One was that Struve's field was too close to that of several first-rate astronomers Adams already had on his staff—Merrill, Alfred H. Joy, and Roscoe F. Sanford. Even though Struve was good, Mount Wilson Observatory did not really need another stellar spectroscopist. The second reason was that Struve was a foreigner by birth, and Adams felt "it would be undesirable to have too large a proportion of Europeans on [an observatory's] staff, even [ones who] are thoroughly competent. The backbone of the staff of an American observatory should, I believe, be American, partly because of the essential differences in viewpoint in nationalities, and partly because . . . [a] European can never represent an American observatory and have the influence that an equally able American would have." When Adams wrote that, he was advising Frost not to hire the Swede Bertil Lindblad on the Yerkes staff, but he was also thinking about his own Mount Wilson staff members Adriaan van Maanen, from Holland, and Gustaf Strömberg, from Sweden.[71]

Although he did not have a staff position for Struve in 1929, Adams did hire the German Walter Baade to the Mount Wilson staff just two years later. He was not even a naturalized American citizen (as Struve had been careful to let Adams know that he was) and had spent only one year in the United States previously, as an International Education Board fellow. Baade had two advantages over Struve; the young German worked on globular clusters and galaxies, an underrepresented specialty at Mount Wilson, and he had a warm, outgoing, extremely pleasant personality.[72] Everyone, even the quiet, reserved Adams, was attracted to Baade, while Struve always remained somewhat stiff, formal, and aloof.

Although there was no job for him at Mount Wilson, Struve wanted to come back for more observing. He asked if he might spend six weeks there in the summer of 1930, as a "volunteer" on his "vacation," at his own expense. The Yerkes spectroscopist wished to extend his work on spectral-line profiles, especially of the helium lines in B stars, to study the Stark effect in more detail. Mount Wilson was clearly the place to do it. Its one-hundred-inch reflector could collect much more light than the forty-inch refractor, and its coudé spectrograph could spread the light out to produce higher dispersion and consequent improved wavelength resolution, to measure the absorption-line profiles in detail. Struve's requests for observing time were always models of politeness, stressing that he did not want to infringe on whatever work the staff members on the scene were doing, and at the same time providing a mass of well-thought-out plans and the reasons his program was important. Adams granted Struve the observing time he requested, saying that the research he had proposed would be an "excellent addition" to the work Mount Wilson astronomer Theodore Dunham was doing on line profiles.[73] Struve and Mary went west for six weeks in June and July, and he got the data at the telescope. By early August they were back in Williams Bay, and he had composed a "minor contribution" for publication in the *Astrophysical Journal* on the interstellar calcium line in a spectroscopic binary, based on spectrograms he had obtained with the sixty-inch just a few weeks before. This had been only a small part of his program; the main work on the line profiles with the coudé spectrograph at the one-hundred-inch required many computations, he informed Adams, and he would not be able to finish them and write the paper until September.[74] The contrast between the energetic Struve and his own staff members, particularly the perfectionist experimenter Dunham, whose projects seemed to drag on forever, could not escape the notice of the Mount Wilson director.

The first Ph.D. student whose thesis Struve supervised was Elvey, who had been a fellow graduate student at Yerkes in the summers of 1922 and 1923. Elvey had then been completing his master's degree at Kansas. He stayed there as an instructor until 1925, then was at Northwestern for three more years before finally coming to Yerkes in 1928. Struve put him to work in all kinds of stellar spectroscopy. At first Elvey planned to do his Ph.D. thesis on the spectra of novae ("new" stars), using existing spectrograms taken over the years at

Yerkes and some he had borrowed from other observatories. But under Struve's tutelage he switched to a detailed study of line profiles in normal stars, interpreting them in terms of the Stark effect and rotation, using the methods his young supervisor had pioneered to learn more about the physical properties of the stars.[75] After completing this thesis in 1930 and receiving his Ph.D., Elvey remained on the Yerkes staff as an instructor.

William W. Morgan, as we have seen, was Struve's second Ph.D., in 1931. As an undergraduate for three years at Washington and Lee, Morgan had majored in English literature. Although he had been interested in astronomy and had taken mathematics and physics courses there, he had to spend the full 1929–30 academic year on the Chicago campus to remedy the gaps in his education. Morgan did not really like the required courses in celestial mechanics and advanced physics; his forte was taking direct plates and spectroheliograms of the sun, and spectrograms of stars, and studying them for patterns— similarities and dissimilarities. He wanted to do his thesis with Struve, who suggested he work on the "peculiar A stars."[76] These stars, a minority of the "normal" A stars, show spectral lines of various heavier elements in their spectra, including some rare-earth elements. Morgan and Struve were both familiar with several of them from the A-star radial-velocity program. Morgan's thesis was an excellent one, in which he classified the brightest peculiar A stars into several groups and studied a few of them in detail.[77] Struve recommended Morgan very strongly for a Yerkes faculty appointment when he completed his Ph.D. He was "a young man of great promise . . . [who had] already done work in stellar spectroscopy that is of outstanding merit." Furthermore, Morgan worked hard and liked to observe, always highly desirable qualities in Struve's eyes. Salary, however, was another matter. Morgan had been an assistant for several years and was making $1,500 a year. Struve coupled his praise of Morgan's "independent work of great originality" with the suggestion that this young paragon be given a $300 raise, to $1,800 a year, effective when he got the promotion—six months after his Ph.D.[78]

Elvey and Morgan were both to play important roles later in the regeneration of the University of Chicago Astronomy Department under Struve, Elvey at McDonald Observatory and Morgan at Yerkes. In that same year, 1931, another young man, almost exactly Morgan's age, first came to Yerkes Observatory. He was the Belgian Pol Swings, destined to become Struve's closest collaborator and to return to Yerkes and McDonald time after time to work with him.

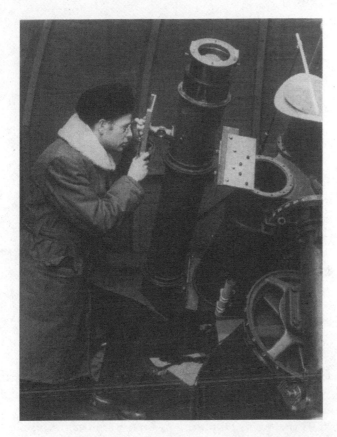

Figure 21
William W.
Morgan at the
Bruce photo-
graphic telescope
(1935). Courtesy
of Yerkes
Observatory.

Swings was in many ways almost the exact antithesis of Morgan. Born in a small town in Belgium, he had been a brilliant student of physical sciences and mathematics at the University of Liège. He earned his doctorate at age twenty-one, with a highly mathematical thesis on planetary orbits in general relativity and also in a modification of classical mechanics with generalized, velocity-dependent potentials. Then he switched to astrophysics and became an expert in spectroscopy, just as quantum mechanics was making it a true theory-based science. Swings spent 1927–28 at Paris and Meudon Observatory, the French center of astrophysics, under a Belgian government fellowship. Back at the University of Liège he had helped set up a spectroscopy laboratory, then had spent nearly two years at the Institute of Physics at the University of Warsaw, Poland, a world center of research in laboratory spectroscopy. The experimental data he obtained in Warsaw earned him a second "special" doctorate at Liège in

1931 and a fellowship to spend four months in the United States working in astrophysics.[79]

Swings wanted to use the fellowship at Mount Wilson. He knew theoretical physics and laboratory spectroscopy well, but he had very little experience in astrophysics. Pasadena, where he could use existing spectrograms for research, measuring, analyzing, and interpreting them, seemed to him the ideal place. However, when he and the secretary of the Belgian Research Council wrote Adams for permission for Swings to go there, completely financed by his fellowship, the Mount Wilson director rebuffed him. Adams was already committed to having three other visiting astronomers at the same time Swings wanted to be there. The observing program would be "exceptionally full." Adams seemed to believe that Swings would need more spectrograms than were on hand and would want to use the telescopes, although his application had clearly stated that he did not. At any rate, the answer was a firm though diplomatically phrased no.[80] The reasons Adams gave were certainly real ones. But in addition, his antipathy to foreigners, the fact that Swings's letter, although written in English, was completely unidiomatic and practically a literal translation from French, and especially the reality that the young Belgian had no observing experience were undoubtedly also factors in the Mount Wilson director's decision.

Once he had learned that his first choice was out, Swings tried Yerkes Observatory as the next best place to gain experience in astrophysics in the United States. Foreign visitors with their own funds were few and far between there (apparently Baade in 1926 had been the most recent one), and Frost was happy to accept him.[81] Van Biesbroeck undoubtedly had put in a good word for his young compatriot, if any was necessary. Swings came to Yerkes and worked with Struve for three months in the fall of 1931. The result was two published papers, written jointly by the two of them, one on emission lines in the spectra of hot stars, the other on molecular bands in the spectra of cool stars.[82] Swings supplied the impetus and the theoretical ideas; Struve provided the astronomical insights and access to the observational material.

Swings and the Belgian Research Council official who had arranged his fellowship seized on a phrase in one of Adams's letters as an invitation for him to come to Pasadena for a brief visit at the end of his stay in America. By then Swings, a keen linguist, could speak English much better than he had written it from Belgium seven months before, and he charmed Adams with his Old World courtesy, his passionate work ethic, and his keen interest in observational astrophysics. In subsequent visits to the United States, Swings was always welcome at Mount Wilson.[83]

Struve too had learned how good a collaborator Swings was, strongest in exactly the subjects where he himself was weakest—theoretical physics and laboratory spectroscopy. He appreciated the Belgian's willingness and even eagerness to work long hours. Struve kept in close touch with Swings, answering his observational questions, suggesting new topics for him to study, sending him spectrograms to measure, and requesting more experimental data on spectra from him.[84]

Thus by the end of 1931 Struve's career was well under way. He had started all his pathways into the future. He was a survivor, a doer, a hard worker. With only a poor, fragmentary training, almost entirely in the older positional astronomy and in the "old new" astronomy of radial velocities, he had made himself an observational astrophysicist. He had worked at all the major American observatories and had visited England and the Continent on a research leave. He probably knew more astronomers abroad than nine-tenths of contemporary American astronomers did. He had trained Elvey and Morgan, two of the young astronomers who would work with him for years, and had helped another, Swings, become a productive astrophysical research worker. Now Struve was ready to become a director.

CHAPTER FIVE

# The Boy President, 1929–1932

In May 1929, as Otto Struve's year at Cambridge University was drawing to a close, he received a letter from his director, Edwin B. Frost, informing him that the University of Chicago had found its new president, Robert M. Hutchins, a "young man of thirty who has achieved distinction as dean of the law school at Yale."[1] It was high time for a new chief executive; the University of Chicago had suffered through two short presidencies, those of the elderly Ernest D. Burton (1923–25) and the mercurial Max Mason (1925–28), leaving long-time vice president Frederic Woodward in charge as acting president until a new president could be selected and brought in.

Woodward himself had been a professor in the Law School, and as vice president he had interested himself in strengthening the educational program of the College, the undergraduate part of the University of Chicago. This made him suspect to the senior professors, who were far more interested in research in their own fields and in graduate students. They feared that if Woodward were named president, he would divert resources from their departments to the Law School and undergraduate teaching. Hence they vetoed him, and all other potential candidates from inside the university were eliminated on much the same grounds. The search then shifted to outside candidates, and Hutchins emerged victorious.

Immediately dubbed the "boy president" by the Chicago newspapers, he had graduated from Yale only eight years before. Voted by his fellow students of the class of 1921 the man most likely to succeed, he had become an English and history teacher at Lake Placid School, located in upstate New York in the fall and spring and in Coconut Grove, Florida, during the winter. It catered to wealthy boys

Figure 22    Frederic Woodward, vice president of the University of Chicago.
Courtesy of the Department of Special Collections, University of Chicago
Library.

who had flunked out of more prestigious prep schools. This was a job
far below the talents of the brilliant, charming, tall and strikingly
handsome Hutchins, and a year later James R. Angell, the new president of Yale, hired him as secretary of the university. His duties included working with the Yale Corporation (the equivalent of the
University of Chicago's Board of Trustees), public and alumni relations, raising money, public speaking, and supervising student regis-
tration and commencement exercises. Nevertheless Hutchins found
time to take courses in the Yale Law School in summers, late after-
noons, and early mornings, and he earned his LL.B. degree in 1925.
Angell was his mentor; Hutchins watched him closely and learned
from his mistakes as well as his triumphs.

Immediately on graduating magna cum laude from the Law School, Hutchins was appointed to its faculty as an instructor. He was a brilliant success as a teacher, and after only two years, in 1927 he was promoted to acting dean and associate professor. Less than a year later, in early 1928, he became a full professor and the "permanent" dean. A vigorous activist, Hutchins made things happen at the Yale Law School and attracted wide attention. He was a natural if daring choice for the presidency of the University of Chicago at the end of the booming 1920s.

Hutchins's father, William J. Hutchins, was an ordained Presbyterian minister, a liberal churchman. He had preached the social gospel and "applied Christianity" in Brooklyn, New York, where Robert Hutchins, one of three brothers, was born. In 1907 William Hutchins took his family to Ohio to become professor of homiletics (preaching) at Oberlin College, a center of liberal evangelical Protestantism. The college had been committed to education for women and blacks from its founding, long before the Civil War, and had sent many of its sons and daughters to China as missionaries. Robert Hutchins was an undergraduate at Oberlin for two years, 1915–17, then served in the United States Army Ambulance Corps during World War I, with his brother Will. Their unit had been recruited at Oberlin, and Robert returned there for classes in the summer of 1919 before entering Yale. His father left Oberlin in 1920 to become president of Berea College in Berea, Kentucky, a liberal Christian institution with close ties to Oberlin. There he worked actively to promote interracial understanding and to prepare the students, many of them quite poor and all of them white (in his time colleges and universities in Kentucky were segregated by law), for practical, healthy lives close to the land. Robert's brother Francis was to succeed their father as president of Berea in 1939, and their brother Will was a longtime teacher and then headmaster at the Asheville School in North Carolina. The "boy president" of the University of Chicago came from a family with a strong educational tradition of service, reform, and morality. His minister father had publicly announced that he did not believe in the literal truth of the Bible; Robert Hutchins himself had learned to smoke, drink, and swear in the army and to deal with professors as well as with the rich and famous at Yale. At thirty he was no longer a boy, and he was ready to be a university president.[2]

Hutchins came to Chicago in 1929 as the choice of the Board of Trustees and the special favorite of its chairman, Harold H. Swift,

Figure 23 Robert M. Hutchins, when he became president of the University of Chicago (1929). Courtesy of the Department of Special Collections, University of Chicago Library.

head of the giant meatpacking corporation that helped make Chicago famous as "hog butcher for the world." Everyone predicted great success for the brilliant new boy president, with so many analogies to the first brilliant boy president, William Rainey Harper, who had also come to Chicago from New Haven. But between April 1929, when Hutchins accepted the presidency, and November, when he was formally inaugurated in Rockefeller Chapel four months after starting on the job, the whole world had changed. The stock market crash in October set off the Great Depression. The University of Chicago's endowment shrank rapidly. Initially it seemed a perturbation from which the economy would soon recover, but that was not to be. Once-wealthy trustees and donors were to find themselves increasingly impoverished in the coming years. Students and their families would have to dig deeper and deeper in their pockets for tuition, dormitory fees, and the cost of books. Hutchins was to become the manager of an educational institution that, like all others, had to cut costs to survive. Many of the decisions that were to affect Yerkes Observatory in the next decade were made in that frame of reference.

By January 1931, in the middle of Hutchins's second year as president, Dean Henry G. Gale was passing down the word to Frost that Yerkes had to "save" (return) some money from its budget, to help the Physical Sciences Division as a whole meet its quota of $15,000. Otherwise the accountants foresaw a substantial deficit. Still optimistic, Gale assured Frost that whatever saving he could make this year "is not to be undertaken as indicating a permanent reduction in the budget." Frost was proud that "we have always operated with extreme economy here," but he promised to do all he could to meet the president's and dean's demands for economy.[3]

The trustees had long felt that the University of Chicago administration needed strengthening, and as a step in this direction Hutchins, with the approval of the faculty, had formally set up four divisions, with deans, through which the departments communicated with the central administration.[4] This made little real difference to Yerkes Observatory, which as a research and graduate-student unit had previously dealt mostly with Gale in his capacity as dean of the Ogden Graduate School of Science; now they dealt with him in his new post of dean of the Physical Sciences Division. Frost continued to write directly to Woodward, who remained vice president, on many administrative matters. In May 1931, near the end of the new president's second year, Frost had still not met Hutchins and was trying to get him to visit the observatory. Although the director's immediate aim was to persuade the president not to be so free and easy in giving astronomy clubs and classes from competing colleges permission to visit the observatory at night and look through the forty-inch (thus disrupting research, from the professional astronomer's viewpoint), he also had a much more important matter to discuss. This was the question of his successor as director of Yerkes Observatory.[5]

Frost was due to retire on June 30, 1932, at age sixty-five. He did not want to do so, and when an item appeared in the *Chicago Daily News* in the summer of 1931, suggesting that the trustees might waive the normal retirement age and keep him on for a few more years, he eagerly wrote Woodward that he would be happy to continue if asked. "No one likes to retire if he is still keenly interested in his work," he wrote. Frost dictated this letter to his wife, who wrote it by hand to preserve its confidentiality; he could not write it himself because of his blindness.[6] But there was no thought in the central administration of keeping him on. Gale, Woodward, and Hutchins were all quite conscious of the highly negative evaluations of the Yerkes director that Walter S. Adams and Harlow Shapley had registered.[7] The *Daily News* story in fact was a garbled account of an appeal that a wealthy Lake Geneva resident had made to her friends Hutchins (her houseguest one weekend) and Laird Bell, another member of the Board of Trustees, that Frost and his wife be allowed to continue occupying the director's house at Yerkes after his retirement, until his death. Woodward politely but firmly told Frost that everyone was "conscious of his distinguished service" but that it was the administration's duty "to keep our eyes turned to the future." Though he was old enough to know how Frost felt, the vice president wrote, his experience had taught him "the importance of new blood and new ideas and the consequent advantage of keeping the management of affairs largely in the hands of younger men who still have a career before them." Privately Woodward and Swift agreed that both the new director, whoever he might be, and Frost would probably be happier if the latter moved far away, but in any case he could not expect to keep the house. "It will be difficulty enough to find a first-rate man who is willing to accept the directorship without complicating the matter by telling him that the most suitable house on the grounds will not be available for his use," the vice president advised. Frost would have to move out.[8]

Frost's own choice for his successor as director was Joel Stebbins of the University of Wisconsin. Stebbins was an outstanding research scientist and a pillar of the astronomical establishment. Born in 1878, he had earned his undergraduate degree and done a year of graduate work at the University of Nebraska, his native state. Then he had spent another year as a graduate student at the University of Wisconsin, moved on to Lick Observatory and the University of California for his Ph.D., and in 1903 joined the faculty of the University of Illinois. There in 1907, with its twelve-inch refractor, he became one of the earliest pioneers in photoelectric photometry, measuring quantitatively the brightness of stars by the weak electrical currents their light, collected by a telescope, produced. From the first it was much more accurate than the old visual photometry. Initially Stebbins used a selenium photoresistive cell, but by 1913 he had switched to a more sensitive photoelectric cell. He moved on to the University of Wisconsin in 1922, continuing his photoelectric work with its 15-1/2-inch Washburn Observatory refractor on the campus in Madison. Stebbins was a protégé of William Wallace Campbell, the powerful director of Lick Observatory from 1901 until 1930; he became secretary of the American Astronomical Society in 1918, was a member of the American delegation to the meeting at which the International Astronomical Union was organized at Brussels in 1919, and was elected to the National Academy of Sciences in 1920. Through these activities, Stebbins was closely allied with Adams and George Ellery Hale, and in 1931 he was named a research associate of the Carnegie Institution of Washington, which allowed him to use the sixty-inch and one-hundred-inch reflectors for photoelectric work on fainter, more distant stars on summer observing trips to Mount Wilson.[9]

In 1926 Frost had first broached the idea of bringing Stebbins to Yerkes Observatory as assistant director, presumably to succeed him as director when he retired.[10] By the fall of 1928 Frost had evidently sounded Stebbins out unofficially and had satisfied himself that he was interested. The Yerkes director proposed that Stebbins build a photoelectric photometer in Madison, to be paid for by the University of Chicago, bring it to Yerkes, install it on the forty-inch refractor, and spend the summer of 1929 using the instrument there as a visiting professor. Madison was close enough to Williams Bay that Stebbins could get back and forth by car in two hours or so and spend an occasional night at Yerkes during the school year testing the photometer. Stebbins agreed to this proposal and provided well thought out time and cost estimates for building the photoelectric equipment and careful suggestions, based on his own experience, on how it might best be used. In particular he emphasized that photoelectric photometry required clear skies and that in the uncertain midwestern weather it would be best to keep it mounted on the smaller twelve-inch refractor at Yerkes, so that it could be used at a moment's notice but could also be quickly transferred to the forty-inch if the sky conditions improved.[11]

By January 1929 Frost had the funds to pay for the photometer and approval to bring Stebbins to Yerkes for the summer. The Wisconsin professor had emphasized the importance of having one person at Yerkes who could work with him, learn his techniques, and carry on the photoelectric observing. Frost, looking backward, thought in terms of a new faculty member to replace the recently deceased John A. Parkhurst, who had done visual and photographic photometry for years with little apparent result. However, the Chicago administration did not want to make any new long-term appointments at Yerkes until a new director had come on board. Hence Christian T. Elvey, then a graduate student working on his thesis, whom Stebbins had suggested—apparently because he had the most knowledge of what later came to be called electronics—was assigned to work with him. Stebbins kept close watch on the photometer as his instrument maker, Oscar Romare, built it in Madison and was in constant touch with Frost about the project.[12]

In April the photometer was far enough along for Stebbins and Romare to take it to Yerkes and check that it would fit on the telescopes; on the same trip Stebbins found a room he could rent in Williams Bay for the summer. By mid-June Romare had completed the photometer, and Stebbins brought it to the observatory.[13] Stebbins and Elvey worked with it that summer and began a program of measuring the colors and magnitudes of stars in galactic clusters. It was a program well suited to the forty-inch refractor, undoubtedly initiated by Stebbins for its astrophysical importance. At the end of the summer he went back to Madison, however, leaving Elvey, who was working on his Ph.D. thesis on stellar spectroscopy under Struve, to complete the project. Under these conditions he could not accomplish much, and though Stebbins came back to Yerkes part time on weekends in the spring of 1930, the results were never published. Elvey, who remained at Yerkes as an instructor after receiving his degree in 1930, continued the photoelectric work, but it did not have a major impact there.[14] Stebbins succeeded brilliantly at Madison in spite of the poor midwestern climate because he dedicated his smaller telescope and his own time, drive, and scientific effort completely to photoelectric photometry, along with all the time and effort of fellow faculty member C. Morse Huffer. In contrast, at Yerkes Elvey had a larger telescope but was not his own master; he had to work on other projects and could use the photometer only on nights assigned by the director.

By 1929 Frost had made up his mind that Stebbins should succeed him as director. He wrote in confidence to Hale and Adams, the letters typed by his wife, notifying them that under the university regulations he would be retired in 1932 and asking whom they would recommend to succeed him. The University of Chicago "would like to get a young man, with a future before him, if he has had enough experience to justify such an appointment," Frost wrote. Such a man was "not easy to find," but Frost went on to state that he believed the best candidate was Stebbins, even though he was then fifty-one. The Yerkes director did not know of anyone "in his early forties in whom [he] would have confidence." Stebbins was a good scientist with research ability, tact, and administrative experience. He could be counted on to do the *Astrophysical Journal* editorial work that went

Figure 24   Edwin B.
Frost as an older man.
Courtesy of Yerkes
Observatory.

with the Yerkes directorship, and he was a member of the National
Academy of Sciences to boot. Frost stated that he would quote their
views to Woodward, but he made it clear that what he really wanted
was their recommendations for Stebbins. Furthermore, he wrote, he
did not intend to consult anyone but the two of them and was "particularly desirous that no clique of Eastern Astronomers should meddle
in our affairs." Frost clearly was referring especially to Shapley at Harvard and Henry Norris Russell at Princeton, who did not respect him
at all and would almost certainly recommend favorites of their own
rather than a midwestern astronomer with West Coast training and
connections. Hale, and no doubt Adams as well, strongly approved of
Frost's choice, for exactly the reasons he gave, and said so.[15]

Frost's opinion was interesting, but the decision was not up to him.
In the 1930s at the University of Chicago (and essentially all other
universities) there was no open recruitment for a new director, no
search committee, no formal recommendation from the department

faculty. It was simply a management decision, much like appointing the head of a branch bank today. Gale, as dean, would recommend a new director, generally after consulting Woodward, and Hutchins would make the appointment. Both the dean and the vice president were familiar with Frost's views, but the new president had still not visited Yerkes Observatory by the spring of 1931, and the director had never had a serious conversation with him.[16]

Gale took his time reaching his decision. No doubt he talked informally with all the senior faculty members at Yerkes and with many on the campus, particularly in physics. He apparently consulted only one outside adviser, his close friend Adams. In May 1931 Gale wrote the Mount Wilson director, asking for his recommendation. The dean started somewhat facetiously, stating that his own first choice was Adams and his second Edwin Hubble. Both were outstanding research astronomers and former Yerkes graduate students. But as Gale suggested to Adams, he knew there was little chance that either of them would leave Mount Wilson for the Yerkes directorship. After those two, he wrote, nearly all the Yerkes staff were strong for Stebbins. But Gale himself preferred Struve. The dean's only stated reservation was that making him director would cut into his research activities. After those two the prospects were dim, though Gale mentioned a few possibilities, none of them very attractive. He asked Adams for his frank opinion.[17]

The Mount Wilson director confirmed that he would not come and firmly stated that Hubble "ought to remain here with a view to the completion of the 200-inch." (At that time it was widely believed that Hubble would become the director at Palomar when the big telescope was completed and went into operation.) Adams strongly recommended Stebbins for the Yerkes director, an excellent scientist with a keen interest in research, excellent training, good judgment, good knowledge of Yerkes, administrative experience, and "a high standing among American astronomers." Stebbins's only drawback was that he was "easy going and perhaps would not push along all departments of the work as vigorously as some other man might do." On the other hand Adams did not think Struve would make a good director. He greatly admired Struve's research but questioned his administrative ability. Perhaps more important, "it would be difficult for him as a foreigner to represent the Observatory adequately on many occasions."[18] Many American astronomers, almost all of them from families whose backgrounds stretched back to New England two centuries or more before, had a strong antiforeign, anti-immigrant bias, and Adams was one of them.[19] Gale replied that most astronomers he had consulted favored Stebbins but that he himself "wonder[ed] . . . if we ought not to consider Struve more seriously." He was "quite lacking in social qualifications" and might not be a good administrator, but those were of "secondary importance." Struve's scientific qualifications and drive made him a real possibility, Gale insisted. He wrote Adams that if Stebbins would not accept the Yerkes directorship, he intended to recommend Struve for it.[20] Russell, who was much more open and welcoming to foreign-born astronomers than Adams, was to pass through Chicago that summer on his way west, and Gale intended to consult him. No record of their conversation has been preserved, but judging from his later attitude Russell very probably did not recommend Struve for the post, believing he was too young and inexperienced and should be allowed to spend full time on his research. Adams had warned his friend not to take Russell's advice, but in fact Gale was even more positive toward Struve than he had admitted. Actually, at that point the dean was undecided between him and Stebbins.[21]

By the late fall of 1931 the dean had made up his mind. During Frost's serious illness and hospitalization, Gale recommended Struve for immediate appointment as assistant director, with full responsibility for the administrative and scientific work on the observatory.[22] During Frost's previous absences from Yerkes for long winter vacations in the South, he had left George Van Biesbroeck, now fifty-one years old, in charge, but this time Gale explicitly named the younger Struve instead. He would not have done so had Stebbins been about to become director. Struve took hold well during his acting directorship. He certainly kept the research going on, and in addition he submitted a proposal for a research grant to the American Academy of Arts and Sciences on behalf of the observatory and also made it a point to invite Willem de Sitter, an important astronomer from the Netherlands who was traveling through the United States, to visit Yerkes Observatory.[23] Thus Struve was explicitly demonstrating both his administrative and his "social" qualifications, the two areas in which Gale had thought he was deficient. He had easily passed whatever test his temporary elevation provided, and in mid-December the dean recommended him for promotion to full professor, with a $750 salary raise to $5,000 a year beginning July 1, 1932. Gale had certainly made up his mind by then to name Struve as director. At the end of February 1932, with Frost back on duty, the dean notified him that he formally recommended Struve's promotion to full professor and his appointment as director.[24]

Struve was then only thirty-four years old and had still been a mere assistant professor only two years before. His appointment was a great surprise. The senior faculty members at Yerkes had originally expected Stebbins to be named director, while most of the graduate students had thought Van Biesbroeck would get the job.[25] Frank E. Ross's friends at Lick Observatory thought he deserved it and were surprised that he had not been offered it. Ross himself insisted that he did not want the position. "Uneasy lies the head that wears a crown, and we only pass this way but once" was his breezy way of expressing it. He had favored Stebbins and agreed with the Lick astronomers and Adams that an American should have been appointed director, but he thought that Struve would probably work out all right. Ross's analysis was that the University of Chicago saved several thousand dollars a year in salary by appointing Struve rather than Stebbins.[26] It certainly was true and probably was a factor in Gale's decision, made in that terrible depression year. He was pressuring Frost (and all the other heads of departments in the university) to cut expenses and put off capital purchases.[27]

Also, the dean's enmity toward Hale made him skeptical and even faintly hostile toward Stebbins, with his National Academy of Sciences membership and American Astronomical Society leadership position. But surely the most important reasons for Struve's appointment were his great research productivity and his age. Hutchins, the boy president, had made it very clear that he wanted to lead a youth movement at Chicago. In this frame of reference Gale selected the thirty-four-year-old Struve rather than the fifty-three-year-old Stebbins, the fifty-two-year-old Van Biesbroeck, or the fifty-eight-year-old Ross.

However, although the dean had chosen his "boy director," Struve did not leap to accept the offered post. He wanted a large reflecting telescope in a good observing climate, with which he could take the kind of high-dispersion spectrograms he needed for his own research. And, it turned out, he had the leverage to get it. In early February, soon after Gale had first "tentatively" offered him the directorship, Struve opened a letter from Shapley asking if he was interested in coming to Harvard as assistant director. Harry H. Plaskett, the son of Struve's nemesis at the Dominion Astrophysical Observatory, had been Shapley's assistant director but was moving on to a professorship and directorship at Oxford University in England. Shapley, impressed by Struve's prodigious research productivity, wanted to bring him to Harvard College Observatory as Plaskett's successor. Struve was definitely interested. He wrote to tell Shapley so but added the news of Gale's tentative offer of the Yerkes directorship. Clearly Struve, the son, grandson, and great-grandson of observatory directors, had learned how to play off one offer against another. He asked Shapley for time to consider the offer and for permission to tell Gale about it.[28]

The Harvard director immediately wrote back to raise the offer and to grant the requested permission, which he was bound to do by the code of the time. It was only a formality in any case. Now Shapley defined the position, about which he had been rather vague in his previous letter. If he came to Harvard, Struve would be a full professor as well as assistant director. Shapley had talked it over with Russell, and they agreed that Struve would be far better off there than at Yerkes. The only fly in the ointment was that Shapley now announced that Plaskett would retain a connection with Harvard College Observatory as a research associate and would come back from England from time to time to give Struve the benefit of his presumably more advanced knowledge of astrophysics.[29] No doubt the Harvard director did not realize how condescending his letter must have seemed to the sensitive Struve. Nevertheless Struve wrote back that he was glad Shapley had discussed him with Russell and that he would prefer to accept the Harvard position. But, he went on, Gale had formally recommended him for the Yerkes directorship, and President Hutchins supported the recommendation. Now the final decision was up to the Board of Trustees at Chicago. He wanted to wait and see if he would get the directorship, he wrote, but he was maddeningly imprecise about which of the competing offers he would take in that case.[30]

Shapley seized on the first part of Struve's letter—that he would "prefer" the Harvard position—as an acceptance, no doubt for tactical reasons. He fired back a telegram and a letter stating that Harvard president A. Lawrence Lowell that very day had promised Struve the professorship he had accepted.[31] That same day, in early March, Shapley wrote to Gale and Frost to notify them that he had formally offered Struve the professorship and assistant directorship at Harvard and that he believed the young spectroscopist would take it. In all three letters, but particularly the two to Gale and Frost, Shapley was extremely patronizing about how much more Struve would be able to do at Harvard, with its superior resources, than at Yerkes, which he strongly implied was finished as a research center. Harvard College Observatory, with its two sixty-inch reflectors outfitted with "modern" spectroscopes, needed Struve, a man "of energy, originality, and experience" who was "versed in modern astrophysics." Yerkes Observatory could get along without him. The best future Shapley could see for it was to give up on astrophysics and try to persuade S. A. Mitchell, the elderly director of the University of Virginia's McCormick Observatory, to leave it and turn the Yerkes forty-inch into a full-time parallax factory. Struve and astronomy would both be advanced when he moved to Harvard. Shapley would have advised Struve to stay at Yerkes, he gravely proclaimed, were there any hope of improving it, but this could not be done without money, and he clearly thought there was no possibility of that at the University of Chicago. He finished by inviting Struve to come to Cambridge later that month to attend a meeting that would celebrate the completion of a new "astrographic building at the Harvard College Observatory." [32]

Shapley thought he had won the battle, but he exulted too soon. Gale had just begun to fight. Undoubtedly he took Shapley's letter, so derogatory toward the University of Chicago and by implication its president, straight to Hutchins. It convinced the Chicago president, as nothing else could have, that he did not want to lose his putative "boy director," whose worth had now been authenticated by the very attractive Harvard offer. He called Struve to his office on the campus in Chicago and unleashed all his considerable personal charm on the impressionable astronomer. Hutchins was "very strongly opposed to my leaving, and he practically refused to consider my resignation at this time, saying that it is the wish of the University [of Chicago] to give me the best possible conditions for my work," Struve reported to Shapley. Struve "regret[ted] this unexpected complication," he said, because he had fully intended to take the Harvard offer, but now he would have to listen to what Hutchins had to say, which would take time. Struve would come to Harvard and talk with Shapley, Russell, and Frank Schlesinger of Yale, another power in the eastern establishment, while Hutchins and Gale came up with a plan for a large reflector in a good climate, although of course Struve did not explicitly mention this last part to Shapley.[33] But the Harvard director surely understood it. Gale, for his part, wrote Shapley that he was "gratified that you have so strikingly confirmed my estimate of Struve." The dean wrote the Harvard director that he was convinced that Struve was the one available man who could do "something radical" and "put Yerkes Observatory on its feet again." [34]

Shapley wrote back to Struve that Hutchins "just" wanted to keep him at Chicago, by implication to prove his power, while he himself wanted to bring him to Harvard so he could do better scientific work there. Russell, who claimed he had had no idea that Shapley wanted to hire Struve away from Yerkes until two weeks earlier, now chimed in strongly on the Harvard side. The interests of astronomy in America demanded that Struve go there. "The opportunities for work at Harvard, with the two great reflectors available, will be much greater than at Yerkes." Another factor was that Shapley's health might break down from overwork unless Struve came to bear some of the administrative load as assistant director, Russell claimed. At Harvard Shapley had "secured a large endowment and collected a numerous staff," while Yerkes "to be frank, . . . appears to have been kept on a maintenance budget for many years." Russell made it very clear to Struve that "it appears to me that you could do more, both for science and yourself, at Harvard, and if I were in your place I would go there." [35] It was very clear to the great Princeton astrophysicist, revered by all, who had been Shapley's major professor and thesis adviser.

But it was not so clear to Struve. He certainly knew that the "two great reflectors" were cheaply built sixty-inch telescopes, one near Cambridge at a site with even worse observing weather than Yerkes Observatory, the other in the Southern Hemisphere, operated by routine observers who were very difficult to control effectively from so far away. The coudé spectrograph for the Oak Ridge Station (in Massachusetts), which Russell had held up as just the instrument for Struve's research, was still to "be provided," as he had admitted. And Struve had observed Shapley's dominant personality close up; he surely recognized that being his assistant director would be no bed of roses. Struve obviously preferred to be a director himself; that was the norm at the time, and his own driving personality demanded it.

Gale did his best to refute Shapley's arguments. The University of Chicago fully intended to offer Struve distinctly improved facilities. It had "a certain responsibility to the science of astronomy in America." It was "quite unwilling to have Yerkes Observatory quietly slip back into a second class institution." To prevent this, Gale said, Chicago would have "to provide a large reflector especially adapted for spectrographic work. It will be located at a favorable site, probably not at Williams Bay." In fact the dean knew that Williams Bay was not a good site, and his use of the word "probably" indicates that he had no firm plan at the time. The "first steps [toward a new larger telescope in the West] had been taken before the offer came from Harvard," he maintained, and he certainly had begun steps to nail down a good site for it.[36]

Struve went to Harvard toward the end of March, stopping off at Princeton on the way to confer with Russell. He, Shapley, and Schlesinger, who came to the meeting at Cambridge, all tried to persuade Struve to take the Harvard position. The young Yerkes spectroscopist was pleasant but noncommittal. He still had not made up his mind.[37]

Meanwhile at Chicago Gale and Hutchins had been working to get what it would take to keep him at Yerkes. Some of the University of Chicago trustees were interested in astronomy, particularly Swift, the chairman of the board. His heart was in the right place, even though he persisted in believing that the main optical element in a large reflecting telescope was a "lens."[38] He could be counted on for support. Gale and Hutchins agreed that, somehow or other, they would build a sixty-inch telescope in west Texas or Arizona, the areas Struve preferred for their good weather and accessibility by train from Chicago. Gale played his cards carefully, asking his friend Adams if he could let Struve use the Mount Wilson telescopes as a guest while the Chicago reflector was being built. Adams, who hated Shapley, was glad to do whatever he could to help keep Struve at Yerkes. He was willing to be "fairly liberal" and promised to let Struve observe three nights a month, as much telescope time as the regular Mount Wilson astronomers were assigned. Adams also rushed Gale the advice he had requested on suppliers of glass for the mirror and opticians who could grind, figure, and polish the disk.[39]

Then, just at that time, almost miraculously, the opportunity opened up by which the university could get the telescope in those dark depression years. Very probably Struve suggested it—he always said that he had. Ironically enough, it probably came to the surface through a letter from Shapley himself to Clifford C. Crump, Frost's protégé and the least serious astronomer on the Yerkes staff. Certainly the opportunity existed because of Frost's own pro bono activities for astronomy years before.

The opportunity was that the University of Texas had $840,000 to build a large telescope but no astronomers on its faculty to use it, or even to advise on how and where it should be built. The money was the bequest of William J. McDonald, a banker who had lived out his life in Paris, Texas. He was a bachelor who, on his death in 1926, left most of his fortune to the state university to build an astronomical observatory. McDonald left much smaller bequests to various relatives. They sued to break the will, claiming that he had been "not of sound mind and disposing memory" when he wrote it. The university feared one of their main points would be that he must have been insane to leave his money for research in astronomy![40]

McDonald's bequest was completely unexpected in Austin. The university authorities first heard of it from a reporter when the will was probated in Paris. Harry Y. Benedict, dean of the College of Arts and Sciences, soon emerged as the leader of the University of Texas side of the case. Born in Louisville, Kentucky, he had done his undergraduate and master's degree work at Texas before earning his Ph.D. in mathematics at Harvard in 1898. In his time mathematics was closely associated with astronomy, and Benedict had worked for two years as an assistant at the McCormick Observatory of the University of Virginia before entering the Harvard graduate school. Asaph Hall, the discoverer of the two moons of Mars, was among his teachers there. After earning his Ph.D. and spending just one year on the faculty at Vanderbilt University, Benedict joined the Texas faculty, rising through the ranks to professor of applied mathematics (meaning for engineering students) and astronomy in 1906 and to dean in 1911. He considered himself primarily a mathematician, but he joined the American Astronomical Society in 1925, a year before McDonald's death. Benedict had never done any research after his Ph.D. thesis, devoting himself entirely to teaching and administration, but he was well acquainted with astronomers from his early days.[41]

Very soon after learning of McDonald's bequest, estimated to be over $1 million, Benedict wrote to several observatory directors, including Frost, to notify them of it. The Texas dean intended to use it to establish "the greatest observatory in the South." Frost, who spent more time and effort on letters like this than most other directors, sent Benedict (whom he apparently did not know) some sound advice on how to use the money effectively. He recommended building a students' observatory on the campus in Austin with a ten-inch or twelve-inch refractor and a thirty-inch or thirty-six-inch reflector, at a cost of perhaps $200,000, and spending the rest on a research observatory in the mountains of west Texas, near New Mexico. Frost offered to help Benedict any way he could.[42]

A month later the Texas dean wrote Frost again, as well as many other directors, telling them that the will was being contested by McDonald's nieces and a nephew. He asked for references to bolster the university's effort to prove that leaving gifts to support astronomy was not a sign of insanity. Nearly all of the sympathetic but busy astronomers sent Benedict one- or two-page responses. In contrast, Frost threw himself into answering this plea for help. In just a few days he produced a very powerful five-page statement citing Leander McCormick, James Lick, Charles T. Yerkes, Andrew Carnegie, and many other donors whose gifts to science had made their names live forever. As Frost himself admitted, his hasty brief was not at all well organized, but it was just the kind of raw material the university lawyers needed for ammunition.[43] This letter undoubtedly fixed Frost and Yerkes Observatory strongly in Benedict's mind and made him receptive to the Chicago offer of cooperation that was to come in 1932. Frost, though himself an ineffective scientist (in his later years) and leader, nevertheless brought both Struve and McDonald Observatory to the University of Chicago.

As the case dragged on through the courts in a series of appeals and retrials, Benedict succeeded to the presidency of the University of Texas in 1927. Finally in 1929 the university settled the case out of court, giving $250,000 to the purported heirs and leaving $840,000 for the telescope. In 1927 Benedict had already sent physics professor John M. Kuehne to visit a number of leading western research observatories to collect information and ideas. After the case was settled, Benedict was determined to use the money to build a first-class research observatory, but Texas had absolutely no tradition in that direction. It was purely a teaching institution, and Benedict and the regents who controlled it knew they would never be able to get money from the legislature to operate an astronomical research center. The depression, which bore down hard on faculty salaries, compounded their problem. Nevertheless, Benedict hoped to solve it during his presidency.[44]

Planning ahead for the day when the William J. McDonald Observatory would be built and would need an astronomical library, Benedict began collecting books for it. In February 1932 he learned from Shapley that the Yerkes Observatory library had a duplicate set of Harvard College Observatory publications it did not need and would be willing to give them to another observatory. Benedict wrote Crump, the librarian and secretary at Yerkes, asking him to ship the duplicate publications to Austin. The University of Texas would pay the shipping charges; Crump should send them by the "cheapest mode of transportation," since there was "no haste." Crump packed and shipped them immediately, with a graceful compliment to Benedict, whom he apparently had met previously. This exchange of letters, occurring just at the time Struve received the Harvard offer from Shapley, undoubtedly brought the Texas money available for a big telescope for a university with no astronomers back to the attention of the close-knit group at Yerkes, who had no money for a big telescope in Texas or Arizona. From the timing, and Struve's later statements that he had initiated the idea of the cooperative arrangement with Texas, it seems almost certain that it originally came to him as a result of the letter from Benedict to Crump, written at Shapley's suggestion.[45] Very probably Struve had suggested working out a cooperative arrangement with the University of Texas at his meeting with Hutchins and Gale in Chicago on March 8.[46] Certainly on March 22, while Struve was en route to or in the East being wooed by Shapley and "advised" by Russell, Hutchins telephoned Benedict directly and proposed that their two universities work together in building and staffing a McDonald Observatory in Texas. Benedict quickly seized the idea and rushed copies of the relevant documents to Hutchins by airmail special delivery. All the initiative came from Chicago; only a month earlier Benedict had assured a correspondent that he had no intention of even beginning to plan "the McDonald Observatory" within the next year.[47]

On March 28, immediately after Struve's return, Gale telephoned him from the campus asking him to come to Chicago at the end of the week and meet with him and Hutchins. They had "plans for the Observatory which they ha[d] been working on." At this conference on April 1 Hutchins assured Struve that he would "carry through the project of a large reflector," but he did not yet have a definite schedule for it. Struve disclosed this in a letter to Shapley but did not reveal that confidential negotiations were under way with the University of Texas. He was hopeful, but not completely won over. "My own attitude is that if the President could do something now, or at least set a definite time-limit, it would be hard for me to refuse to stay here. I think I should regard it almost a duty to put the observatory in a position where it could do useful work in astronomy."[48] Shapley, who went off on a short vacation at this crucial moment, remonstrated with Struve, pointing out all the advantages of Harvard and the disadvantages of Yerkes, but it was too late.[49]

Just at this time Mary Lanning Struve was hospitalized and underwent an operation for what her husband described as "an abdominal tumor and appendix." In the restrained, coded language of the time this probably meant she had suffered a miscarriage and had undergone a hysterectomy. There are indications, mentioned earlier, that she may have had previous miscarriages in 1927 (at Harvard) and in 1929 (in England). Expressed in the very guarded circumlocutions of that era, these vague hints are impossible to verify or deny today, but it seems probable that from April 1932 on Struve knew he would not have a fifth-generation observatory director in his family.[50]

Meanwhile Hutchins, Benedict, and their underlings were working toward an agreement along the lines proposed by Struve. Whatever he felt at the time, he left his wife, who was "making a good recovery in hospital" and flew to Austin—his first airplane trip—to be inspected by Benedict and receive his assurances that the deal would go forward. Hutchins had telegraphed Benedict on April 4 to arrange the meeting. Struve arrived in Austin on April 13, and on April 18 Hutchins signed a formal proposal, prepared by Struve, that was presented to the University of Texas regents five days later. They accepted it in principle, and Struve's plan for a cooperative observatory became a reality.[51] It was not the first time Struve had put his responsibilities to his wife second to what he considered his duties to Yerkes Observatory and astronomy, nor would it be the last.

While Struve was gone, a formal letter from Harvard University arrived at Williams Bay, appointing him professor of astrophysics and assistant director of Harvard College Observatory.[52] By then he was sure he would stay at Yerkes. He hastily wrote Shapley to tell him so. He attributed his decision to "a number of consultations" he had had with Hutchins and Gale but did not even hint at the agreement with Texas, which had been formally proposed only the previous day and was still in the works and hence confidential. To Shapley, Struve stated that Hutchins had "promise[d] me full support in the modernization of the Observatory, and I believe that I can now feel reasonably certain that satisfactory conditions for research will be provided here." The "enthusiastic interest" of Hutchins and Gale would be important assets to Yerkes Observatory and to astronomy in general." At the end of his letter Struve wrote Shapley that he felt that under the changed circumstances it would be "improper" for him to accept the $150 traveling expenses for his trip East that Harvard had paid him, and he returned it.[53]

That same day Dean Gale wrote Frost and Adams, whose discretion he trusted, and gave them the general outline of the plan. Struve would stay, and the University of Chicago would join in a cooperative plan with the University of Texas to build "their reflector." Texas would pay all the costs of building the observatory and telescope, and subsequent capital expenses, from the McDonald bequest. Chicago would pay all the salaries, and the observatory would be operated jointly. They hoped to get an eighty-inch reflector (which would be the second largest in the world until the two-hundred-inch, then in the planning stage, was completed) with the available funds. Adams received a much fuller account than Frost, because Gale wanted him to write Benedict, endorsing the plan and emphasizing that the observatory should be located at a good site in the mountains of west Texas, rather than near Austin, a cloudier region.[54]

The Mount Wilson director came through handsomely, writing a very positive assessment of the advantages of the cooperative plan for both Texas and Chicago. Struve, Adams said, was "very well known among astronomers and recognized as a man of rare ability"; the University of Texas would profit greatly from his experience "from the outset more than it could hope to if it attempted to construct and operate the observatory quite independently." The telescope should definitely be built in the mountains of west Texas, with its high percentage of clear nights, not in Austin.[55]

President Hutchins himself telephoned Hale, who claimed he was "as much interested as ever in the University of Chicago and the Yerkes Observatory," to ask for his support. The old founding director responded to Benedict with a strong endorsement of Struve and the cooperative observatory plan. However, Hale emphasized that the building of McDonald Observatory should not mean the closing down of Yerkes and its forty-inch refractor. It was "invaluable for solar work, especially with the Rumford spectroheliograph," the instrument he had raised the money for, built, and used more than thirty years earlier. When he congratulated Struve on the directorship two months later, Hale made the same point again, after reminiscing about the old days at Yerkes. He was pleased by the cooperative agreement, which Benedict sent for his comments, and compared it with the "somewhat similar scheme of cooperation" between Caltech and the Carnegie Institution of Washington to operate Palomar Observatory when the two-hundred-inch was completed.[56]

Struve drew up the parts of the draft agreement dealing with the observatory and the directorship himself, and they were eventually adopted in toto. Under it the director of Yerkes Observatory would automatically also be the director of McDonald Observatory and would report directly to both university presidents. It also provided that the observatory would be built and outfitted "according to the plans of the Director and under his supervision."[57]

By mid-May it was clear that only the details of the agreement remained to be worked out, and Struve began making his peace with Shapley, Russell, and Schlesinger. He knew their support would be essential in making his directorship a success. To each of them he now revealed for the first time the negotiations that had been going on with the University of Texas since March. Struve emphasized what an opportunity the Texas arrangement was for Yerkes Observatory and how important it was for the future of astronomy at the University of Chicago. The McDonald bequest would give its astronomers the use of a modern eighty-inch reflector at a good western site, rather than an old forty-inch refractor at a poor midwestern site. He sent each of them a copy of the draft agreement and asked for their comments and suggestions.

To Russell, Struve confided that he planned to have the telescope equipped with a coudé spectrograph, with which he himself would obtain high-dispersion spectra of normal and peculiar stars for detailed study. This was exactly the program Struve knew would most appeal to the older Princeton astrophysicist. Russell was somewhat mollified but did not trust the University of Chicago to provide adequate support. He had warned Robert L. Moore, a mathematics professor from Texas who had visited Princeton and told him of the plan, against Chicago and Hutchins. Maybe it would work out, but Struve should insist on more money and should not overcommit himself to administration. If he followed this advice, he could make "very efficient and valuable use of the new telescope." [58]

To Shapley Struve wrote expressing his thanks for the Harvard director's "most kind letter" of April, in which he had accepted the younger man's decision to stay at Yerkes. Struve had now written to President Lowell, explaining his reasons for not accepting the Harvard offer, and had received "a very gracious note from him in reply." The rest of Struve's letter was much the same as the one to Russell, but with less emphasis on coudé spectroscopy. Shapley, the most cheerful of the three, replied that he was "very much pleased" with the prospects of the new Texas telescope. After a series of humorous, cutting remarks about state universities, their failure to support research, and the "notorious" Texas legislative bodies, he gave Struve some advice on ways to save money in having the telescope made. This was a favorite subject of Shapley's, and the Harvard telescopes of his time showed it. They were widely known for being poorly built, and they never proved satisfactory. Struve was wise enough not to pay any attention to these well-meant words of wisdom. At the end of his letter Shapley sent his ironic "congratulations" to Hutchins, who had been "very wise" in keeping Struve, avoiding spending any money on an observatory, forming a connection with the Southwest, and maintaining Chicago's supremacy in the Midwest. None of the other eastern astronomers understood Struve's decision not to cast his lot with them, but Shapley would bring them around. Two months later the new Yerkes director wrote the longtime Harvard director to congratulate him on "secur[ing]" Donald H. Menzel," an excellent choice" to fill the vacant position he himself had turned down.[59] Menzel was not named assistant director then, but twenty-one years later he was to succeed Shapley in the top job at Harvard, after Struve had declined it.

Schlesinger, who received a letter similar to the other two, was the most negative about the Texas plan. He believed it hurt astronomy, but his reasoning was hard to follow. What he disliked was that the University of Chicago, committed to keeping Struve on its faculty, had gotten off without using any of its own money to build a new telescope or observatory, at the same time as it also "virtually can-cel[led] the McDonald bequest so far as astronomy is concerned." What he seemed to mean was that Chicago got it rather than Yale! It seemed a "questionable arrangement" from the Texas standpoint, and he himself would have advised against it. Furthermore, what Schlesinger really objected to was the "major premise in this transac-tion . . . that large instruments are necessary for astronomical research. This is, of course, not so." He was a confirmed small-telescope user, for astrometric work.[60] Thus the three important eastern directors saw Struve's situation with eyes heavily conditioned by their own in-dividual situations.

Adams already knew about the Chicago-Texas agreement from Gale, but Struve wrote him also, to arrange to borrow one or two small visual telescopes for the site survey he had quickly begun orga-nizing. He told the Mount Wilson director of his plan to shift all the stellar spectroscopy to the new telescope when it was completed, re-serving the forty-inch for solar work and astrometry. Adams, who was not emotionally involved in Struve's decision, but favored the Midwest over the East, congratulated him, arranged for one of the small tele-scopes to be shipped to Yerkes, and gave the new director-to-be some much better advice on telescope building than Shapley had.[61]

Struve had Christian T. Elvey, his former Ph.D. thesis student, now an instructor on the Yerkes staff, prepare a quick climatological study of Texas from existing weather data. It showed that the amount of clear weather increased steadily from east to west.[62] Struve had earlier favored a site in the Amarillo region, but after the discussions he had in Austin in April, he began to think in terms of the Davis Mountains

Figure 25
Otto Struve,
when he be-
came director
of Yerkes
Observatory
(1932). Cour-
tesy of Yerkes
Observatory.

in west Texas instead.[63] He sent Elvey and assistant Theodore G. Mehlin on a lightning site survey of the Lone Star State. They traveled by car, taking the telescope Adams had lent to Yerkes from the earlier Palomar site survey to test the "seeing," or image quality, wherever they stopped. They made measurements from Austin in the east to the Davis and Hueco Mountains (near El Paso) in the west and to Amarillo in the north. By modern standards the survey was far too rushed; the results did not average over even a single season. But Struve was in a hurry. Based on the climatological data and the quick survey, he decided on the Davis Mountains area. He drove to Texas himself with his wife Mary so they could camp at several of the possible sites and he could test them himself. After a few nights observing with a small telescope, he decided on the mountain now called Mount Locke, and that is where McDonald Observatory was built.

Struve succeeded Frost as director of Yerkes Observatory on July 1, 1932, just before his trip to Texas. On that morning Franklin E. Roach, a graduate student, was seated in an office measuring a spectrogram. The door to the hall was open, as was customary there except when confidential discussions were in progress. Roach heard Struve's measured tread as he walked down the hall, making the first of the daily inspections to see who was working that were to be a feature of his reign. He came into the office where Roach was at the measuring machine. "Good morning, Mr. Struve," Roach said, using the obligatory form of address to all faculty members at Yerkes at that time. As Roach remembered it, years later, this conversation followed:

> Struve: Good morning. What are you doing?
> Roach: I'm measuring a spectrogram.
> Struve: What star is it?
> Roach: β Lyrae.
> Struve: Who told you to measure that spectrogram?
> Roach: Mr. Frost did.
> Struve: From now on I'll tell you what stars to measure![64]

Whether those were the exact words they exchanged may be highly debatable, but they convey the picture every survivor of Struve's years as Yerkes director has of him. From the start he got right down to business, was deeply immersed in research at the detailed level, and never minced words with underlings in making his directives known. Those tactics were to pay off brilliantly in scientific results and in the training and development of a host of top-notch research scientists at Yerkes Observatory over the next decade and a half, but they were to end in Struve's deep disappointment and bitter departure from the institution he had revivified.

In 1932 the Great Depression, Struve's amazing research productivity and drive, and the Harvard offer had all coincided to bring him to Hutchins's attention. The young president, a product of Yale, had quickly determined to fight to keep his Russian-born astronomer. He promised he would get him a telescope somehow. Hutchins was far more interested in law, undergraduate education, philosophy, and debating than in any of the physical sciences, but the stern, serious research astronomer had captured his attention. Then Struve's brilliant creative idea of combining the Yerkes staff with the Texas bequest and building a great observatory showed Hutchins how he could actually follow through on his promise without bankrupting the University of Chicago. He charmed Benedict, but the Texas president wanted desperately to build a research observatory in his state and was ready to be charmed. Everything fitted together perfectly, and Struve not only got the observatory but simultaneously became Hutchins's favorite scientist. The new Yerkes director could always go right to the president in Chicago if he needed help.

Struve always talked straight, never wasted time, and had no other agenda than making Yerkes Observatory and the University of Chicago the most powerful astronomical research center in the world. Hutchins learned to trust and believe in him. Struve never crossed the president and supported him to the faculty without question. Hutchins clearly enjoyed talking with him much more than with the elderly, stuffy chairmen of the other science departments, who spent so much of their time in Chicago defending their turf and criticizing the flamboyant boy president. For fifteen years Struve and Hutchins were made for each other, but as they both grew older and their responsibilities became larger and more complicated, their relationship was bound to sour. Then they both departed, in quick succession. But right up to that sad ending, the eighteen years of Struve's directorship were eventful ones indeed.

# CHAPTER SIX

# The Boy Director, 1932–1936

Otto Struve started as director of Yerkes Observatory on July 1, 1932. He became a full professor, and his salary went up from the $4,000 a year he had been earning as an associate professor (to which he had been promoted just two years earlier) to $6,000. Edwin B. Frost retired, and the $6,500 salary he had received for years as director and professor was reduced to $3,000, to be paid from the regular university funds.[1] He had no pension, nor did any Chicago professor of his age. Those came only years later. Frost had not been sure that Struve would accept the directorship until April, and only then did the older man begin work on the home he now knew he would have to move into to make way for the new occupant of the director's house.[2]

Frost owned a large parcel of land adjoining the observatory property, left to him by George S. Isham, doctor and longtime friend of the observatory. Frost and his wife had subdivided it into large lots, saving the choicest one with a view of nearby Lake Geneva for themselves and selling a few others to University of Chicago faculty members who wanted Williams Bay homes. He named the area University Heights, but it soon came to be known as Frost's Woods. He and his wife—confirmed New Englanders who had come to Williams Bay a third of a century earlier somewhat fearful and contemptuous of what they regarded as half-civilized Wisconsin—had made many friends in the Lake Geneva area and had been converted by the beauty of its woods and hills. They had decided to stay, and Mary H. Frost had a plan waiting for their home in retirement when the fateful moment came. She and Frost called in the builders, who completed the house in seventy days. He named it Brantwood (for his mother's maiden name), but to the locals it was always "the Frost house." They moved in on June 28, 1932, leaving the director's house open for Otto and Mary Struve and his mother, Elizabeth, who lived with them.[3] She had retired from the position as part-time computer at the observatory that she had held in her first years there.

Struve was then not quite thirty-five years old, five years older than Robert M. Hutchins had been when he became the "boy president" of the University of Chicago in 1929. Yet compared with fifty-five-year-old Walter S. Adams, director of Mount Wilson Observatory, sixty-seven-year-old Robert G. Aitken of Lick Observatory, forty-six-year-old Harlow Shapley of Harvard College Observatory, and fifty-four-year-old Henry Norris Russell of Princeton University, Struve was by far the youngest head of a major American research observatory. From the start he was stern, wanting to establish that he was in complete charge. When he learned that his initial appointment as director was to be for only one year, he exploded. His enemies at Yerkes, particularly Clifford C. Crump, whom Frost had brought in as secretary and associate professor two years before but who had never done any research, were sure to spread the word that the new director was on trial and to work against him. Struve protested vehemently to Hutchins but, always correct, did so through Dean Henry G. Gale. As director he had to take the long view and plan for the future. How could he do that if he were on a one-year appointment? How could the University of Texas take seriously its agreement with the University of Chicago to build and operate McDonald Observatory over a thirty-year period if Chicago could not trust the man who was to carry it out for more than one year? Harvard University had offered Struve the assistant directorship of its observatory without any time limit, but he had declined it to stay at Chicago, on Hutchins's promise to build a new observatory with a large telescope at a good site in the West. If he were appointed for only a one-year term as director, Struve said, he could not "feel free to proceed with the Texas project or with my own personal plans." Hutchins got the message and took the threatened resignation very seriously. He called Struve to Chicago for a personal conference. The president reassured the new director that it was all a mere formality. He promised that Struve would certainly be reappointed, evidently for as long as he would keep the job, and that no word of the initial short term would leak out. It did not, and no one at Yerkes Observatory knew that the director was formally on approval for his first year. Probably Struve

Figure 26   Henry G. Gale, dean of physical sciences at the University of Chicago. Courtesy of Yerkes Observatory.

had overreacted and there would have been no rebellion in any case, but he certainly had established his direct access to the president and verified that Hutchins would support him when the chips were down.[4]

As soon as he became director, Struve started receiving stern notices from Dean Gale that expenses must be reduced. Struve realized they were not aimed at him alone; the depression was fierce, and all department chairmen were under severe pressure to cut out all nonessential costs and purchases. Gale sent him a personalized appeal for additional savings, and Struve reported that there was a "real spirit of cooperation among all the employees at the Observatory," which enabled him to put in place his own "economy program." It included terminating the contract of Arthur S. Fairley, a young instructor without tenure who had been a favorite of Frost. Fairley had produced almost no research results, and getting rid of him saved not only his salary, but the cost of the photographic plates he used for the photometric measurements he never published. Struve also stopped the daily direct photographs and spectroheliograms of the sun, another routine program Frost had continued long after its research value had ended. The new director knew how to use required economy measures to make the changes he had long felt were necessary to restore Yerkes Observatory to its former high position as a research center by cutting away deadwood.[5]

Struve proudly reported to the dean than he had no time to go to the upcoming solar eclipse in New England himself, but that he was sending a Yerkes Observatory expedition, at no cost to the university. Philip Fox, once a Yerkes assistant, later a longtime professor at Northwestern University, and now the director of the Adler Planetarium in Chicago, was sending a large group and a truckload of equipment to the eclipse; he had offered to take up to five hundred pounds from Yerkes as well. The observatory party consisted of George Van Biesbroeck, who would be with a group from Mount Wilson at the eclipse, Ralph Van Arnam, an instructor at Lehigh University and one of the perennial summertime graduate students at Williams Bay, and Frank R. Sullivan, who had begun work as the night assistant at the forty-inch refractor in 1900. In thirty-two years he had never had more than a one-week vacation. Struve was giving him slightly more than two weeks to get to the eclipse site at Island Pond, Vermont, in Van Arnam's car, set up the instruments, see the eclipse, and return to Williams Bay by bus. The director would personally pay his assistant's bus fare home, in recognition of the many long nights at the telescope, but Sullivan "of course" would have to pay for his own room and board. Gale, who knew Sullivan well, agreed that he deserved the time off to go to the eclipse. In spite of the economy drive, the dean instructed Struve to pay for Sullivan's bus ticket from observatory funds instead of out of his own pocket, but he did not go so far as to pay the night assistant's other travel expenses.[6]

Alas, it was all in vain: on August 31, the date of the solar eclipse, the Adler Planetarium and Yerkes group at Island Pond, Van Biesbroeck and the Mount Wilson group at Lancaster, New Hampshire, and almost all the other astronomers who had sought the best locations along the eclipse track were clouded out. Only the Lick Observatory expedition at Fryeburg, Maine, got some fairly good data through thin cirrus clouds. Frost, on the other hand, who had led the fruitless eclipse expedition to Catalina Island in 1923, accompanied his wife on a visit to their son, who lived in the outskirts of Portland, Maine. There they enjoyed perfectly clear skies, and the old director's grandchildren described the entire ninety-two seconds of totality to him.[7] The one sightless, retired astronomer along the eclipse path, without any telescope, camera, or spectrograph, turned out to be the only one who had picked the right site to observe it![8]

Soon after his return to Williams Bay, Frost reentered Billings Hospital in Chicago for the gallbladder operation he had been too weak to undergo the previous fall. He survived it and returned to his home. He did not keep an office in the observatory or do any postretirement work in astronomy. Frost took his exercise walking alone on the path through the woods from his house up to the observatory, feeling his way lightly along a wire his wife had had strung waist-high between the trees.

The old director made one last pro bono contribution to astronomy and the observatory he had served so long. In the fall of 1931, just before he was hospitalized, Frost had suggested to Fox that the Century of Progress Exposition, more popularly known as the Chicago World's Fair of 1933, be opened by an electrical signal, triggered by the light from a star captured by the forty-inch refractor and focused by it onto the photocell in the photometer Joel Stebbins had built and Christian T. Elvey was using.[9] Frost's idea echoed the theme of progress through science and technical development that the Fair celebrated. The 1930s were the decade when "electric eyes" came into vogue to open doors or count customers, and using the giant refractor to collect the faint light from a star and turn on the lights of the Fair on the shore of Lake Michigan had tremendous popular appeal. Furthermore, the star Frost proposed to use, Arcturus, is one of the brightest objects in the spring sky, easily visible to anyone, even from the well-lighted sidewalks of Chicago. Its distance from the sun is approximately forty light years, so the photons that opened the Century of Progress Exposition in 1933 could be said to have started on their journey through space when the same forty-inch telescope (without its lens) was on display at the World Columbian Exposition of 1893, just three miles farther south along the same Lake Michigan shore. Henry Crew, the Northwestern University physics professor and former Lick Observatory astrophysicist who was in charge of science exhibits of the Fair, quickly accepted Frost's idea.

By late 1932, when detailed planning began, Frost had retired and Struve, quite critical of Fox as a failed research scientist who had become a showman (his view of almost everyone engaged in raising public awareness of science), was initially skeptical. When Fox invited him to a conference in Chicago to make final arrangements for sending the amplified current from the photocell over the telephone lines from the forty-inch dome in Williams Bay to the Hall of Science on the Exposition grounds in Chicago, Struve had just returned from an observing trip. He could not miss a colloquium to be given on the campus by the visiting young Dutch astronomer Jan H. Oort, who had "caused quite a revolution in our ideas of the structure of the galaxy and its rotation." He sent Elvey to the conference in his place, to help draw up the tentative plans. When Fox informed Struve that he had recommended Yerkes Observatory be listed as one of the beneficiaries of any potential surplus of the receipts from the Fair, to the extent of 1 or 2 percent, but that the general manager had raised the observatory's share to 3 percent, Struve's dedication to popular education suddenly increased tremendously. Yerkes Observatory would "cooperate in every way to make the opening of the World's Fair interesting and attractive to the general public." He even promised to attend the next meeting of the Chicago Astronomical Society, the amateur group Fox sponsored at the Adler Planetarium, and to join the Society (and pay his dues) himself.[10]

The Fair opened on May 27, 1933. Over 30,000 people were present in the great open-air court of the Hall of Science; more than twice as many attendees heard the ceremonies from loudspeakers scattered through the grounds, and hundreds of thousands more throughout the United States listened to the broadcast on the major radio networks. The noted Frederick Stock conducted a symphony orchestra of 125 and a chorus of 2,500 in a concert; the famous baritone Lawrence Tibbett was the soloist. Frost, well known in Chicago as the blind astronomer, delivered a short talk extolling the spin-off values of astronomy, exemplified by the photocell, developed by "scientific men in Germany" (at the time a recent enemy of the United States). Then Fox gave an equally brief technical explanation of telescopes and electric eyes. It was cloudy in Chicago that night, but luckily it was clear at Williams Bay (or perhaps it was cloudy there but Elvey shone a flashlight up on the dome). Although backup photometers were on-line at Harvard, Allegheny Observatory in Pittsburgh, and the University of Illinois Observatory in Urbana, the signal from Yerkes actually turned on the lights that officially opened the Fair.[11]

By then Frost was well into writing his autobiography, *An Astronomer's Life,* a disappointing volume that told more about his New England boyhood, his Dartmouth College friends, his travels in Europe, and his love of nature in Wisconsin than it did about his

Figure 27   Edwin B. Frost, Christian T. Elvey, and Otto Struve at the forty-inch refractor with the photocell that converted the light from Arcturus to an electrical signal to start the Chicago World's Fair (1933). Courtesy of Yerkes Observatory.

work in astronomy and at Yerkes Observatory. He died two years later, on May 14, 1935, following a second serious operation, and was genuinely mourned in the little village as a great astronomer. On the main street the citizens of Williams Bay dedicated a small park to his memory. A massive boulder is its centerpiece; Frost's ashes are interred below a headstone near it. To this day he is revered by the inhabitants of Williams Bay, whose parents and grandparents loved the "blind astronomer's" human attributes, while Hale and Struve are practically forgotten there.[12]

Van Biesbroeck and Frank E. Ross were the two main senior research workers, besides Struve, on the Yerkes faculty in his early years as director. Both were highly productive in spite of the obstacles imposed by the poor Wisconsin observing conditions. Ross, repeating Barnard's exposures with the Bruce photographic telescope, continued to find many "high proper-motion stars" with relatively large angular motions in the sky. Some were high-velocity stars, some were white dwarfs or faint red dwarfs; all were worthy of further astrophysical studies. He spent as much of his time as he could in the Southwest, taking direct photographs of Milky Way fields with his five-inch Ross lens, the optical system he had designed, mounted on a camera he could use most effectively in the clear skies and stable atmospheres of Lowell, Mount Wilson, and Lick Observatories. Subsidized by a grant of $500 from the National Academy of Sciences, the University of Chicago Press was able to sell these atlases, containing twenty large photographic prints, at $10 a copy.[13] Ross spent half of each year in Pasadena as an optical designer on the two-hundred-inch project headed by Hale and John A. Anderson and was not much help to Struve.[14]

Van Biesbroeck, on the other hand, worked self-effacingly as his former student's aide and confidant. Trained as an engineer, Van B (as he was universally called) was in charge of the instrument shop and designed whatever auxiliary apparatus could be made there, areas in which Struve had no training or competence. When the director was gone from Williams Bay on extended observing trips, he left the older man in charge.[15] Van Biesbroeck observed visual binary stars, reserving the close pairs for the all too rare nights of good seeing, the only occasions on which anyone could resolve them, and measured the positions of comets and asteroids.[16]

Struve was determined to keep Elvey, William W. Morgan, and Philip C. Keenan, the three young Ph.D.'s he had done so much to train, on the Yerkes staff in spite of the drive for economy. They were all productive research workers, as he frequently reminded Gale. Struve made the elderly dean feel he was a member of the Yerkes team by appealing to him for spectroscopic gratings from the ruling machine in the Ryerson Physical Laboratory that Gale supervised as his sole remaining research activity. The young director would outline exactly what sort of grating he needed and explain its scientific purpose, then diplomatically praise the final product.[17] Struve also assured George Ellery Hale, whose influence was still tremendous, that he would have Keenan continue solar research with the forty-inch. It was a program dear to the old first director's heart, and he reciprocated with a strong letter of support to Gale.[18]

The next to go, to make room in the staff list and the budget for bright new research workers, was Crump. Frost had succeeded in pushing through his appointment as an associate professor in 1930, but only for a three-year term. Struve had no intention of renewing it. Crump had published only two actual research papers in his whole life, one in 1916 on his Ph.D. thesis at Michigan, the other in 1921 on his work at Yerkes immediately after it. He was aware that his splendid teaching record at Ohio Wesleyan, and in his one year at Minnesota, meant nothing to the new director. Crump knew that some of his friends, marginal astronomers like himself, were losing their jobs as the Great Depression worsened.[19] He tried to justify his own research inactivity by working up the data he had on two variable radial-velocity stars he had been desultorily observing for years. One of them, $\beta$ Cephei, had been the subject of his Michigan thesis; the other, $\delta$ Ceti, had been discovered to have a variable radial velocity by Frost and Adams in 1902. Although Crump had measured over 1,100 spectrograms of $\beta$ Cephei, taken by many observers, his final conclusion was that more data were needed. Both stars in fact are short-period, small-amplitude pulsating variables, but Crump interpreted them as spectroscopic binaries even though his own results showed this was impossible—the sizes of the orbits he determined for the "pairs" were smaller than the sizes of the stars themselves. Crump ignored this discrepancy, finally completed both papers in August 1932, and submitted them to the *Astrophysical Journal*. They were not published until spring of the following year, indicating that the new managing editor, Struve, had had serious doubts about them (as he did about everything Crump did) and had returned them for major revisions.[20]

Undoubtedly the moralistic young director had made sure Gale was well aware of the real reason Crump had so abruptly left Ohio Wesleyan in 1929, which would make him quite unsuitable for a permanent faculty position at Chicago (unless he were a real research star). As late as 1940, long after Crump had left Chicago, when he was under consideration for the presidency of Grinnell College, Struve gave him a very negative evaluation as administrator, scientist, and secretary. In his letter to the chairman of the trustees' search committee, Struve concluded that Crump's "record, especially at Ohio Wesleyan University, was not of a kind that would inspire confidence in him as a leader of young men and women." Needless to say, Crump did not get the job.[21]

Half a century later one young staff member of the time said that he had hated to watch Struve get rid of Crump.[22] One of the crueler ways he did it was to assign the middle-aged secretary to observe with the forty-inch on Saturday nights. He knew that Crump, whose father had recently died on his farm near Greens Fork, Indiana, liked to go there to check on his mother on weekends and help her manage it. If he left the observatory on Friday afternoon and could not get back in time to take his turn at the telescope, Struve carefully kept a record of his absence.[23] By April 1933 Crump knew there was absolutely no hope he would be retained at Yerkes. He tried to find a teaching job "on some University or College campus," but there were none to be had.[24] On September 30, when his appointment ended, Crump moved back to Greens Fork and the family farm. From there he traveled as an itinerant lecturer, probably speaking to church groups, until he landed a position as professor of mathematics and astronomy at Ripon College in central Wisconsin. There Crump had an excellent teaching record until his retirement twenty-one years later at age sixty-six. He was then immediately appointed professor at his alma mater, Earlham College in east-central Indiana, where he continued teaching right up to his death in 1969. At least six of his Earlham graduates went on to outstanding graduate schools of astronomy, one of them to Chicago, and to professional careers in astronomy; every one of them idolized him.[25]

With Crump gone, Struve did not appoint a new secretary to succeed him. He preferred to run everything himself and to keep the position for a full-time research astronomer. Struve appointed Keenan "library adviser," a very part-time responsibility, and insisted that he himself must control the Yerkes library rather than leaving it to the university librarian on campus.[26] Likewise, he tried to handle publicity for the observatory himself through the campus administration.[27] One of his recurrent struggles was to keep control of the maintenance and repair work on the buildings at Yerkes. He had only a skeleton staff of workmen at the observatory, and any major job was therefore the responsibility of the Buildings and Grounds Department in Chicago. Lyman R. Flook, its superintendent, was Struve's nemesis. Frost had jollied him along, but Struve tried to dictate to him. Flook fought back hard, insisting that his work crews must do the tasks he had assigned them and not be diverted to other jobs Struve felt were more important. Struve was completely unrealistic in wanting Flook to clear every job with him before starting it. He argued, for instance, that putting up scaffolds to paint the dome of the forty-inch refractor could hamper observing, but he left no one in charge who could speak for him during his frequent absences on observing trips. William B. Harrell, the business manager in Chicago and Flook's superior, under orders from Hutchins to get the jobs done but keep Struve happy, was constantly caught in the middle in the guerrilla warfare between the two strong-minded chieftains.[28] Somehow he kept the peace and research flourished, but at a tremendous cost to Struve's sense of well-being.

As he pruned the deadwood from the Yerkes staff, supported the best young research workers, and fought to maintain the physical plant as he thought it should be maintained, Struve was simultaneously turning out research results at an only slightly slower rate than before he became director. He and Elvey put together a long paper, of which Frost was officially the first author, reporting all their observational results on the spectrum and variable radial velocity of the unusual F-type supergiant $\epsilon$ Aurigae. Frost had begun obtaining spectrograms of it in 1899 and had published some of the data in 1921; Struve and Elvey had published a preliminary report on their more astrophysical results in 1930. This 1932 paper was almost entirely a collection of their data; Struve did not have the theoretical equipment to analyze it. Half a decade later, with two bright new young theorists he had recruited to the Yerkes staff, Struve was able to understand and explain these observational results in terms of the structure of the supergiant and of its even larger, cooler, nearly invisible companion.[29]

In 1932 he began the work to which he was to devote much of his life, the study of B stars with emission lines in their spectra, showing that they have greatly extended outer structures, or shells. In most cases they resulted from rapid rotation of the star, which was losing or throwing off mass around its equator, forming a large disk or ring of low-density gas. In some cases the mass loss was steady and the resulting "shell" was quiescent; in other cases mass was thrown off in bursts, and the shell's structure varied more rapidly. Struve used his observational data on the widths and strengths of the absorption and emission lines and simple physical reasoning to determine the dimensions and rotational velocity of the shell and its dimensions, density, ionization, and temperature. He had decided that the basic nature of the great majority of stars, with "normal" spectra, was already known or soon would be; Struve was pushing on to study the rarer stars with "peculiar" spectra, which he and a few other astronomers were beginning to realize represented short-lived, transitory stages of stars, evolving from one configuration to another. Most of Struve's research at this time was based on existing spectrograms in the Yerkes collection, supplemented by additional ones that he, Elvey, and graduate students working with him had obtained specifically for his program.[30]

However, Struve knew that he needed a larger telescope soon. He could not wait until the McDonald reflector was completed. As an interim measure, he worked out another agreement to use the recently completed, nearly defunct sixty-nine-inch reflector of Perkins Observatory. This was the Ohio Wesleyan telescope Crump had started. After he left in 1928, the Methodist university had brought in Harlan T. Stetson, a Yerkes Ph.D. of 1915, to replace him. Stetson had been a very good teacher at Harvard University for thirteen years before accepting the Perkins Observatory directorship, but he had no research experience at all, and Shapley had pushed him out of Harvard. The Ohio Wesleyan administration had found Stetson through a teachers' agency rather than by recommendations from astronomers and observatory directors, which they did not seek.[31]

In 1930 Stetson hired Nicholas T. Bobrovnikoff, the more recent Yerkes Ph.D. who had then been a postdoctoral fellow at Lick Observatory, as an assistant professor at Ohio Wesleyan, but they both had full-time heavy teaching loads and no time for research. James W. Fecker, who had bought the former John A. Brashear Company in 1926, a few years after its founder's death, made the sixty-nine-inch primary mirror for the Perkins telescope and its Cassegrain secondary. He finally completed and installed them in the mounting in the dome in December 1931, with the Great Depression already in full force. Ohio Wesleyan University had no research tradition whatever, and no money to spend on astronomy. By April 1932 Stetson had succeeded in getting the telescope into focus and was able to send its crucial dimensions for mounting a spectrograph to Struve, who was then still debating whether to stay at Yerkes. He had already conceived the idea of a cooperative arrangement to use the existing sixty-nine-inch reflector at its poor midwestern site until the big reflector he really did want was ready at a good western site.[32]

Gale and Hutchins encouraged Struve to go ahead with an agreement with Stetson for the use of Perkins Observatory. Struve's plan was simple; he would have a spectrograph built in the Yerkes shop and send it to Delaware, Ohio, in exchange for half the observing time with the sixty-nine-inch reflector. He would station an assistant there to make sure that the Yerkes share of the time was always fully utilized. Stetson, Bobrovnikoff, and anyone else at Ohio Wesleyan could use the telescope and spectrograph for the other half of the time. In addition, Hutchins used his influence with Max Mason, his predecessor as president of the University of Chicago, now head of the Rockefeller Foundation, to have it make a grant to Ohio Wesleyan to keep Perkins Observatory in operation. The grant, $10,000 a year for two years, provided Ohio Wesleyan would match it dollar for dollar, came through on July 1, just as Struve took over the directorship.[33]

Struve began working very fast. He had already decided that the Perkins Observatory spectrograph would be made so that it could be used interchangeably with either a prism or a grating. The latter would be especially effective for work in the red spectral region and hence for observing the Hα line of hydrogen, a strong emission feature in the "peculiar" Be stars in which he was becoming so interested. Struve had machinist Carl Ridell start working on this spectrograph in the shop even before the agreement was signed. The grating would come from Gale's ruling machine. Struve had even picked out the assistant he would send to Perkins Observatory to do the Yerkes observing. He was Franklin E. Roach, an unusually mature, well-motivated graduate student who, after earning his bachelor's degree at Michigan, had worked two years, then had taken his master's degree in 1930, after one year at Yerkes Observatory, and learned to observe at the telescope.[34] Roach, married and with a young child, had had to drop out of school then to get a full-time job, as the industrial spectroscopist for a large machine-tool company in Milwaukee. But in 1932, with the depression in full force, he was laid off and could not find another job. His only option was to come back to an assistantship at Yerkes. His salary dropped from $200 to $62.50 a month, and he wanted to do a good thesis and earn his Ph.D. as soon as he could. Struve knew that he was self-reliant, intelligent, skilled, and experienced.[35]

Stetson was glad to have Roach assigned to Perkins Observatory, and at Struve's request he arranged for him to be allowed to take a graduate course at nearby Ohio State University, free of tuition. Hutchins and President Edmund D. Soper at Ohio Wesleyan University were glad to approve the cooperative agreement Struve had worked out. Basically it provided that Yerkes would leave the spectrograph at Perkins but retain ownership of it, while Ohio Wesleyan would keep the telescope in operation and pay its operating expenses, such as maintenance, heat, and electricity. Either party could terminate the agreement on six months' notice. Struve sent Roach to Delaware in October to look over the telescope and find a place for his little family to live.[36]

Ridell had the spectrograph nearly enough finished to be used in early November. Struve took it to Delaware even though he knew it would have to be returned to the shop for further work in a month or two. He wanted some results to show as soon as he could get them. He had just fought off a weak attempt by Gale to take back $1,000 as part of the economy drive, from a fund Struve had earmarked for the Perkins spectrograph and Roach's salary. Stetson, though in principle wanting to help, spent most of his time organizing dinners, arranging publicity, and giving talks to teachers' groups on "our research at the Perkins Observatory," but Roach, with long-distance coaching from Struve, pushed through the preparations to mount the spectrograph on the telescope.[37] After seeing it in place, Struve hurried on to Cleveland for a conference on the contract for the big Texas reflector, leaving Roach to get the Perkins instrument into operation and obtain some spectrograms whenever the Ohio skies cleared. The first results were disappointing, with the focus poor and very long exposure times required even for bright first-magnitude stars. Struve, who was soon off to Texas for a two-week conference in Austin and decision making at the McDonald Observatory site, fired off a series of terse suggestions to help track down the trouble. He enclosed a list of B stars for which he wanted spectrograms. Roach had some good ideas of his own and was soon reporting, equally tersely, on the results of the tests. Within a month he was getting much better results, including a good spectrogram of $\gamma$ Cassiopeiae, one of the Be stars Struve had asked him to observe. Two weeks later, just before Christmas, Roach had improved the system so much that he overexposed several plates, a welcome change, which indicated that they could get spectrograms of faint stars with the Perkins reflector and spectrograph. Struve was pleased. He had been "anxious to begin measurements"; now he could do so. The spectrograms Roach had sent him of Be stars were "very good," and Struve wished him, his wife, and their baby a Merry Christmas.[38]

By that time Stetson had worked out a regular observing schedule for the sixty-nine-inch reflector. Under it, Roach was assigned Monday, Wednesday, and Friday nights all night, as well as half of Sunday night. Bobrovnikoff had started observing cool red stars with molecular bands in their spectra, a program well suited to the grating spectrograph and his own good knowledge of molecular spectra. He took two half nights a week, and Marvin Cobb, the one Ohio Wesleyan graduate student, took all the rest. Stetson himself did not observe at all with the telescope; he did show it to visitors and otherwise saved himself for teaching, administration, and conferences. He was a very different type of director from Struve.[39]

Over the Christmas "vacation," Struve wrote two short papers on the H$\alpha$ emission line in $\gamma$ Cassiopeiae and $\beta$ Orionis, based on the spectrograms Roach had obtained at Perkins Observatory. Roach was already at work on his own thesis, on a topic Struve had suggested. The Yerkes director was "exceedingly enthusiastic" about the whole arrangement with Ohio Wesleyan and was sure that many more papers would come out of it.[40] All was not that simple, however. When Roach inserted the prism in place of the grating and began testing the new spectrograph, he could compare it directly with the spectrograph on the forty-inch refractor, also a prism system used in the blue spectral region. Roach immediately learned that the Perkins reflector required just as long an exposure time as the Yerkes refractor to obtain a similar spectrogram of the same star, although the sixty-nine-inch should have been at least three times faster than the forty-inch. Struve was incredulous, but it was true.[41] It meant that they were wasting much of the light that fell on the sixty-nine-inch primary mirror, and if they would correct this fault, they could work on even fainter stars with the prism or the grating. Struve's optical experts, Ross and George W. Moffitt, eventually tracked the problem down to the Cassegrain secondary mirror of the telescope, which Fecker had not figured correctly. Stetson resisted making the tests that led to this analysis, and in spite of constant prodding by Struve to send the mirror back to its maker to be reworked, he temporized and did nothing to correct the problem. He left for his "summer place in Maine" as soon as classes ended at Ohio Wesleyan in June and took an apparently relaxed, distant view of the defects in his telescope. In reality, Stetson's problem was that he had trumpeted the sixty-nine-inch as a great success; now he could not afford to admit that he had been wrong and the optics needed further work.[42] Ohio Wesleyan University was suffering even more from the depression than the University of Chicago, and to survive it imposed salary cuts (which reached 40 percent in 1933) on its faculty. Bobrovnikoff sent his wife back to Berkeley where she could get a job teaching school so they could both eat regularly. All the Ohio Wesleyan administrators and faculty members now resented the observatory as a white elephant that "old Perky" had foisted on them and opposed spending any more money on it.[43]

By fall Roach feared he would be unable to finish his thesis because of the faulty secondary mirror, which even Stetson now admitted was the main problem. The Perkins director continued to procrastinate and hardly seemed to understand the problem.[44]

By then President Soper had lost confidence in Stetson. He asked Struve for his opinion of him, of Bobrovnikoff, and of the future of astronomy at Perkins Observatory. Struve's reply was absolutely true, but devastating. He was always completely fair and objective in his evaluations, but his standards were very high and he never minced

Figure 28   Otto
Struve, already a
successful Yerkes
director (1934).
Courtesy of Yerkes
Observatory.

words. Bobrovnikoff was far more important scientifically than Stetson, Struve reported. Bobrovnikoff had produced results and could turn Perkins Observatory into an important center of research; Stetson could not. The director was pleasant, friendly, and gentlemanly, but he could not cope with the problems of his observatory. Struve was sympathetic to Stetson's plight, but he thought it would probably be best to turn Perkins Observatory over to "a more powerful observatory" and let it move the telescope and dome to another site. Soper accepted Struve's judgment and withdrew whatever moral support he had been giving Stetson.[45] Arthur H. Compton, America's and the University of Chicago's third Nobel Prize winner in physics, was the chairman of the Perkins Observatory Board of Visitors. Like Soper, he trusted Struve completely, and Stetson's days were numbered.[46] With no real progress by December, Struve announced a plan of terminating the agreement with Ohio Wesleyan and withdrawing Roach, the spectrograph, and all University of Chicago support from Perkins Observatory. This would certainly mean the end of the Rockefeller Foundation grant, due for renewal the following spring. Soper had "one or two very frank talks" with Stetson and made him "realize that he was not doing work which satisfied us here." The president was now openly supporting Bobrovnikoff and communicating with Struve through him. Just a few days before Christmas 1933 Stetson abruptly resigned his position, apologized to Struve, and decamped for Harvard, where he had secretly obtained a temporary appointment.[47]

Soper put Bobrovnikoff, the only astronomer left on his faculty, in charge of the observatory. Bobrovnikoff pushed hard on testing the Cassegrain secondary and returned it to Fecker's shop in Pittsburgh for further work. When it came back to the observatory not much improved, he returned it again. Finally, after three tries, Fecker succeeded in getting approximately the right figure on it. He did not have a large flat mirror to test the secondary mirror properly with the primary, so his attempts to improve its figure were all hit-or-miss affairs.[48]

Struve had already decided in January that there was little hope that the Rockefeller Foundation would renew its grant to support Perkins Observatory. He warned Roach to plan to finish all the observing for his thesis (on the near-infrared spectra of stars, the subject Struve had suggested to him) by summer. The Yerkes director sent another graduate student, J. Allen Hynek, to Delaware in May to take over from Roach. The Rockefeller grant did not come through, and Bobrovnikoff thought the observatory would have to close, but Struve had encouraged him to apply for smaller grants and helped him get them. By the end of summer Bobrovnikoff was sure that Perkins Observatory would stay in business and he would keep his job.[49]

Roach returned to Williams Bay in early June, completed his thesis, and passed his final examination in August. Evidently he did not do well with the questions on celestial mechanics, the old, preastrophysics astronomy that was still part of the Ph.D. ritual even at Yerkes Observatory. Nevertheless he got his degree and was soon on his way to McDonald Observatory, as the first astronomer Struve stationed there. Part of his plan when he sent Roach to Delaware in 1932 had been to train him, and test him, in working effectively on research on his own, in difficult circumstances. Roach had passed the test with flying colors. He was Struve's first Ph.D. student after he had become director.[50]

Struve and Yerkes Observatory also profited from the short-lived cooperative agreement with Ohio Wesleyan University. The director's first two short papers based on spectrograms Roach had obtained at Perkins Observatory were worth the price alone. One had confirmed the idea that the emission lines in rotating Be stars arise in an outer extended shell; the other had enabled Struve to put forward the hypothesis that most supergiant stars are losing material into space in radial, outward flows from their surfaces.[51] He had obtained Perkins material for several more papers of his own, and Roach had done his thesis there, and after him Hynek, whom Struve had sent specifically to get as much data as he could before Ohio Wesleyan closed its observatory.[52] Probably most important of all, Struve had seen at first hand the heartbreaking problems of trying to squeeze research results out of a badly underfunded observatory that he did not himself control and could visit only rarely. These experiences strengthened his resolve to deal with McDonald Observatory and the University of Texas in a completely different way.

At Perkins Observatory Bobrovnikoff worked tirelessly to bring about a cooperative agreement between Ohio Wesleyan and nearby Ohio State University to operate the observatory as an ongoing research institution. The Russian emigré, whom conservative, nativist American astronomers like Joel Stebbins and Heber D. Curtis thought would prove a complete failure, in fact turned out to be an extremely effective diplomat. He succeeded in his negotiations, hired Hynek as the first new staff member at the now jointly operated observatory, and preserved it for science. Even Curtis applauded.[53]

Many other things were going on at Yerkes Observatory as the Perkins cooperative arrangement played out. In November 1933 Struve received a nasty shock when he learned of rumors drifting around the campus that President Hutchins was working on a plan for a merger with Northwestern University in nearby Evanston, Illinois. The idea was anathema to Struve, who saw everything in terms of his Astronomy Department. He was doing his best to purge the nonproductive deadwood from his faculty and bring in bright young researchers. He shot off a letter to Dean Gale in Chicago, making it clear that although he would be glad to take over Northwestern's Dearborn Observatory for student teaching, its director, Oliver J. Lee, would be a "deplorable" new faculty member. Lee, a Yerkes Ph.D., had been a singularly unproductive member of its faculty for twelve years before leaving when he was not promoted to associate professor. Struve wanted no part of him in the proposed joint university. Gale, the veteran of countless bright new ideas that had petered out against the entrenched status quo, reassured Struve. Probably nothing would happen, and if it did he thought he and Struve would be able to head off Lee's appointment. Gale was right; as soon as

Figure 29   Mary R. Calvert at eyepiece of twelve-inch refractor. Courtesy of Yerkes Observatory.

Hutchins's merger plan hit the newspapers, bitter opposition from Chicago and Northwestern alumni, traditional rivals, killed it. Struve was saved from Lee.[54]

A much more pleasant episode was Sir Arthur S. Eddington's visit to the University of Chicago and Yerkes Observatory in April 1934. The great English theorist had written several popular and semipopular books on astronomy, physics, and the universe, of which the most famous was *The Nature of the Physical World,* published in 1929, and his *The Expanding Universe* had just appeared in 1933. President Hutchins invited Eddington, as part of his lecture tour of the United States, to give two popular talks on the campus. In accepting, the English theorist expressed his hope of visiting Yerkes Observatory as well. Struve was very pleased to welcome his Cambridge mentor whenever he could come, for as long as he wanted to stay. Eddington could spend a weekend at Williams Bay, so Struve drove to Chicago to hear his Thursday evening lecture, brought him to the observatory on Friday afternoon, entertained him as a houseguest, arranged for him to give a colloquium on Saturday, and held a dinner for him that evening. Undoubtedly the famous Eddington's visit solidified Struve's importance in the minds of many faculty members on the campus who had not previously known him.[55]

Two new members joined Struve's team in 1934. One was secretary Lillian Ness, who was to work for him until he left Yerkes sixteen years later, then follow him to Berkeley and to the National Radio Astronomy Observatory in 1959. Frost's secretary for the last four years of his directorship had been Gertrude Enders, whose work he evidently found satisfactory. When Struve took over, however, he had a much greater workload for her. He had much more correspondence with Hutchins, Gale, and other administrators on campus, the Press, which published the *Astrophysical Journal,* the Buildings and Grounds officials he thought were not maintaining Yerkes Observatory adequately, and all the many observatory directors and scientists with whom he had so much business to transact and so many research ideas to discuss. He expected all his letters to be typed quickly, accurately, and neatly. Enders did not measure up to his exacting standards, and no doubt he told her so. Finally, after a year and nine months with him she left for Milwaukee "because of a reorganization in the office work at the Observatory, which demands a greater proportion of purely typographical work," according to the lukewarm recommendation Struve gave her. There was nothing wrong with her; she had worked "satisfactorily," displayed "considerable initiative and efficiency," and "her honesty. . . ha[d] been above reproach." Leaving a steady university job with a low-rent observatory house was a hard fate in the cruelest part of the Great Depression, but Struve knew he needed a better secretary. Frost would never have done what he did.[56]

Ness arrived at Yerkes Observatory and began typing her first letters for Struve in May. She was thirty-nine years old and had worked as a secretary for years. He soon found that she could take whatever dictation he gave her and turn it into neat, well-typed letters by the next day, or by that night if there was a crisis. Her work ethic exactly matched his own. Struve was not long in learning that he could count on her to take any responsibility he gave her and fulfill it well.[57]

The other new addition to Yerkes Observatory was Louis G. Henyey, who came as a young graduate student and assistant in June 1934. Then twenty-four, he had been an undergraduate and then graduate student at the Case Institute of Technology in Cleveland, where he had earned his bachelor's degree in 1932 and his master's a year later. At Yerkes his starting salary as an assistant was the standard $900 a year. In the late winter and spring of 1935, when Struve brought Hynek back to Chicago to help with the teaching on the campus, he sent Henyey to Perkins Observatory to take his place. The new assistant turned out to be well trained as an observer and a brilliant theorist as well. At the end of that summer Struve raised Henyey's salary to $1,050. He was to work closely with Struve as a graduate student and then as a faculty member at Yerkes and later at Berkeley.[58]

Struve also protected those who worked well for him. In July 1934, when the annual letters of appointment for the staff employees arrived from the campus, Mary R. Calvert learned that she had been demoted from assistant to computer. She was Edward E. Barnard's niece, who had been his assistant until his death and then stayed at Yerkes as a general high-level assistant, particularly skilled in photographic work. She could handle any assignment Struve gave her. The younger women on the staff looked up to her as an example; her $1,500 annual salary, though well below what married male assistants with Ph.D.'s made, was comparable to the younger single men's. Struve immediately protested Calvert's change in status forcefully to Dean Gale. The dean lamely admitted there had been a "slip" in his office, and she was restored to her assistantship. Fifteen or more years Struve's senior, the bright southern spinster enjoyed working with him. She was the one Yerkes employee who mothered him, sending him chatty, vivid reports on the state of the observatory and on his mother's health when he was away from Williams Bay on long trips.[59]

For all his harshness, Struve did have a soft spot for old, loyal staff members. At the end of 1933 he recommended that George C. Blakslee, who was to be retired at age seventy-two, be granted a "retiring allowance" of $45 a month. He was no longer capable of doing the photographic work he had been handling since 1916 and had practically no other source of income except what he could make from occasional lectures on the wonders of the universe. There were no pensions as a matter of right for employees like Blakslee at that time, and there was no social security system. Families were expected to take care of their parents when they could no longer work. Blakslee had three married daughters; two of his sons-in-law had been unemployed for over a year, and the third, Roach, was then receiving $750 a year under his assistantship. That was the reality many families faced in the depression. Gale approved retiring Blakslee and paying him $45 a month, but he stipulated it should be a salary, with no duties, rather than a retiring allowance. This meant it would come out of the astronomy budget.[60]

Struve continued to insist on the necessity of raises for Elvey, "one of the most promising astrophysicists in this country," and Morgan, "now doing most of our spectroscopic work; [i]n this field he is the best investigator in America." On July 1, 1935, Elvey received a $300 raise to $3,000 a year, and Morgan got $400, which after a conference between Struve, Gale, and Hutchins was raised another $200 effective October 1, to an annual salary of $2,400.[61] In part Morgan's large raise may have been because Struve feared he would receive a job offer from Harvard. It also certainly reflected the fact that Struve, at Gale's urging, made Morgan his de facto assistant director at Yerkes and left him in charge during his frequent trips to Texas. Struve had been embarrassed by a letter written by Van Biesbroeck's wife, published in a Belgian astronomical magazine, in which the two of them were described as "the two Europeans" who had founded McDonald Observatory themselves and would use it entirely for their own research. All the Americans on the Yerkes staff apparently would serve only as observing assistants for them, in her fond eyes. She had intended the letter only for a family friend, but in print it aroused unfavorable comment from American astronomers who did not like the concept of a "foreigner" as director and would dislike having another as assistant director even more. Hutchins and Gale wanted Struve to have someone who could stand in for him, and Morgan was the available American, whom Struve may have thought had some of the secretarial talents of his father-in-law, Storrs B. Barrett.[62]

Whatever Morgan's administrative abilities, he was a dedicated astronomical spectroscopist. When Nova Herculis flared up in December 1934 as one of the brightest "new" stars in years, Struve and Morgan mounted an all-out spectroscopic campaign with the forty-inch refractor. At the American Astronomical Society meeting in Philadelphia on December 29, the Yerkes director presented their paper, in which they described the changes in the star's emission and absorption lines between December 18 and 27. One of them, probably Struve, had observed it on Christmas Eve, and Morgan must have telephoned a description of the December 27 spectrogram to Struve in Philadelphia just before he gave the paper.[63]

Struve was not content with observing with only the Perkins sixty-nine-inch reflector and the Yerkes forty-inch refractor. What he really most wanted to do was to obtain high-dispersion spectrograms with the McDonald Observatory reflector, designed as an eighty-inch. It would be years before it would be ready; in the meantime he sought observational material where he could. In 1932 Struve persuaded Theodore Dunham, at Mount Wilson, to obtain a high-dispersion spectrogram of $\tau$ Scorpii with the one-hundred-inch telescope and its coudé spectrograph. This was the B star with the narrowest spectral lines Struve knew, ideal for measuring and identifying many previously unobserved lines. Struve could also study its line profiles, unperturbed by the rotational broadening most B stars exhibited so strongly. Dunham's contribution apparently was only to take the spectrogram at the telescope; Struve seems to have done all the measurements, identifications, analysis, and profile studies.[64]

Measuring and analyzing spectrograms were easy compared with the human problems Struve faced. In the 1930s, as Adolf Hitler rose and solidified his power, German astronomers he defined as Jewish were persecuted and driven from their jobs. The United States was one safe haven that many of them tried to reach, but the number of positions there was small, and capable young American astronomers were having difficulty finding places themselves. Harlow Shapley, at Harvard, and Frank Schlesinger, at Yale, were leaders in trying to help these refugees from Nazism obtain positions in American observatories.[65]

Struve, with his family's German background and his many international contacts, was highly aware of the problem. He was not prejudiced against Jews and detested "the mad activities of the present German government in regard to scientists of non-Ar[y]an origin." He believed that "the history of the world knows nothing equally despicable and equally foolish as the expulsion of the best men Germany had had."[66] Yet when he had to make a choice, saving astronomical research won out over saving a victim of Nazi oppression.

The victim was Hans Rosenberg, one of the earliest pioneers in photoelectric photometry. During World War I he had served at the front as an artillery officer in the German army. Nevertheless he was forced out of his position as director of the Kiel Observatory in 1934, officially on leave of absence, but with no hope of returning as long as the Nazis were in power or, as it later turned out, of even getting his pension. The Emergency Committee in Aid of Displaced German Scholars and the Rockefeller Foundation provided $6,000 to pay Rosenberg's salary for a year and a half as a visiting professor at Yerkes Observatory. Fifty-five years old, Rosenberg arrived in Williams Bay with his family in early May.[67] He was glad to be there, and Struve was glad to have him, as long as someone else was paying him and he was not filling a position a younger, more active researcher might take. Though Struve believed that "the persecution of Jewish scientists in Germany" was the worst thing that had happened since the end of World War I, "probably worse than any of the excesses in Russia," he made it clear from the beginning that Rosenberg should not expect to be hired permanently.[68]

Rosenberg was especially welcome because he brought with him a number of photocells, photometers, electrical laboratory instruments, and other research equipment, much of it quite difficult to obtain in the United States.[69] At Yerkes he began a program of photographic photometry, helped design a photoelectric photometer for McDonald Observatory, and gave a series of lectures on astronomical photography. All were admirable efforts, but none was the kind of creative, highly productive research Struve wanted from his faculty. The Emergency Committee and the Rockefeller Foundation would continue Rosenberg's funding for another year if the University of Chicago would appoint him to a permanent faculty position at the end of that period, but Struve was adamantly opposed. Rosenberg himself believed for a time that if he returned to Germany he could retire, draw his pension, and live out his life safely there, but Hitler's actions soon convinced him that going back to his homeland would be a dangerous mistake. Hutchins's assistant then arranged for the Jewish Charities of Chicago to continue Rosenberg's salary for another six months.[70]

As the end of that period approached, the Emergency Committee in New York came through with the money for another four-month extension, which the United Jewish Charities of Chicago matched. Each time the foundation executives pressured the University of Chicago administration to give Rosenberg a permanent appointment, but each time Struve stood firm against it. Poor Rosenberg sold a set of his prized Zeiss optical filters and a slide projector he had brought with him to finance his family's travel to he knew not where.[71] They had to move out of their house and sold the fine European furniture, china, and hangings they had brought with them from Germany.[72] Struve had tried hard to find a job for him in America, but no university wanted to hire the fifty-seven-year-old refugee, who still preferred to express himself in German.[73] In two and a half years he had published only one paper. That was not enough for a Yerkes staff member under Struve.[74]

In 1937 Rosenberg left Williams Bay with his family, and in 1938 he made his way to Turkey, where he had been appointed director of the Istanbul University Observatory. It was a fortunate outcome of his desperate search for an opportunity to go on in astronomy long term. Years before then, in fact even before Rosenberg had sold his photo-cells and furniture and given up his job, Struve had begun appointing the brilliant team of young astrophysicists for whom he had been saving the precious research positions.[75]

# CHAPTER SEVEN

# Resurrection on the Campus and at Yerkes, 1893–1937

Otto Struve resurrected Yerkes Observatory. His arrival as a graduate student, his appointment as director, and his insistence on keeping the best young, productive Ph.D. students Christian T. Elvey, William W. Morgan, and Philip C. Keenan on the staff by getting rid of the deadwood Edwin B. Frost had accumulated were all steps toward that resurrection. But it really began in earnest when Struve added three brilliant young astronomer-astrophysicists from abroad to the Yerkes faculty, Gerard P. Kuiper from Holland, Bengt Strömgren from Denmark, and Subrahmanyan Chandrasekhar from India. In later years Struve liked to recount how he had gone to Robert M. Hutchins and told him that these men were the best astronomical research workers in the world and should be hired; the University of Chicago president, convinced, simply told his director to sign them up. What Struve did not mention was that all three were his second choices; in each case he had first failed to persuade another candidate to join the University of Chicago faculty, then had successfully wooed a substitute.[1]

The story begins on the University of Chicago campus, however, long before Kuiper, Chandrasekhar, and Strömgren were even born. The first astronomy faculty member there, Thomas Jefferson Jackson See, began his appointment as an instructor in astronomy on January 1, 1893, at a salary of $800 a year. He was to have one of the strangest careers on record in American astronomy. President William Rainey Harper had begun negotiating with See in the early summer of 1892, even before he had finally signed up George Ellery Hale as an associate professor, and before Harper and Hale had persuaded Charles T. Yerkes to provide the money to build "his" observatory.[2] See, a native of Missouri, had done his undergraduate work at its state university and had then entered the University of Berlin as a graduate student in astronomy. As soon as he heard that Harper was recruiting a faculty for the soon-to-open new University of Chicago, See (who had met him briefly in Berlin) wrote to apply for a post. He was scheduled to complete his Ph.D. in November and could come to Chicago as soon after that as Harper needed him. See's training and experience were all in the older astronomy; he was basically a theoretician of celestial mechanics who had also observed and measured double stars visually. This was exactly the kind of astronomy Harper was familiar with from his own schooling and from whatever informal contacts he had with his faculty colleagues at Yale, or in almost any university in America or in Europe. See fitted his idea of a professor much better than Hale; the young Missourian (three years older than the director-to-be) had also studied Latin, Greek, German, and French, and during summer vacations from Berlin he had traveled in Italy, Egypt, Greece, and England, visiting ancient churches and castles as well as art and historical museums. Best of all, he had come to know Eri B. Hurlbert, a Bible scholar, longtime friend of Harper, and professor of divinity at Chicago. It added up to a perfect picture for the president, and he sent See a contract.[3]

Very soon thereafter, Harper began to learn what the real See was like. George D. Purinton, president of the University of Missouri, hearing of the appointment, warned Harper that this alumnus, although intellectually brilliant, was "thoroughly unscrupulous, an intriguer . . . of a dangerous type, a genius in prosecuting his own claims to preferment, and to sum up, largely devoid of moral principle." Worst of all, according to Purinton, See had spread a story that his salary at Chicago would be $6,000 a year, well above what any professor was getting at Missouri.[4] Actually, Harper had made Chicago famous by paying that amount to full professors, but he was much more economical, even stingy, with instructors, and it is highly unlikely that See himself had told anyone this. The Chicago president demanded an explanation of the charges, and See defended himself in a sixteen-page letter denying them all but admitting that he had been one of the student leaders in a successful campaign to oust the previous president of the University of Missouri, who was "dictatorial" and had "gathered on the faculty a crowd of servile supporters." See himself had taken a stand of the side of "all the ablest and most independent members of the faculty" and had used his influence "in favor of decency and reform and progress." Harper received a deluge of letters of support from See's friends in Missouri and elsewhere. The young astronomer, still in Germany, wrote Hurlbert denouncing the "blackmailing intrigue" of the former Missouri president whose career he had ended. See urged the divinity professor to assure Harper that the charges were "not only a gross misrepresentation and injustice but a deliberate and willful falsehood designed for the express purpose of revenge." Evidently Hurlbert was not sure where the truth lay, for he mildly endorsed this letter to Harper. "If See is innocent, I hope lies will not kill his chances—Hurlbert." [5] The Chicago president decided to go through with the appointment, but he was to regret it.

Harper had been negotiating with Hale and See simultaneously but did not name either one to the other. Part of the agreement under which he hired Hale provided that the Yerkes director would not supervise the work of any classical astronomers who might join the faculty, no doubt because of See's Ph.D. and three-year age advantage. Even before See arrived on the campus, Hale learned that his new colleague's fellow students in Berlin considered him "a very capable fellow in some directions, but . . . very peculiar." The Kenwood director had hoped to the last that Harper would not go through with the appointment, but he did. With See lecturing on the campus and barred from observing at Kenwood, where Hale was doing research and planning Yerkes Observatory, explicitly free of any teaching duties, naturally the two quickly became rivals. In Berlin himself in 1894, Hale questioned Wilhelm Förster, See's thesis adviser, and heard what he interpreted as far from a vote of confidence in the young Missourian. Hale hastened to report the bad news to Harper. See was not skilled with instruments, Hale said he had learned. He could be positively dangerous if he was allowed to use the forty-inch refractor when it was completed. Hale did not want See at Yerkes; he should be confined to the campus, preferably under some senior professor who could direct his work. He had "absolutely no personal feeling" against See, Hale told the president, but he had to write this letter to protect the telescope, much as he hated to do so. [6]

From the moment he arrived in Chicago, See was embroiled in a bitter, legalistic argument with President Purinton of the University of Missouri, whom he charged with malicious slander for informing Harper of his activities during his student days. Although See clamored for recognition, he was notably uncreative and, except for one early paper on spectroscopic binaries, tied to the past in his research. [7] His new methods in celestial mechanics proved to be little more than minor variants on old, well-tested ones, and he made no new discoveries that brought him the fame he desired. See published a long paper giving his views on education in astronomy, in which he derided "the New Astronomy" as simply fact collecting while glorifying the "Old Astronomy," which he also called "physical astronomy, or celestial mechanics," as the explanation of those facts in terms of gravitational forces and dynamics. He deplored "the neglect of astronomical teaching and the placing of the centre of astronomical effort almost entirely in observational and photographic work [which] has undoubtedly had much to do with bringing about the present state of affairs." This was a direct shot at Hale, whom he did not mention by name in the article, though he praised Sherburne W. Burnham and Edward E. Barnard for their visual observations. See scoffed at the idea that "nothing is required for the cultivation of astronomy but a big telescope." [8]

In 1895 See demanded a promotion to associate professor, and when Harper refused, he went on leave to the recently founded Lowell Observatory in Flagstaff, Arizona Territory. He had given up his dream of becoming the head of the Chicago Astronomy Department and now hoped to take over the scientific leadership of the new observatory from its director and source of funds, Percival Lowell. Once again, however, See was soon at loggerheads with all his coworkers. They accused him of plagiarism, of mistreating his subordinates, and of denigrating Lowell in his absence. In less than three years See lost his job at Lowell Observatory, unceremoniously fired in the summer of 1898. [9]

During all the time he was on the Lowell Observatory staff, See had been carrying on guerrilla warfare against Harper, alternately demanding a promotion with inflated claims of his own research and denouncing the president in letters to the University of Chicago trustees. See's appointment and leave expired on October 1, 1898, just a year after Yerkes Observatory had gone into operation, and Harper sternly notified him in advance that it would not be renewed. The trustees supported the president fully, and See had argued himself out of two jobs within just a few months. [10] However, he managed to get an appointment at the Naval Observatory, where he used its twenty-six-inch refractor for double-star measurements, as he had used Lowell's twenty-four-inch in Flagstaff and in Mexico. See was no Burnham, though, and some of his more spectacular purported observations turned out to be outright falsifications. He was banished from the Naval Observatory after only three years, but since his position was "professor," then a commissioned officer in the navy, it was difficult if not impossible to fire him. Instead he was sent to Annapolis to teach mathematics to midshipmen. Once again See proved a complete misfit, and just one year later, in 1903, he was reassigned to the Mare Island Navy Yard, near San Francisco. There he observed, calculated, wrote and published papers and books (which no one but he himself took seriously), and fulminated against the entire American astronomical establishment until long after his retirement in 1930.[11]

See had met and corresponded with Harry Y. Benedict, the University of Texas mathematics and mathematical astronomy professor, very early in his career. When Benedict, who had been dean of the College of Arts and Sciences since 1911, succeeded to the presidency in 1927, See sent him a flowery letter of congratulations, comparing his situation to "that of Marcus Aurelius on the throne of the Roman Empire—Plato's dream come true." In 1930, the year he retired, the captain sent his old "friend" a more disturbed letter, mixing recommendations of George Willis Ritchey as a telescope builder with ravings against Hale. Two years later, See suddenly learned that Benedict had signed the agreement with the University of Chicago under which Otto Struve, Hale's successor as director at Yerkes and a "foreigner" to boot, was going to build McDonald Observatory around a large reflecting telescope, to be dedicated to astrophysics. His fury knew no bounds. Now retired, with time on his hands, See dashed off a fifty-page diatribe to Benedict, denouncing Chicago, Hale, Struve, reflectors, astrophysics, and every bit of the sketchy plans he had read.[12] Benedict ignored See's hysterical letter and went right ahead with Struve. None of the Texas trustees paid any attention to See either; Struve later reported that one of them, responding to a complaint against the concept of a "foreigner" as director, replied, "Better a Russian than a damn Yankee."[13]

See's successor on the faculty on the campus in Chicago was Kurt Laves, a German who received his Ph.D. in astronomy at Berlin in 1891. He came to Chicago in August 1893, at the time of the World Congress of Astronomy and Mathematics, and stayed at the new university as a docent, the first step in Harper's academic hierarchy. In 1897, with See on leave at Lowell Observatory and not expected to return, Laves was promoted to instructor in astronomy. He taught mathematics as well and slowly went partway up the ladder, to assistant professor in 1901 and associate professor in 1908. Laves did only a little published research, almost all of it on the motion of the earth in the earth-moon system, and was never promoted to full professor before he retired in 1932, just as Struve became director at Yerkes.

The dominant astronomy faculty member on the campus for many years was Forest Ray Moulton. Born in a log cabin in Michigan, he was the first boy in his hometown of LeRoy to get a college education, at nearby Albion College. He graduated with a bachelor's degree in 1894, worked for a year to earn some money, and then entered the University of Chicago as a graduate student in 1895. See and Laves were two of his teachers, and Moulton received his Ph.D. in mathematics and astronomy summa cum laude in 1899. He was the first doctor the University of Chicago astronomy program produced. Moulton was a hardworking, highly intelligent go-getter who was hired immediately as an instructor. He rose rapidly through the academic ranks, becoming a full professor in 1912.

Even before he had completed his thesis in celestial mechanics, Moulton began working with Thomas C. Chamberlin, the former president of the University of Wisconsin whom Harper had lured to Chicago as head professor in the Geology Department with a salary of $7,000 and the promise of more time for research. Geological evidence had convinced Chamberlin that the earth was much older than the ten- to fifty-million-year age of the sun deduced on theoretical grounds by the English physicist Lord Rayleigh. Going on to study the origin of the solar system, Chamberlin had further come to believe that the then current idea of its formation, proposed a century earlier by Pierre-Simon Laplace, could not be right either. Laplace's "nebular hypothesis" traced the formation of the planets to the condensation of protoplanets from a rotating gaseous disk, composed of atoms and molecules affected by gas-dynamical forces and orbiting the sun. Chamberlin showed that the earth had much more probably not condensed from a gas, but arisen from cold, solid "planetesimals." Parts of his argument depended on the properties of orbits, gravitational forces, and angular momentum, of which he had generalized intuitive concepts but no detailed mathematical knowledge. Chamberlin needed a celestial mechanics theorist, and in Moulton he found his man. Moulton was glad to explain the basic ideas of Newtonian dynamics to the older Chamberlin and to work out the specific calculations he needed to make concrete predictions from his theory and test them against existing observational data. The two worked very well together, and the Chamberlin-Moulton planetesimal hypothesis became the most widely accepted theory of the origin of the earth and

Figure 30   Forest
Ray Moulton.
Courtesy of Yerkes
Observatory.

the solar system. On this picture the swarm of "planetesimals" orbiting the sun owed its origin to the gravitational interaction between the sun and another star that passed near it, drawing out a tidal streamer from each, which condensed into the objects in orbit.[14] Only after World War II did this picture give way to a newer form of the Laplace idea, under which every star naturally developed a rotating nebula, which then condensed into solid planetesimals, at least in its inner parts, before forming the planets.

Moulton became a famous theoretical astronomer. A prolific writer, he published numerous books, including elementary and advanced astronomy texts, popular books on astronomy, and research monographs on orbital theory and tidal interactions between stars. A good teacher, he considered Edwin Hubble one of his prize undergraduate students on the campus and encouraged him to go on as a graduate student in astronomy at Yerkes after his return from his

Rhodes scholarship in England.[15] During World War I Moulton went to the Aberdeen Proving Ground in Aberdeen, Maryland, as a major in the Ordnance Department and very quickly developed mathematical methods for computing accurate firing tables for long-range artillery shells. To him it was essentially an exercise in orbital theory in a resisting medium (the atmosphere).[16]

In his scientific work, Moulton was highly combative. In 1899, the same year he received his Ph.D., he severely criticized a paper by See, in which his former teacher claimed to have detected the presence of an unseen third body (a faint star or a massive planet) in a visual binary system, from its gravitational effect on the other two stars. Moulton proved that See had been wrong and seemed to relish doing so. See dashed off a reply, claiming that in his paper he had not actually stated that there was a third body but had only mentioned it as a theoretical possibility. This was false, as Moulton politely pointed out to the editor of the *Astronomical Journal*. The editor inserted an editorial note after See's published reply, coldly and unmistakably declaring that it was "not in accord with the facts." Therefore, the editorial note concluded, See could expect to have any papers he submitted very carefully scrutinized in the future. He never again published a paper in the *Journal*.[17] In 1909 Moulton engaged in a different, three-cornered quarrel with See and Lowell over priority for the planetesimal hypothesis and "errors" they claimed to have found in it. Once again Moulton emerged the clear winner.[18]

Then, just three years later, See, writing on the conditions under which a satellite could be "captured" gravitationally by a planet, criticized some of Moulton's work on the problem. The younger man composed a long, biting reply in which he showed conclusively that See, in his book *The Capture Theory of Cosmical Evolution*, had copied sentence by sentence and equation by equation almost verbatim from Moulton's much earlier textbook on celestial mechanics, without ever mentioning it as a source. Furthermore, wherever See had deviated from Moulton's book, he had gone wrong, the injured author claimed. The editor of *Popular Astronomy* published Moulton's bitter attack on See in full, and the Naval Observatory professor could not or did not reply. See never published another scientific paper in *Popular Astronomy* either. His former student had finished him off as a serious figure in American astronomy.

Still later, in 1929, Moulton angrily charged Sir James Jeans and his younger collaborator Harold Jeffreys, who were developing what they called the tidal theory of the origin of the solar system, with not giving Chamberlin and himself credit for its main ideas, which he considered identical with those of their planetesimal theory.[19] By then Moulton had become restless and dissatisfied. One of a large family, he had six brothers, several of them outstanding successes in the business world and one a famous economist. Though Moulton was well known for the planetesimal hypothesis, it was hard to imagine that he could do much more in astronomy. Celestial mechanics seemed a relatively stagnant field; most of the big problems that could be handled analytically were already solved, and the high-speed, huge-memory computing machines that today make it possible to explore the long-term evolution of planetary orbits were not yet invented. He could not even hope to be chairman of the Astronomy Department; that went with the directorship of Yerkes. Frost was not sympathetic to Moulton's aspirations; the theorist was only the straw boss of a little astronomy outpost on the campus. In his spare time Moulton had managed a farm and had acted as architect and contractor for an apartment building in Chicago. In 1927, in the heyday of Coolidge prosperity, holding companies, and unchecked stock speculation, Moulton took the final plunge. He resigned his professorship, left academe, and became a director of the Utilities Power and Light Company of Chicago. As an investor and scientist, Moulton tried to combine the roles of business, financial, and technical adviser to its president. The day of reckoning arrived in the early 1930s, as many holding companies came crashing down in the Great Depression. Samuel Insull, the utilities kingpin of America, with headquarters in Chicago, fled the country in 1932. He was extradited and tried three times; although acquitted each time, he died a broken man. Moulton only lost his paper fortune and his job in 1933. But he survived as director of concessions of the Chicago World's Fair, and then in 1937 he moved to Washington as the "permanent secretary" or executive officer of the American Association for the Advancement of Science. He continued at its helm until he retired in 1948 at age seventy-six. In his eleven years he had built up its membership, its financial base, and its publishing operations and thus played an important part in strengthening the role of science in America.

Two of Moulton's Ph.D. students at Chicago who stayed on as faculty members in mathematics and astronomy were William D. MacMillan and Walter Bartky. MacMillan, born in Wisconsin, did his undergraduate work in a small college in Texas and then went to Chicago as a graduate student, receiving his doctor's degree in 1908. He listed one of his specialties as "real and implicit celestial mechanics," far more a mathematician's term than an astronomer's. Bartky, born in Chicago, did all his undergraduate and graduate work there, completing his Ph.D. and becoming an instructor in 1926. He was promoted to assistant professor the next year when Moulton resigned. MacMillan did a little research; Bartky did practically none, although in 1924 Struve did get him to look into the mathematics of the "curve of growth," the relation between the strength of a spectral line and the abundance of the atoms or ions that produce it.[20] Both were basically teachers, and Bartky became heavily engaged in the physical sciences courses in the College, the undergraduate general program for students in their first two years at the university. Struve counted on both of them to examine the graduate students in celestial mechanics, which he still believed every real astronomer, no matter how astrophysically inclined, should learn, just as he had done himself.[21] Otherwise Struve had little use for either of them, and he was eager to replace MacMillan, due to retire in 1936, with an astrophysicist on the campus.[22] Bartky had expressed interest in a split appointment, spending part of his time in mathematical statistics, and Struve, who abhorred divided responsibility, would have been glad to get rid of him altogether.[23]

The Yerkes director had decided that the time had come to begin building up the Yerkes staff. So far he had promoted and hired from within, but now he resolved to go outside the University of Chicago, find the best astrophysicists in the world, and bring them to Yerkes or the campus. Early in the summer of 1935 he met with Hutchins and Dean Henry G. Gale and secured the president's approval to find the best young research workers he could, no matter what country they came from.[24] Later that summer Harlow Shapley organized the first of a series of annual summer schools in astronomy at Harvard, which continued through 1942.[25] Struve participated in it, along with several other visiting faculty members, who lectured on their specialties to the Harvard faculty, research staff, and graduate students and visitors from other eastern universities. There Struve had opportunities for many confidential discussions with Shapley, Henry Norris Russell, and other top research workers. At the time there was no such thing as a search committee, a departmental vote, or an open job-application process for research positions in astronomy. Instead, a director like Struve made up his mind whom to hire based on whatever advice he wanted to seek, convinced his own superiors (Hutchins and

Gale, in Struve's case), and with their approval asked the prospect if he wanted the job. Struve did consult Frank E. Ross, his senior staff member who was now spending much of his time in California and therefore was well placed to learn about bright young prospects. But Ross's criteria were very different from Struve's; some of his suggested candidates seem ludicrous in retrospect while others, like Kuiper, were excellent. Neither Ross nor Struve even considered young astronomers who already had regular staff positions at Mount Wilson or Lick Observatory; they both knew there was no hope of attracting them to Yerkes.[26]

The first person Struve tried to hire for Yerkes Observatory was Olin C. Wilson, just twenty-six years old and the California Institute of Technology's first Ph.D. in astrophysics. He had done his observational thesis at Mount Wilson Observatory under Paul W. Merrill, one of Struve's mentors, and now had a lowly temporary computer's job there.[27] Wilson had been at the Harvard summer school and had made a strong impression on Struve. He invited the young man to make a side trip to Yerkes Observatory for further discussions on his way back to Pasadena. Following the directors' code of that day, and anxious to keep on good terms with Walter S. Adams, Struve did not directly offer Wilson a job, though he probably more than hinted that he would like to do so. Instead he first wrote the Mount Wilson director to ask him if his young computer "w[ould] be available for a position at the Yerkes Observatory." Adams replied that he hoped to keep Wilson on the permanent staff and had recommended him for a regular position just before Struve's letter had come. Nevertheless, Adams wrote, Struve should not hesitate to write Wilson; the young man "would greatly appreciate hearing from you." Wilson's salary had been "very low," but Adams was recommending him for a raise to $2,500 a year. However, the Mount Wilson director "d[id] not think that the question of a salary is a very serious consideration for him at the present time." This was a common directors' statement that was almost always false, but in fact a dedicated research worker like Wilson was resigned to accepting a lower salary in order to stay at Mount Wilson and observe with its large telescopes. This is just what happened; Wilson declined Struve's offer but appreciated the raise and permanent appointment at Mount Wilson it had brought him, even though he obviously would have preferred a still larger salary.[28]

In place of Wilson, the observational astrophysicist Struve did hire was Kuiper, whom Ross had recommended so strongly. The director no doubt discounted his praise for the young Dutch astronomer (which was largely based on his ability to resolve close double stars visually at the telescope, which Ross claimed he shared), but it reinforced the strong recommendation that Bart J. Bok had also made for Kuiper. Bok, Kuiper's fellow student in Holland, had come to America a few years before him and was an assistant professor at Harvard. Already highly Americanized, enthusiastic and talkative, he was a font of information for Struve. Kuiper, not quite thirty years old in the summer of 1935, had been aiming for a career in astronomy from early childhood. As a student in Leiden he worked under the great Ejnar Hertzsprung and resolved to become the world's astrophysical expert on double stars. In 1933, immediately after receiving his Ph.D., he went to Lick Observatory on a two-year postdoctoral fellowship. Kuiper was a demon of energy and began making important discoveries of close binary stars and nearby low-luminosity stars. He greatly increased astronomers' knowledge of white dwarf stars, the amazingly small, dense objects which could be fully understood only in terms of the then quite new concepts of electron degeneracy and general relativity. Kuiper had hoped to remain on the Lick staff with a permanent position, since its director, Robert G. Aitken, the great classical double-star observer, was to retire in 1935. Aitken also wanted Kuiper to stay as his successor in the double-star field, but it was not to be. Lick Observatory had only two openings, and the new director, William H. Wright, preferred to appoint Arthur B. Wyse and Nicholas U. Mayall, who like himself were excellent astronomical spectroscopists of stars and nebulae and were also Americans, married, and products of the Lick graduate program. Kuiper, a Dutch bachelor who worked on binary stars and nearby dwarfs, was odd man out.[29]

He had wanted badly to stay at Lick, but now he had no choice but to accept a job at the Bosscha Observatory at Lembang, Java. However, Kuiper obtained a temporary, one-year position at Harvard to tide him over before making the long voyage halfway around the world. On Bok's recommendation Struve invited him to stop at Yerkes on his way east to Cambridge.[30] Kuiper came, announcing that he had just discovered two more white dwarfs before leaving Lick. Struve was impressed by his energy, knowledge, and enthusiasm for research, and after a little further checking he wrote Kuiper, who had gone on to Harvard, asking if he would be interested in a position on the Yerkes and McDonald staff. Kuiper was definitely inter-

Figure 31   Gerard P. Kuiper (ca. 1936). Courtesy of Yerkes Observatory.

ested. He wanted to stay in America and observe with big telescopes, not fritter his life away in Java, waiting for a professorship to open up back in Holland. Furthermore, although he had "postponed the great step as long as [he could] . . . the right person came at the right time" in Cambridge, and he had fallen in love and was to be married to an American, Sarah P. Fuller. Struve was a master at communicating his own enthusiasm for the wonderful opportunities a bright young research worker would enjoy at McDonald as soon as the telescope was completed, and Kuiper, through Bok, who acted as his agent in the negotiations, made it clear that he would accept.[31]

Thus armed, Struve arranged for a conference with Hutchins and Gale in Chicago on October 8, 1935. At this meeting he outlined his plans for the future. He reported that he had approached the American, Wilson, who had declined the informal job offer, and then the Dutchman, Kuiper, who would accept. Hutchins authorized him to go ahead and appoint Kuiper as an assistant professor at the salary of $3,000 a year. When he first heard of him from Bok, Struve had considered making him assistant director at Yerkes, to lighten his own administrative burden. Both Bok and Shapley had warned him against it, telling Struve it would bring even more criticism from the "real" American zealots like Joel Stebbins and Wright, who were particularly opposed to putting "foreigners" in even minor positions of power. Struve himself had seen this in the "two Europeans" furor the previous year, when Julia Van Biesbroeck's enthusiastic letter about her husband and Struve appeared in print. After meeting the supremely self-centered Kuiper in person, Struve knew that although he was an outstanding research scientist, he could not possibly fill the role of assistant director. Adams, who had seen Kuiper several times, described him confidentially as "a little over-enthusiastic about his results, but doubtless this characteristic would be tempered with time." Hence it was at this same conference that Struve decided to name Morgan to this post, although it was not to be officially announced or even formalized. It brought Morgan his salary raise to $2,400. As a local product who had not yet received any firm outside offers, he did not qualify for a larger raise in Struve's calculus of salaries.[32]

As soon as he got back to Williams Bay, Struve wrote Kuiper and offered him the job. The double-star astrophysicist quickly accepted, informing his new director at the same time of the paper on white dwarfs he was then composing. Kuiper was to come to Williams Bay the following fall, with his young bride, at the end of his year at Harvard. Morgan got his raise too and was promoted to assistant professor on Hutchins's initiative rather than Struve's, just as Kuiper arrived.[33]

At the same conference in Chicago, Struve had made his recommendation for MacMillan's successor on the campus. The candidate Struve wanted was Svein Rosseland, professor at Oslo and director of its astronomical institute. He was one of the outstanding theoretical astrophysicists in the world, and his book was the bible of the subject. Rosseland had been a visiting lecturer at Harvard and had worked with Arthur S. Eddington in Cambridge, so he knew English well. Struve had written him in August to see if he would be interested in the professorship on the campus, perhaps first trying it for a year as a visitor. Rosseland had replied that it was a difficult decision for him, which would depend on the "financial and scientific conditions in Chicago." Probably he would be agreeable to a visiting appointment for a year. Armed with this information, Struve recommended him to Hutchins, who approved offering him $6,000 as a visiting professor for one year and letting him know that he could expect more if he accepted the permanent appointment. On campus Rosseland would have the "regular" teaching load, two courses each quarter, but Hutchins promised to reduce it if Rosseland needed more time for important research. However, the president was a little disappointed that the Norwegian professor was so old (he was forty-one); Hutchins would have preferred a younger man.[34]

Since all the campus astronomy faculty positions were joint ones with the Mathematics Department, its chairman, the elderly Gilbert Ames Bliss, had veto power over MacMillan's replacement. Struve took a copy of Rosseland's book and a pile of his papers to the campus and dropped them off in Bliss's office before his conference with Hutchins and Gale. The next day the impatient Struve wrote Bliss that the president was "very favorable" to appointing Rosseland and asked if the calculus of variations expert had yet read the material he had had for only two days when he received the note. Struve told Bliss that the matter was "urgent" because Rosseland was "contemplating another possibility," although the Norwegian's letter had in fact said nothing like this. Bliss wearily replied that he had been examining Rosseland's work, and though the time was far too short, his equations looked all right and his writing style was excellent, so Struve could go ahead and send him the offer.[35]

Struve did so immediately, explaining the salary and teaching load carefully. He emphasized that there would be no courses in the summer, which he hoped Rosseland would spend at Williams Bay, working and discussing research with the astronomers in residence there. Struve had a cottage waiting for him near the observatory, as Hutchins had authorized. All this was in the official letter; Struve was duty-bound to send a copy of it to the president. However, in a second, unofficial letter he sent with it, he told Rosseland that MacMillan's current salary was $5,500, while Ross and Van Biesbroeck, the two senior astronomy professors, were each making $5,000 a year, all less than the $6,000 of the offer. These salaries were lower than at Harvard (where Rosseland had declined a professorship in 1929), but the living expenses were lower in Chicago than in Cambridge, he explained. Hutchins wanted to get Rosseland, Struve wrote, and he broadly hinted that if the Oslo astrophysicist demanded more money, the president would very probably give it to him. The director emphasized that he himself very much wanted Rosseland to come, not only to do research, but to guide and inspire the entire Yerkes staff in their astrophysical efforts. Struve was highly conscious of his own deficient training in astrophysical theory. No copy of this second letter went to the president's office. However, Rosseland, after a month's consideration, declined the offer. He had his own institute in Oslo, where he was working with scientists in related fields, including hydrodynamics and planetary atmospheric physics, whom he knew well. He had already started using the new Bush differential analyzers, a very early form of electromechanical analog computer, for astrophysical calculations. Rosseland really had no desire to leave his native land for a new and unknown situation.[36]

Struve had foreseen that this might well be the outcome of his negotiations and had begun thinking about alternatives long before he received the Oslo professor's refusal. A very important source of advice was Russell, America's leading theoretical astrophysicist. He agreed to stop in Williams Bay in October to consult with Struve, and to return to Chicago with him for dinner with Gale, before continuing his trip to California. Russell, like Shapley, was not very satisfactory as an adviser. Both tended to recommend their own protégés of the moment, recent Ph.D.'s or refugees from Europe who needed jobs, rather than thinking deeply about the whole field of possible candidates and advising Struve on the best for his observatory and university. They were both committed to internationalism in hiring, although in Russell's case this commitment was mostly verbal and seldom extended to his own institution. Neither Russell nor Shapley yet took Yerkes seriously or believed that Struve would be able to rebuild it into a first-class research institution. Russell ranged over a large number of possibilities but concentrated on Rupert Wildt, a young German theoretician whose wife had a Jewish grandmother and who therefore had had to flee with her to America. Russell also advised Struve that Donald H. Menzel "has such a good place at Harvard that I doubt you'll get him—or if he ought to go." Likewise Shapley could find faults in every candidate Struve was considering from the Midwest or California, but he enthusiastically recommended "six, possibly seven young men" who seemed to be "extraordinarily good," all of them from Harvard and looking for jobs.[37] Struve though, was looking for the best. He considered it "a great pity that there are no theoretical astrophysicists in America besides Professor Russell and Dr. Menzel. It is imperative that we start a new school of thought at one of our universities and gradually develop a group of competent astrophysicists."[38] In fact, when he wrote "we" he often

Figure 32   Bengt Strömgren with electrical calculating machine (ca. 1935).
Courtesy of Copenhagen University Observatory.

meant "I," and he had just one university in mind, the University of
Chicago.

By the time Rosseland declined his offer, Struve's thoughts had
turned firmly to Bengt Strömgren. Bok had suggested him and Russell
had approved him as "good" but added that his work was "rather
abstract." Strömgren was the nearest thing to a child prodigy that
theoretical astrophysics had seen. The son of the director of Copenhagen Observatory, Young Bengt had learned calculus at eleven,
started a regular observing program at thirteen, and published his
first paper, on the orbit of a comet, as fifteen. He graduated from
Copenhagen University at nineteen and completed his Ph.D. thesis in
1929, at age twenty-one. Trained by his father in celestial mechanics
and positional astronomy, he made himself a theoretical astrophysicist, partly through his contacts at Niels Bohr's Institute of Theoretical Physics. In 1932, using the latest quantum mechanical results on
the opacity of matter at high temperatures, Strömgren proved that
there is a high hydrogen content in stellar interiors and thus, by implication, in the universe.[39] After ascertaining that Kuiper was enthusiastic about Strömgren, Struve arranged to see Hutchins, got his

approval, and cabled an offer to Copenhagen in early January 1936. Possibly because this time the president was getting the young candidate he wanted (Strömgren was just a few days short of twenty-eight), Struve did not have to clear this offer with Bliss. It was for an assistant professorship at $4,000 a year, the higher salary than Kuiper's recognizing Strömgren's faculty position in Copenhagen and his demonstrated accomplishments. As in the earlier offer to Rosseland, Struve specified that Strömgren was to teach and do research on the campus during the three quarters of the academic year and spend his summers at Williams Bay.[40]

This offer came to Strömgren out of the blue; Struve had not had time to explore the idea with him or with his father in advance but wanted to fill the position immediately on MacMillan's retirement that summer. No doubt he feared the mathematicians would come up with a candidate if he did not have one. Two weeks after receiving Struve's cablegram, Strömgren cabled his reply. He "hope[d]" to be able to accept it for a fifteen- or eighteen-month stay but could not commit himself beyond that. In a follow-up letter he expressed his gratitude for the wonderful opportunity but wrote that it would be difficult for him to decide immediately to remain in America for all the future. He asked if he could get the traveling expenses Struve had offered if he stayed only that long; married and with two small children, he had to consider financial matters, he said. Struve treated this letter as an acceptance and told Strömgren he was recommending his appointment for eighteen months but that he could leave earlier if he wished. He would get his travel funds in any case. An appointment was never considered as binding the faculty member to stay, Struve wrote, but he emphasized that the University of Chicago hoped that Strömgren would decide to remain permanently. On those terms the young Danish astrophysicist was glad to confirm his acceptance. Struve instructed him to go easy when he began teaching astrophysics on the campus, since there had never been any previous competent instruction in it there. But above all, Strömgren should not waste time preparing lectures; he should use the "notes and so on" he already had and "devote [your] energy primarily to research and not so much to teaching.[41]

Under Struve's complicated method of counting and keeping track of faculty positions and vacancies, which Hutchins and Gale accepted, he still had one more opening on the Yerkes staff left to fill. He wanted a theoretical astrophysicist who would complement Strömgren. Struve's first choice was Marcel Minnaert, the hardworking

Dutch theoretician of the solar and stellar atmospheres and their physical properties as revealed on high-dispersion spectrograms. Stellar atmospheres were exactly the field of Struve's greatest interest and the central problem of observational astrophysics of the 1930s, so Minnaert seemed an ideal choice. Russell advised Struve that Minnaert was "definitely the best," and Kuiper was highly enthusiastic about him. In their conference in early January, however, Hutchins objected that Minnaert was too old. Struve, who did not know him personally, had asked Van Biesbroeck Minnaert's age and had relayed his guess, fifty, to the president. In fact Minnaert was forty-two, as Kuiper ascertained from a friend in Holland, but by the time he got the answer (letters went to and from Europe on fast steamships then, requiring about one week each way) it was too late. What Hutchins wanted was "another brilliant young man" like Strömgren.[42]

This brought Struve's attention back to Chandrasekhar, who exactly fitted the president's desired profile. Born in 1910 in India to a high-caste Brahman family, he had demonstrated outstanding mathematical ability as a physics student at the University of Madras. On his own Chandrasekhar had studied Eddington's book *The Internal Constitution of the Stars,* and on graduation in 1930 had received a Government of India scholarship for graduate work in England. At Cambridge he studied with Ralph Fowler, attended Eddington's lectures, met Paul A. M. Dirac and Edward A. Milne, and quickly mastered the theory of stellar interiors as it was known at the time. He had already developed, almost completely on his own, the theory of the internal structure of white dwarf stars. This brought a clash with Eddington, who did not accept the newly derived relativistically degenerate equation of state that applies in the densest, most massive, and therefore paradoxically smallest white dwarfs. All the top theoretical physicists of the time knew it was correct, but Eddington could not accept it or the conclusions Chandrasekhar drew from it. At astronomical meetings Eddington used rhetorical arguments and his great prestige to uphold his own point of view; the shy young Indian could not and would not stoop to such tactics, but he was deeply wounded by them.[43]

In the summer of 1935 Chandrasekhar took part in the International Astronomical Union meeting in Paris and renewed his acquaintance with Russell and Shapley, both of whom he had met at meetings of the Royal Astronomical Society in London. Shapley invited him to come to Harvard as a visiting lecturer for a few months and followed it up with a formal letter a few months latter. Thus on December 8,

1935, Chandrasekhar first saw the New World from Boston Harbor, disembarked, met Shapley at the pier, and settled in for a three-month stay at Harvard College Observatory. This gave Struve, who had met him only once, in London in 1934, a chance to learn more about him.[44]

Russell considered Chandrasekhar "the most brilliant" of the possible candidates Struve was investigating and thought that "he bids fair to one of the best men of the coming generation of theoretical astrophysics." However, Russell warned Struve, Chandrasekhar's "political views are pretty radical, but I don't imagine that would prejudice him with President Hutchins." The last part was a shot at Hutchins's supposed left-wing tendencies; actually he was a strong defender of academic freedom and a crusader for equal rights for all, but at the same time an intimate friend of some of the most conservative capitalists in Chicago, such as Harold H. Swift, chairman of the Board of Trustees. Shapley was even more negative, thinking Chandrasekhar "more of a mathematical physicist than an astronomer" and questioning the idea of appointing him to a faculty position in a department of astronomy. Furthermore, Shapley wrote, Chandrasekhar was "a Communist, as is pretty well known, and does not hesitate to talk politics very vigorously." In fact he was not in the least a Communist but was an Indian who was proud of his heritage and objected to the discrimination he suffered in England. Shapley thought that a little radicalism would not hurt Chandrasekhar's chances at Harvard, where there were several much more radical faculty members, but might be dangerous at Chicago, which he evidently considered more conservative. The Harvard director clearly was not a reader of the *Chicago Tribune,* which was constantly fulminating against the many supposed Communists and other so-called radicals on the university's faculty. When Hutchins pressed Struve to invite Chandrasekhar to visit Yerkes and Chicago, the director passed Russell's and Shapley's views on to the president, in suitably edited form. He himself had no interest in the political views of his fellow faculty members, Struve said, but he wanted Hutchins to know about any possible objections. Furthermore, he himself was "not fully satisfied that [Chandrasekhar's] scientific work is of a character that we should want to promote in Chicago." Struve still considered the internal structure of white dwarfs too theoretical a subject for a Yerkes Observatory faculty member's research and would have preferred more of an applied theorist of stellar atmospheres, like Minnaert.

Hutchins's response was quick and to the point. He wrote Struve: "The only consideration which should be permitted to affect Chandrasekhar's appointment is his distinction and promise as a scholar. I am not interested in his political views except as they might get him into trouble with the police. I suppose he could be made to understand that the advocacy of the overthrow of the government by violence is a felony in Illinois."

This flippant style was typical of Hutchins; in it he communicated his deeply felt position and at the same time referred to the hearings held before the Illinois state senate the previous spring, in which he had brilliantly defended the university against unfounded charges of teaching subversion in Social Sciences 1. He concluded his letter to Struve by telling him that of course he should not recommend Chandrasekhar unless he thought he would be a valuable addition to the department, but he made it clear that he himself would like to see the Indian theorist on the campus with Strömgren.[45]

Quite independently Kuiper, now that he had met Chandrasekhar at Harvard and begun discussing research with him, enthusiastically recommended the young Indian to Struve. Kuiper could see that Chandrasekhar was an expert in all kinds of stellar interiors and a driven research worker. His lectures at Harvard were excellent. Furthermore, Chandrasekhar had spent a year in Copenhagen and knew Strömgren well. They got along well together. Kuiper could visualize himself, Chandrasekhar, and Strömgren taking over the leadership of the study of the astrophysics of stars from Eddington, Milne, and similarly older, unnamed observers (probably, in Kuiper's mind, including Struve).[46] Struve, who had gone to McDonald Observatory immediately after writing Hutchins, received the president's reply there. He at once assured Hutchins that he would invite Chandrasekhar to come to Williams Bay and Chicago to give a few lectures. Struve wrote him as soon as he got back to his office at Yerkes, and Chandrasekhar gladly accepted, saying he would be delighted to visit both places as soon as he had completed his lectures at Harvard in March.[47]

The problem, as Shapley, Russell, and even Kuiper had all warned Struve—who knew it all too well himself—was that Chandrasekhar was what most of them called "of the Oriental race."[48] Aristocratic and handsome, he had quite dark skin and in many parts of the United States would be considered black. This could all too easily lead to his being made to feel uncomfortable by strangers or even being insulted. Worst of all, Dean Gale was a notorious bigot. He belonged to a generation of scientists who had never met a black man as a social equal, had attended completely segregated schools, and had

Figure 33
Subrahmanyan
Chandrasekhar
(1936). Courtesy of
Yerkes Observatory.

been at universities that had at most a handful of blacks as graduate students in the social sciences, none at all in the physical sciences, and none on the faculty. In their daily life they were accustomed to see "colored men" (the polite term of their time for blacks) in subservient, poorly paid, labor-intensive jobs in which few could ever give an order to a white man, and never to a white woman. Struve and Kuiper were free of the prejudice that American society reinforced from generation to generation in this way; Hutchins had been brought up to hate and resist it; and Shapley especially and Russell to a lesser extent had freed themselves from it. But Gale, born near Chicago, a resident of the city all his life, had grown up within this system and now was a self-chosen enforcer of it. He considered anyone with a dark skin a black and did not want even one on his faculty. Chicago was a de facto segregated city in the 1930s, and the university campus was a few blocks away from Cottage Grove Avenue, the dividing line between white and black on the South Side. The university itself owned large amounts of property in Hyde Park, and its Board of Trustees and their agents were determined to keep the area lily white to protect these investments; most of the faculty members who lived there felt exactly the same way. The dean encouraged and fostered this attitude; his views were well known.

Struve realized all this when he carefully wrote Gale in early February, stating his own earlier doubts but emphasizing that they had evaporated when he met Chandrasekhar, "a very pleasant young man of great refinement and of excellent manners. His complexion is of course quite dark but his features are quite different from those of the American Negro." Chandrasekhar was a fellow of Trinity College, Cambridge, who had worked with Eddington and Milne but had now far surpassed his teachers. He was an extraordinarily brilliant man, who "promises to become one of the most successful astrophysicists in the world." Kuiper, who knew him, was enthusiastic about him; Strömgren, who had known him well in Copenhagen, was on excellent terms with him. Hutchins had "suggested" that Struve invite Chandrasekhar to Chicago and Yerkes "so that we could meet him personally and form an opinion of him." The Yerkes director wrote that he would not recommend anyone for an appointment unless Gale was convinced that it would be "an action for the advantage of the university," but that unless the dean objected he was going to invite Chandrasekhar to come.[49]

Gale did not object, no doubt because he knew that Hutchins would overrule him. Struve wrote Kuiper, telling him of Strömgren's cablegram accepting the Chicago faculty position, and asked for his further comments on Chandrasekhar, "especially . . . in his ability as a scholar, although his ability as a teacher would also have to be considered." Kuiper, who was attending Chandrasekhar's lectures at Harvard and discussing research with him daily, was ecstatic about him. The young Indian was an outstanding theorist not only of white dwarfs, but of the internal structure of massive, main sequence stars and of stellar atmospheres and radiative transfer as well. All he needed was more contact with observers (like Kuiper himself) to make his results more understandable (the young Dutchman would "translate" them for him), and his greatness would be recognized. Furthermore, Kuiper assured Struve, Chandrasekhar was no revolutionary; he had visited the Soviet Union and had "a fine time" with the Russian and Armenian astronomers he met, but he was no Communist. The Indian astrophysicist thought that "Russia ha[d] solved the race problem"; Kuiper pointed out to him "that the old Austro-Hungarian monarchy [which he evidently took as an analogy to the United States] had solved this problem pretty well too." But most of all, Kuiper's letters radiated his very strong belief in Chandrasekhar's theoretical abilities and his own great desire to have him at Yerkes, working together with him and Strömgren in solving all the problems of stars. Furthermore, near the end of February Kuiper reported that Shapley and Russell were now planning to propose Chandrasekhar for a three-year membership in the Society of Fellows at Harvard. This was a highly prestigious, well-paid research fellowship, and their recommendations, especially Russell's, would make it almost certain that he would get it. If Chandrasekhar went to Yerkes, Kuiper reported, Shapley "laughingly" feared that it would have the best staff in the United States, including the two best theoretical astronomers (Strömgren and Chandrasekhar).[50]

That was just what Struve wanted to have. By now his mind was made up. Kuiper's letters had convinced him that Chandrasekhar belonged on the Chicago faculty because he was "exceptionally capable and would certainly add to the standing of our department of astronomy and astrophysics." Furthermore, Struve knew he was interested in coming. Recognizing Dean Gale's extremely negative attitude and his ability to make trouble for Chandrasekhar, as well as the general discriminatory situation in Chicago, Struve's idea was to have him spend most of his time at Yerkes but to put in one or two quarters each year on the campus, where Hutchins wanted him. Gale had relented far enough to tell Struve that "in view of [Chandrasekhar's] exceptional ability" he would perhaps agree to his appointment. Keeping President Hutchins fully informed of every stage of the negotiations, the director wrote Chandrasekhar, asking him to consider whether, after his visit, he would be likely to accept an appointment "should a vacancy become available this fall." The position would be either as an assistant professor or a research associate, the latter requiring less teaching, at a salary of $3,000 a year.[51]

Chandrasekhar replied that he would be glad to be considered for a position at Chicago and agreed to come for four days, after he had finished his lectures at Harvard. Shapley and Russell were pressuring him to take the Society of Fellows position, emphasizing the difficulties his dark skin would bring him in Chicago. The one reservation Struve had expressed about Chandrasekhar was that he wished he would do a little observing along with all the theory; when he learned this from Kuiper, the young Indian immediately told his new friend he welcomed the idea and wanted to start some solar work on prominences and the chromosphere. He also hoped to write a book on stellar interiors, based on his Harvard lectures; this idea no doubt attracted Struve even more. He liked astronomers who published papers, but books were still better.[52]

Struve prepared the way very carefully for Chandrasekhar's visit. He wrote to the business manager of International House, the center for foreign students and visiting faculty members on the campus in Chicago, to make a reservation for his visitor. Struve emphasized that he was "one of the most brilliant young scientists from India and . . . holder of one of the most distinguished fellowships of Trinity College of Cambridge." The manager, who had visited Yerkes himself and heard Struve lecture on astronomy, should "instruct the clerks at the desk to extend to [Chandrasekhar] all the courtesies." Struve sent his prime job prospect information on how to get to International House, where he could eat and sleep, and very clear directions on which train to take the following morning from Chicago to Walworth, Wisconsin, where a car from Yerkes would meet him at the station and bring him to the observatory. Chandrasekhar would be his houseguest all the time he was in Williams Bay, and "Mrs. Struve . . . remember[ed] exactly the kind of meals that you prefer" (Chandrasekhar was a vegetarian). Struve would drive him back to Chicago on the last day of his visit, introduce him to some of the physicists and administrators on the campus, then personally put him on the train to Boston.[53]

The Yerkes director also wrote a formal letter to his dean, notifying him of the visit, stating that he had invited Bartky to come from the campus to talk with Chandrasekhar and hear him lecture, and offering to introduce him to Gale himself. Struve told Gale of the Harvard offer and asked permission to make an offer to Chandrasekhar if, after hearing his lectures, "we are all satisfied that he would be a real addition to our department." The director now revealed his plan to have Chandrasekhar, if he was appointed to the faculty, spend most of his time at Yerkes, with no more than one quarter a year on the campus. Struve gave as the only stated reason that he was "informed that [Chandrasekhar] is greatly interested in observational work and would especially welcome an opportunity to use the Yerkes 40-inch telescope for solar observations." Furthermore, Struve wrote, Chandrasekhar had worked with Strömgren under Bohr at Copenhagen, and they were planning to write a book on modern astrophysics. All this was calculated to make the Indian astrophysicist as acceptable as possible to Gale, and there was some reality, plucked from Kuiper's reports, behind each statement. Since Chandrasekhar would be on campus for one quarter at most each year under this plan, Struve would still need someone to do the elementary teaching there. That would be Philip C. Keenan, who had gone to Ohio Wesleyan as an instructor for one year but wanted to get back to the better research opportunities at Yerkes. Struve could hire him for less than $2,500, half of Ross's salary, which was in the astronomy budget but unspent because the optical designer was in Pasadena, working on the two-hundred-inch project for half of each year. The Yerkes director knew how to make a few appointments and a little money go a long way.[54]

Chandrasekhar came, gave his two talks at Yerkes, stayed with the Struves, and went to the campus with the director. Apparently Dean Gale avoided seeing him, but Struve had carefully arranged for Vice President Frederic Woodward and Dean of the Faculty Emery T. Filbey to meet the young prospect instead. They both knew how badly Hutchins and Struve wanted him on the faculty and encouraged Chandrasekhar to come by their warm, friendly attitudes. Before he left Williams Bay, he had told Struve that he would accept the Chicago offer, but after he got back to the East he told Kuiper he was still undecided.[55] Before he could make the official offer, Struve had to meet with Bliss, the elderly mathematics chairman, who "raised various questions concerning the appropriateness of having a Hindu on the faculty." However, the Yerkes director, after "a prolonged discussion" convinced him to accept the appointment.[56] Gale, with no allies left, petulantly approved Struve's recommendation that Chandrasekhar be appointed a research associate but emphasized that he was to be at Yerkes Observatory. The dean explicitly told Struve at this time that he did not want to see the Indian astrophysicist on the campus.[57] That same day Hutchins sent a radiogram to Chandrasekhar, on board the *Berengaria* steaming back to England, telling him that he had made the appointment and asking the young man to accept by a return collect radiogram. The president's direct intervention worked; Chandrasekhar accepted immediately (in exactly ten words).[58] Struve's warm, friendly letter addressed to him at Cambridge forecast "a new chapter in the history of Yerkes Observatory." With Strömgren, Kuiper, and Chandrasekhar on its staff, it would "certainly not be an exaggeration to say that we shall now have the best group of astronomers in the world." He assured the Indian astrophysicist that he and his wife would do everything they could to make his stay in Williams Bay "pleasant in every respect." Hutchins, "completely free of small prejudices so often present in office[r]s of the administration," would guarantee Chandrasekhar a bright future. In particular, the president would allow him to stay at Williams Bay as long as he wished; he would not have to come to the campus at all.[59]

Chandrasekhar and Kuiper were equally ecstatic. The young Indian was extremely positive about the "kind interest" Struve had taken in him and was "looking forward to a very profitable time under your inspiring directorship." Hutchins's personal interest in him had tipped the balance, Chandrasekhar told Kuiper, but the young Dutch astronomer's research drive and ardent desire to collaborate with him had also helped. They both looked forward to scaling the heights of their profession at Yerkes and pushing aside the tired old misconceptions of their elders, even Eddington and Russell.[60]

Old Elis Strömgren, Bengt's father and the director of Copenhagen Observatory, sent Struve a very moving letter, expressing his great pleasure that his son would have the opportunity to work in Chicago, at Yerkes, and at McDonald Observatory, but also saying how much he would miss him and his family. Bengt himself was very pleased that Chandrasekhar, "a very good friend of mine," would be coming too, and also Kuiper, whose papers he knew well. All three of them were working actively and productively on pushing out the limits of astronomers' understanding of the nature and physical properties of stars.[61]

Struve was highly conscious of the potential of the staff he was assembling. When a promising young expert in photoelectric photometry, John S. Hall, wrote to ask about job possibilities, Struve told him he had no opening at Williams Bay but advised him to apply to Shapley. "Harvard is certainly the most active place in modern astrophysics at the present time," the Yerkes director wrote, but in his own mind those last four words were clearly the crucial ones, and he intended, with his young staff, to change that situation in the very near future.[62] In a long, thoughtful letter to Arthur H. Compton, the physicist who had won the Nobel Prize nine years earlier, Struve wrote of the new appointments and outlined the astrophysical qualifications and potential of Strömgren, Kuiper, Chandrasekhar, and Keenan. Compton was pleased to learn of the "real strengthening of our situation in theoretical astronomy." He knew of Strömgren's work and had met Chandrasekhar in India with evident pleasure.[63] The forty-four-year-old Compton was an altogether different type of physicist from Gale. Struve was to become aware of the good and bad aspects of this difference in just a few years, when the Nobel Prize winner succeeded the old spectroscopist as dean.

CHAPTER EIGHT

# Birth of McDonald Observatory, 1933–1939

From 1932 onward, Otto Struve's thoughts were always fixed on completing a large reflecting telescope and getting it into operation. Whatever he did, from sending an observer to work with the Perkins Observatory reflector to hiring new staff members to persuading his dean not to reduce Yerkes Observatory's appropriation for supplies and equipment, was based on that ultimate goal.[1] By December of that year, just six months after sending Christian T. Elvey and Theodore G. Mehlin to Texas on their rapid site survey and three months after his own first trip to the Davis Mountains, Struve had decided the observatory would be on Flat Top Mountain, also called Up and Down Mountain, now named Mount Locke, 6,800 feet above sea level, ten miles northwest of the little town of Fort Davis.[2] He knew that the $375,000 allotted for the telescope and dome would be enough to buy a reflector of about eighty inches aperture. Very early on, Struve decided to give the contract to build it to Warner and Swasey, the Cleveland, Ohio, company that had built nearly all the large American telescopes, starting with the Lick thirty-six-inch and Yerkes forty-inch refractors, except the Mount Wilson sixty-inch and one-hundred-inch reflectors. Warner and Swasey had built the Perkins Observatory sixty-nine-inch reflector and the Dominion Astrophysical Observatory seventy-two-inch in Victoria, British Columbia. They had a large, competent engineering design group and plenty of experience; Struve knew he could count on them to build a satisfactory conventional telescope.

In the past Warner and Swasey had not made the optics for the telescopes it provided, but only the mountings. However, they started their own optical shop, built around C. A. Robert Lundin, who had

Figure 34
Otto and
Mary L.
Struve on
horseback in
Texas during
site-testing
visit (1932).
Courtesy of
the Yerkes
Observatory
Archives.

worked from age sixteen as an optical technician for the firm of Alvan Clark and Sons and had been its manager from 1916 until it dissolved in 1929. Then he had gone into business for himself, unsuccessfully, and now was willing to work for Warner and Swasey. His father, Carl A. R. Lundin, a Swedish immigrant, had been Alvan G. Clark's chief assistant in producing the Yerkes forty-inch lens. Robert Lundin had never made a large mirror, and Struve considered having James W. Fecker, who had done so much optical work for Edwin B. Frost and who had produced the optics for the Perkins Observatory reflector, make the primary mirror for the McDonald telescope. But he knew the sixty-nine-inch produced poor images and decided to give the entire contract to Warner and Swasey, partly to keep the responsibility all in one place. At every step of planning the telescope, Struve sought the advice of Walter S. Adams and John S. Plaskett, the Mount Wilson and Dominion Astrophysical Observatory directors. He also visited the Warner and Swasey offices in Cleveland frequently, and its president and chief designer came to Williams Bay for conferences with him. Struve had no design training or skills and was completely dependent on them and on George Van Biesbroeck, who had a civil engineering degree but no large-reflector experience.[3] Struve also received unsolicited advice on telescope building from Harlow Shapley, mostly along the lines of taking the cheapest possible course and not worrying about the quality of the instrument, but he politely ignored it.[4]

A new element entered the situation soon after the Universities of Texas and Chicago announced their joint agreement to build and operate a large reflecting telescope. George Willis Ritchey, the former Yerkes and Mount Wilson optician, sought the contract to build the McDonald reflector. After completing the Mount Wilson sixty-inch, a tremendous success, Ritchey had come more and more into conflict with George Ellery Hale, who had taken the design responsibility for the one-hundred-inch away from him but had him complete its mirror. He finished it in 1917, during World War I, and soon after the one-hundred-inch went into operation in 1919, Hale fired Ritchey. Banned by the director of the most important observatory in the world, Ritchey could not get any astronomical job in the United States. For five years he could only scrape out a living from his orange groves in Azusa, California. Still highly respected in Europe, Ritchey finally got an opportunity in 1924 to go to Paris, to build a planned French telescope that would be the largest in the world, 102 inches in diameter! However, he had much more grandiose plans—for a 240-inch, a 320-inch, and a 400-inch. The purported donor of the French telescope never came through with the money, and the project dissolved in a sea of acrimony. Nevertheless, before returning to America at the end of 1930, Ritchey did succeed in making and exhibiting his first, small Ritchey-Chrétien telescope. This was a new optical design that he and his assistant, Henri Chrétien, had invented at Mount Wilson in 1910. Part of the origin of his struggle with Hale was that Ritchey had wanted to build the one-hundred-inch as the world's first Ritchey-Chrétien telescope. This was a foolish idea, for the design was as yet quite untested, but Hale and Adams, then the Mount Wilson assistant director, refused even to give Ritchey a real chance to build a smaller one experimentally, then or later.

By 1932 Ritchey had gotten a contract to build a forty-inch Ritchey-Chrétien reflector for the Naval Observatory in Washington and was working on the mirror there, while his workmen assembled and adjusted the mounting. He, James Robertson, his sponsor at the Naval Observatory, and Captain J. F. Hellweg, the naval officer who was its superintendent, were all outsiders to the American astronomy establishment. Hale and especially Adams had vilified Ritchey for years as unstable if not insane, and few American astronomers understood or had even read serious accounts of his concepts, which were actually quite advanced and form in many ways the basis of the large research telescopes of today.[5]

Ritchey not only wrote Struve, trumpeting the virtues of "the many improvements and refinements that I am putting into this revolutionary telescope for the government," but he also met H. J. Lutcher Stark, an important Texas regent, and convinced him of the virtues of the Ritchey-Chrétien design, of his own qualifications to build the McDonald reflector in that form, and of how much better it would be than a conventional, Warner and Swasey–built reflector.[6] Struve was horrified. He wanted a big telescope in a hurry, to do spectroscopic research, not a development project of a new type of system. Ritchey's first experimental twenty-inch Ritchey-Chrétien had good optics but only a cheap, inadequate mounting, and in the cloudy climate of France he had never produced a single even remotely good astronomical photograph with it. The Naval Observatory forty-inch was still years from completion, and in the event it appeared completely unsuccessful also until it was moved to the clear skies of Arizona after World War II. The great advantage of the Ritchey-Chrétien design, a Cassegrain reflector in which both the primary and secondary mirrors are hyperboloids, is that it has a much larger field of good definition than any conventional reflector, making it better for wide-field photography. The compensating disadvantage is that the images away from the center of the field are worse if the Ritchey-Chrétien primary is used with a different secondary mirror, to give a longer focal length for spectroscopy. However, this is not a real disadvantage for spectroscopic work on stars, which until decades later could be observed only singly, on axis, where the image is perfect. Hale and Adams did not understand this latter point and refused to listen to it; most other astronomers (including Struve), who knew little of telescope optics, tended to believe them. Certainly the Ritchey-Chrétien had no advantage in terms of Struve's plans for McDonald Observatory, which were almost entirely spectroscopic, while Ritchey himself, a visionary idea man but a notoriously poor planner, and a perfectionist who frequently "improved" his designs as the work progressed, could be counted on to add years to the project.

Stark suggested waiting to decide on the contractor for the Mc-Donald telescope until Ritchey had completed the Naval Observatory instrument and had it in operation, planned for the summer of 1933. At President Harry Y. Benedict's suggestion, Struve wrote him a letter, to show to Stark and the other regents, outlining all the disadvantages of the Ritchey-Chrétien design. He also wrote Stark himself, assuring him that his one aim was "that the new Observatory will be one of the greatest centers of research in the world" and going through the same arguments in more detail. Struve invited the Texas regent to come to Chicago, to visit Yerkes Observatory, Perkins (to see how not to build an observatory), and the Warner and Swasey plant, and to go over the whole problem with him.[7] Stark did come to Chicago in March and talked with the Yerkes director. After the Texan's departure, Struve sent him a long summary of their conversation, setting down on paper his plans for the future of McDonald Observatory and why a Ritchey-Chrétien telescope would not fit into them.[8]

In addition, Struve asked Adams and Plaskett for testimonial letters supporting his decision to have Warner and Swasey build the telescope as a conventional one rather than going to Ritchey for the job. Both senior directors were pleased to do so, and Struve forwarded their letters to Benedict for the Board of Regents in Austin.[9] If Stark was not completely won over, he was impressed by the earnest young director and accepted with good grace the regents' majority vote to go ahead with Warner and Swasey. They approved the contract in October 1933, with Ritchey's Naval Observatory telescope still far from completion. Stark, a very wealthy man and a great Texas booster, thereafter supported Struve.[10] He visited the Warner and Swasey plant with the Yerkes director in 1934 to inspect the telescope then under construction.[11] Part of Struve's letter to him had pushed his credo that "it is the published work after all that counts and not the mere possession of a large telescope," with the corollary that it would be a mistake to build the observatory as an edifice (as at Perkins Observatory) rather than as a utilitarian research station. Stark accepted this point of view.

With the telescope contract placed, the next step was to order the glass disk for the mirror. Struve decided to have it made of Pyrex, a low-expansion form of glass newly developed by the Corning Glass Company for the Palomar two-hundred-inch project. On his recommendation, the University of Texas placed the order in October 1933, and the glass was cast in late December. The disk had to cool for four months before it could be inspected; it was then found to be fissured and had to be reheated and reannealed, necessitating another four months' delay. The disk pushed out the mold slightly in the process, and the mirror diameter grew to eighty-two inches. By October 1934 the disk was definitely accepted, and Warner and Swasey estimated that the telescope would be in operation in another year. This was wildly optimistic, since the work on the mirror dragged on for years.[12]

Struve did not sit back and wait for the telescope to be completed. That was not his style. He had a good dark-sky site in Texas, and he resolved to use it. It was important to establish a University of Chicago presence there and start getting research results. By early 1933 he was already planning to start photoelectric measurements of the night sky at Mount Locke. In the late summer of 1934 Struve sent Franklin E. Roach, who had just completed his Ph.D. thesis and passed his final examination, to McDonald as the first astronomer stationed there. At Perkins Observatory he had proved his ability to work on his own and get results. With him he took the twelve-inch refractor from Yerkes Observatory and a new photoelectric photometer that he and Christopher T. Elvey had designed and built. Soon Roach had the photometer, a wide-field instrument, set up on Mount Locke and began using it to measure the brightness of the night sky as a function of position, wavelength, and time. Roach's wife and son went with him, and as an assistant earning $2,000 a year, he was the first astronomer to live permanently at McDonald Observatory. They lived first in rented quarters in Fort Davis and later on the mountain, as soon as a house was ready for them.[13]

A year later, in the fall of 1935, Struve sent Elvey to Texas too, as the first University of Chicago faculty member stationed there. He went as Struve's deputy, essentially as assistant director in charge of McDonald Observatory, but without the title. Struve did not like to share authority. Elvey had just married; apparently the Yerkes director had not wanted to send a single man to this isolated western outpost, but a married man would have more of the serious sense of purpose he found so many American astronomers lacked. Struve put Elvey and Roach to work measuring the spectra and surface brightness of galactic nebulae—clouds of gas and dust associated with the stars in the Milky Way. It was an ideal program for the dark sky and small telescopes, the only assets the McDonald Observatory would have until the eight-two-inch reflector was completed. The research they were doing was important, for before they began their work little was known of the physical properties of nebulae. Furthermore, under the agreement with the University of Texas, the salaries of Yerkes astronomers assigned to the McDonald staff and in residence there, like Elvey and Roach, could be paid out of the McDonald endowment fund, freeing money in the Chicago budget for more researchers at Yerkes. President Hutchins, however, insisted that Elvey, as a faculty member, must spend one quarter each year at Yerkes, paid for from Chicago funds, to make it clear he was still on its faculty. Gerard P. Kuiper came to Yerkes in the vacant position created by Elvey's transfer to the McDonald staff.[14]

One of the reasons for Struve's amazing productivity was his drive to do research with whatever telescope or spectrograph he had, rather than waiting for the perfect ones, and his uncanny ability to recognize important new problems, jump into them, and get results. As he wrote Kuiper, the mounting for the twelve-inch refractor that Frost had spent so much time and money "improving," and that Struve had shipped to McDonald, was "quite unsatisfactory," and as a consequence the images of stars on photographic plates he, Elvey, and Roach obtained using a small Ross camera mounted on it were "terrible." However, he continued, "for nebular work accurate guiding is not required, and the plates will be useful, even [though] they do not look nice."[15] The first papers that came out of this work, based on these plates and photoelectric measurements made with the photometer Elvey and Roach had designed and made, gave important new information on the properties of dust in nebulae.[16]

Struve's main interest was in using the eighty-two-inch reflector, when it was completed, for stellar spectroscopy. From the beginning he planned for a coudé spectrograph, a large fixed instrument to be mounted permanently on the floor of the dome, with the light sent to it down the polar axis of the telescope by a series of flat mirrors. Dean Henry G. Gale, a laboratory spectroscopist, had his technician make, with the ruling machine originally built by Albert A. Michelson, the large reflection grating that would disperse the starlight into a spectrum. The Yerkes director asked to have the grating "blazed" (by shaping its grooves in the ruling process) so that it would be especially efficient in the red spectral region and thus useful for his planned studies of the H$\alpha$ emission line in stellar spectra. Struve made several trips to Pasadena, usually after a visit to McDonald Observatory, primarily to consult with Adams on the latest design ideas for the coudé spectrograph. The Mount Wilson director had been an outstanding high-dispersion stellar spectroscopist ever since his early years with the Snow telescope, even before the sixty-inch reflector with its coudé spectrograph went into operation, and he continued to observe and make important discoveries with successive versions of the one-hundred-inch coudé spectrograph until he retired in 1945.[17]

The McDonald coudé spectrograph would be too slow for all but the brightest stars; Struve planned faster, low-dispersion Cassegrain

Figure 35   McDonald Observatory eighty-two-inch reflector dome. U.S. Army Corps photograph.

spectrographs for more general use with the eighty-two-inch. They would be prism instruments, the standard form of spectrograph of the time (because even the best gratings lost considerably more light than prisms). To make a fast spectrograph requires an optically fast camera with a short focal length. Even the best, most expensive camera lenses had bad aberrations for focal ratios faster than about f/3, setting a practical limit to the speed of older astronomical spectrographs. But in 1930 Bernhard Schmidt had invented a much faster form of camera, based on a spherical mirror and a thin, refracting corrector plate. Walter Baade brought the idea from Germany to the United States in 1931 when he joined the Mount Wilson staff, and the optical designers there were soon incorporating fast Schmidt cameras into their low-dispersion spectrographs.[18] Struve was quick to follow this lead. In Chicago a very active amateur telescope-making group was centered at the Adler Planetarium, with members throughout Illinois. Some of the most skilled of them soon taught themselves to figure small Schmidt corrector plates. The Yerkes director placed an order with them, and by 1937 he had f/1 and f/2 Schmidt cameras waiting to be used in the Cassegrain spectrograph.[19]

Struve was hampered throughout his directorship, especially in planning the McDonald Observatory spectrographs, by not having any sound instrument designer on the Yerkes staff. Frank E. Ross would occasionally help out by computing the design parameters for a lens, but he was away in California at least half of each year, preoccupied with his own research at all times, and unwilling to spend any appreciable effort on providing detailed optical advice for the director. As a senior professor with tenure, a national reputation, a half-time position on the two-hundred-inch project, and very close associations with the Mount Wilson and Lick Observatory power brokers Hale, Adams, and William Wallace Campbell, Ross could ignore Struve's requests with impunity.[20]

The person who actually did optical and mechanical design at Yerkes was George W. Moffitt, a Stanford Ph.D. in physics who had been on the Yerkes staff since 1926, when Frost had fired Oliver J. Lee. Ross, who had known Moffitt at the Eastman Kodak Research Laboratory, recommended him to take over the astrometric work Lee had done, and Frost, always anxious to fill a vacancy quickly, had hired him. Moffitt was equally undistinguished in parallax determination and spectrograph design, but he was the only person with talents in that direction whom Struve had on his payroll. In 1935 Moffitt was still a research associate without tenure, but earning $3,000 a year, equivalent to an associate professor's salary. He had done the design work on the Perkins Observatory Cassegrain spectrograph and was working on the plans for the McDonald instruments. Struve coldbloodedly decided to keep Moffitt on the staff for a few more years (without any more salary raises), get what he could out of him in designs, and then get rid of him when the building stage was over at McDonald, to free a position for a more productive research worker.[21]

By 1937, with the spectrographs designed and the eighty-two-inch planned for completion in a year, Struve decided it was time for Moffitt to go. The director recommended against continuing his appointment, maintaining that he could find a job with a commercial lens company without difficulty. President Robert M. Hutchins wrote a formal letter giving Moffitt six months' notice that his appointment would be terminated effective September 1. Moffitt protested vehemently, but Hutchins, Gale, and Struve held firm, and there was no recourse from their decision.[22] Moffitt did find an optical job and soon afterward became a partner with Richard S. Perkin and Charles W. Elmer in forming a new optical design and consulting group in New York. However, the former Yerkes designer left the firm in a messy series of charges and countercharges just before World War II, during which the successor Perkin-Elmer Corporation became an enormously successful precision optical company. Moffitt had lost out once again, just before fortune struck.[23]

In 1936 Struve decided to let Roach go, apparently believing he was not working hard enough. Struve told Shapley that Roach did not want to stay on Mount Locke "for family reasons," but in fact the young research assistant had hoped and expected to get a permanent job there. Roach had taken his wife to Pasadena for the birth of their second child, since they both were apprehensive about the level of medical care in west Texas. To Struve, Roach believed, his consequent absence from his post on the mountain, ready to observe every night, indicated a fatal "lack of obsession" with research.[24] More probably the Yerkes director would have called it a lack of seriousness or of purpose. He may also have decided that it was time to get someone from outside the University of Chicago on the McDonald staff. Roach landed a faculty position at the University of Arizona, primarily a teaching job, and after World War II he went on to become an outstanding pioneer in research on the light of the night sky, the program he had begun at McDonald Observatory.[25]

Struve asked Shapley and Bart J. Bok, who was fast becoming his confidant at Harvard, to suggest candidates for the research position at McDonald. They both recommended Sidney D. McCuskey, who was just finishing his Ph.D. there. Before Struve could make him an offer, however, McCuskey accepted a job in the mathematics department at the Case Institute of Technology in Cleveland. Struve decided not to make a counteroffer, because McCuskey's interest in salary and vacation time "clouded the issue." Bok further disappointed Struve by reporting that McCuskey had wanted to take the McDonald job, but his wife did not want to face the isolation there.[26] Shapley, who had earlier told the Yerkes director he had six or seven "extraordinarily good" young men finishing their Ph.D.'s at Harvard and looking for jobs, next recommended Carl K. Seyfert, as did Bok. Both emphasized that Seyfert's wife had been one of Shapley's research assistants for several years and that she and her husband had plenty to do on their own and would not mind living on a mountain in west Texas. Seyfert was eager for the position and snapped it up as soon as Struve offered it to him, starting work in the fall of 1936. The Yerkes director had also sent Paul Rudnick, who was completing his Ph.D. thesis at Yerkes, to McDonald earlier that year, to do photoelectric observing.[27]

A year later Yerkes and McDonald Observatories got their first National Research Council postdoctoral fellow, Jesse L. Greenstein. He had done his undergraduate work at Harvard, then, during the depression, worked for a few years in his father's business in New York. After that he returned to Harvard, where he completed his Ph.D. in astronomy in 1937. His thesis was on interstellar dust and thus fitted in very well with the nebular research going on at McDonald Observatory. Greenstein was especially interested in astrophysics. He was very good, and Shapley wanted to keep him at Harvard but was willing to let him broaden his horizons for a year or two with Struve, whose papers the young man had read and admired. The Yerkes director was glad to have him if he could get the government-supported postdoctoral fellowship; only about fifteen or twenty were awarded each year in physics, with perhaps one in astronomy. Shapley was a powerful figure in American science and no doubt recommended Greenstein strongly for the fellowship; he did get one and came to Yerkes in October 1937—to stay for a decade.[28]

Not long after Greenstein arrived at Yerkes, he, Louis G. Henyey, and Struve conceived the idea of putting together a nebular spectrograph, using the prisms and the small, fast Schmidt camera, all on hand waiting the completion of the McDonald eighty-two-inch reflector. As Greenstein remembered it nearly half a century later, it all began with Struve asking a question as the three of them sat out a cloudy night in the Yerkes library. The director wanted to know what would be the most efficient spectrograph possible. As they discussed it, they saw that all that is required is a fast camera, a prism, and the fewest possible lenses and mirrors, to hold the light losses to a minimum. The basic design is then simple.[29] A typical complete stellar-spectrograph system consists of a telescope lens, which brings light to a focus, a slit that blocks off the light from other stars and the sky, a collimator lens, which makes the diverging rays of light behind the slit parallel again, a prism to disperse the light into a spectrum, and a camera lens that refocuses the rays or spectrum onto a photographic plate. But because the light rays from a star are all parallel before they enter the telescope, it is possible to do away with the telescope and collimator lenses altogether. Then if the slit is far enough away from the prism and camera (which in the Yerkes system was not a lens, but the small f/1 Schmidt camera), the spectrograph will record the spectra of whatever bright stars shine through it, and of the sky or any other extended object (such as a nebula) in the field as well. Because the camera aperture was so small (approximately four inches in diameter), only the spectra of the brightest stars showed on the plate. However, the spectrum of an extended object, such as the night sky itself, or of a nebula (if its area was large), was recorded, just as the image of a scene on the earth is recorded by a very small camera lens.

The Yerkes nebular spectrograph was mounted on the side of the giant forty-inch refractor, which was used only to hold it. The slit (twenty-five inches long and one inch wide) was mounted at the top end of the tube, alongside the forty-inch lens. Near the eyepiece end, fifty-eight feet away, the spectrograph, consisting of the prism (actually two prisms in series) and the camera, was mounted alongside the tube. Since the slit subtended a length of two degrees and a width of five minutes of arc, the spectrograph recorded the spectrum of the individual stars within this area on the sky as narrow strips and the spectrum of the sky and nebula (if there was one in this area) as a long strip. If emission lines were present in the nebula they could be detected in this way, and those that came from the night sky could be eliminated by comparison with the spectra of blank areas outside the nebula. The whole system was very fast, because it eliminated the inevitable light losses in the telescope lens (or mirror, in a reflecting telescope used at the Cassegrain focus, as the eighty-two-inch would be) and because the f/1 Schmidt camera had such a fast focal ratio. Nevertheless the exposures were long, typically three to four hours for even the brightest nebulae. The spectrograph therefore had to track, or follow, the motion of the nebula in the sky; this is the reason it was mounted alongside the great refractor, which in this arrangement was essentially a gigantic guide telescope, keeping the nebular spectrograph pointed at a nebula during a long exposure. This novel system made very good use of the resources available to Struve at Yerkes—the forty-inch refractor and the prism and Schmidt camera for the McDonald Observatory spectrograph—and applied them to an important research problem, the spectral analysis of faint nebulae.[30]

Greenstein and Henyey (who had just completed his Ph.D. that summer and had been kept on the staff as an instructor) obtained the first results with this nebular spectrograph on the well-known North America Nebula, which has an emission-line spectrum, and on the fainter IC 1318, which has a lower-ionization spectrum of the same type. They quickly published these first results, and Greenstein went to the campus in Chicago in early December to make tracings of their spectrograms, to show the emission lines and continuum in graphic form, for an oral presentation of their paper at the American Astronomical Society meeting in Bloomington, Indiana, just after Christmas. Struve always believed in getting results and publishing them quickly.[31]

Although these first spectra from Yerkes were important, McDonald Observatory, with its much higher number of clear nights and its dark skies with no hint of light pollution, was obviously a better site for the nebular spectrograph. Struve quickly moved it there. Since there was no large telescope to serve as a mounting, Van Biesbroeck designed an arrangement something like a coudé telescope (with overtones of a heliostat, in technical terms) in which a rectangular flat mirror, pointed at the sky and driven at the correct sidereal rate, reflected the light from a nebula to a fixed mirror mounted seventy-five feet up the slope of a convenient rise so that the optical axis was parallel to the axis of rotation of the earth (at an elevation of 31° on Mount Locke). The slit was an arrangement of two curtains, which could be open or closed to any desired width, mounted directly in front of the first, flat mirror. The fixed flat mirror reflected the light back down the hill to the spectrograph, mounted on the same rotating mount as the slit mirror, thus preserving the alignment on the nebula throughout the exposure.[32]

With this system Struve and Elvey amassed data on many faint nebulae, nebulous areas, and even apparently blank areas of the sky. Their observations revealed for the first time the very many faint gaseous emission nebulae and diffuse nebulous areas, far more numerous than astronomers had expected from the early pioneering work of Edwin Hubble in his Yerkes thesis and his later, more extensive work at Mount Wilson. His spectral information had necessarily been restricted to bright nebulae; the McDonald results went down to the faint limit set by a very dark sky and opened astronomers' eyes to the near ubiquity of ionized interstellar gas in the Milky Way.[33]

These papers aroused wide interest and brought Struve many congratulatory messages on the new results. Adams was especially pleased by the "ingenious and extraordinarily efficient spectrograph which you have designed."[34] Even crusty old William H. Wright, now director of Lick Observatory and a firm believer in keeping jobs in American astronomy for "real" Americans, was impressed. In

1936 he had commented very negatively to the news of Struve's appointments of Kuiper, Bengt Strömgren, and Subrahmanyan Chandrasekhar at Yerkes Observatory, writing that there was "no justifica-
tion for loading up American observatories with foreigners, to the
discouragement of our own young men." Nevertheless, the Russian-
born Struve's observational results won Wright over. He had pioneered the study of the spectra of the brighter planetary nebulae years
earlier; now he congratulated Struve on the new results. Wright wondered if the bright Hα λ6563 line Struve and Elvey had identified
might not in fact be the [N II] λλ6548, 6583 lines, close enough to
Hα to be blended with it and indistinguishable from it on the low-
dispersion McDonald spectrograms. This was a possibility that had
not occurred to the two McDonald observers, who were tyros in the
nebular field. Struve handsomely admitted it and added a footnote to
their second paper, still in press, mentioning this possibility that had
come from Wright, but stating that in fact Hα must dominate, even if
the [N II] lines were present. Ira S. Bowen, the world expert on nebular emission lines, confirmed Struve's statement to Wright, who ended
up with a much higher opinion of at least this one "foreigner." [35]

Although these nebular research results made very good use of the
spectrograph and the McDonald Observatory site, Struve and all the
other astronomers were eager to get to work with the eighty-two-inch
reflector. But turning the glass disk, pronounced acceptable in the fall
of 1934, into a finished parabolic mirror seemed to take forever.
Warner and Swasey had estimated then that it would require one year
to complete the job, but Struve had been skeptical. After two years, in
November 1936, he was definitely impatient and thought the company "shamefully slow in their optical work." He had his new staff
of astronomers assembled, but they had no big new reflector to use.[36]

In 1937 Lundin figured the big mirror to nearly the correct parabolic form, but not quite close enough. Then for months he was unable to make more real progress. He would work on the mirror, trying to improve its figure, but each time he retested it he found that it
was not quite good enough. Once, in February 1937, the optician had
the figure nearly within the specifications, but then when he tried to
improve it further, the next test showed that the figure was worse and
deviated more from the true paraboloid. For months Struve made frequent trips to the Warner and Swasey plant in Cleveland to speed
things up, but he could not do the optical work himself. E. P. Burrell,
the company's chief engineer, had died; without his guidance Lundin,
who had never previously made a large parabolic mirror, seemed

helpless. Plaskett, the elderly Dominion Astrophysical Observatory director who acted as the consultant on the job, could not come up with advice that solved the problem. Finally, in March 1938 Struve, with Van Biesbroeck and Kuiper in tow, demanded a full demonstration of Lundin's test procedure. They discovered, to their horror, that he had been misinterpreting the results of the knife-edge tests that were the standard method of monitoring progress. Lundin could tell when the mirror was perfect, but he took the deviations revealed by each test, which in fact showed the slope of the mirror at each radius, as if they gave the height or depth of the mirror. Thus the additional figuring of the glass that he carried out based on his incorrect interpretation did not converge to the required perfect parabolic form. Struve was furious. To a confidant he expressed his "utmost disgust" with Lundin and "the entire Cleveland crowd" and his fear that Plaskett, "this grand old man whom I respect perhaps more than any other spectroscopist in the world," was losing his "former keen insight." [37]

Struve had no choice but to go on with Lundin and the Warner and Swasey Company. They had a contract to complete the mirror and the telescope, and no other firm could replace them now. But the Yerkes director did succeed in his demand to insert his own remaining optical expert into the Warner and Swasey shop. He was E. Lloyd McCarthy, who had been a graduate student and assistant at Yerkes since 1934, a protégé of Ross and Moffitt. McCarthy was a first-class young optical designer, who had provided the detailed plans for a slitless spectrograph to be used by Kuiper on the eighty-two-inch telescope. McCarthy was looking for a temporary job, since he wanted to leave Yerkes and switch to graduate work in optics at the University of Rochester. Warner and Swasey agreed to hire him to work with Lundin in the interim, interpret the knife-edge tests, and suggest more sophisticated ones. McCarthy was permitted to report informally to Struve on how the work was progressing. From then on it went very swiftly. Under McCarthy's tutelage, Lundin completed figuring the eighty-two-inch mirror by mid-October, and Struve, summoned by Plaskett, agreed that it now was well within the specifications. [38] The "grand old man" told him that the eighty-two-inch mirror was better than the sixty-nine-inch Perkins primary, the seventy-two-inch Dominion Astrophysical Observatory instrument, and the one at the Dominion Observatory in Toronto, and that it might be "the best large mirror in the world." However, none of those three telescope mirrors were ever regarded as particularly good by independent opticians, and the one-hundred-inch Mount Wilson primary that Ritchey had completed in 1917 was certainly far superior to them and to the eighty-two-inch paraboloid. Nevertheless the latter was completed and acceptable at last, and Struve expressed his gratitude to McCarthy, now at Rochester, for helping push it through to completion.[39]

With the primary finished, Lundin could go ahead and complete the smaller, hyperboloidal Cassegrain and coudé secondary mirrors for the telescope. Fred Pearson, who had been Michelson's optician in the physics department on the campus and remained there under Gale, accompanied Struve on his visits to Cleveland to inspect them until they were accepted also.[40]

Warner and Swasey shipped the telescope optics to Texas in February 1939. Struve had spent several weeks in January at the McDonald site and then returned briefly to Chicago and Williams Bay before going east to Philadelphia for a meeting of the American Philosophical Society. There, in a symposium on astrophysics, he gave a long, important paper summarizing his research on stars with extended atmospheres, supergiants such as $\alpha$ Cygni, and shell stars such as 17 Leporis. Because of the low density, the physical conditions in these outer envelopes of stars may deviate strikingly from the near-thermodynamic equilibrium conditions in the atmospheres of normal stars. Obtaining spectra of such stars, analyzing them, and understanding them in physical terms had become Struve's chief field of research, and he had made himself an acknowledged world expert in it.[41]

Struve had sent Van Biesbroeck to Texas to take charge of receiving and installing the telescope mirrors. Safely packed in nested boxes, they arrived by train in Alpine on February 21, and by truck on Mount Locke the next day. The telescope mounting had been completed and assembled in the dome long in advance, and the McDonald and Warner and Swasey workmen soon had the primary mirror and the two secondaries unpacked and installed. Plaskett and Struve hurried to the mountain to see the system perform. Both the primary and the Cassegrain secondary mirror gave "perfect" images, and Struve was satisfied, at least for the time being. He had the Cassegrain spectrograph put on the telescope, and by March 5 he had begun taking spectrograms of stars with it. The flat mirrors to divert the light along the declination axis of the telescope and then down the polar axis to the coudé spectrograph had been the last made and had not yet arrived on the mountain, so Struve could not begin using it for another month. He assigned the first extended observing run with the eighty-two-inch telescope to Van Biesbroeck, in recognition of the important role he had taken in all stages of getting it into effective operation.[42]

The large grating ruled in Gale's laboratory for the coudé spectrograph had turned out to be unsatisfactory; it did not concentrate enough of the light into one spectral order to be at all efficient. Instead of it, Struve had turned to a pair of large prisms, mounted in the Littrow arrangement (in which the light, reflected by the rear, silvered face of the second one, goes through them twice), as the dispersing element. The lens (which was both collimator and camera in the Littrow mounting) gave a curved, inclined field, so that glass plates could cover only a small wavelength region in good focus, or film could be used in a shaped holder. It was a technology Hale and Adams had used at Mount Wilson thirty years before, outmoded by 1939, which exhibited graphically Van Biesbroeck's and Moffitt's limitations as instrument designers as well as the need for a better grating and a large Schmidt camera.[43] Struve began using the coudé spectrograph in May and immediately saw its deficiencies, but he was hopeful that they could be corrected.[44]

Struve's first exposures with the new spectrographs were not intended simply as tests; he took spectrograms of stars he needed for his research. On March 5, the first night he used the Cassegrain spectrograph, he obtained a good spectrogram of the shell star 17 Leporis. It was taken in the ultraviolet spectral region, for which the Cassegrain spectrograph was specially optimized with quartz prisms and lenses. Hence this spectrogram showed many new features. Before the end of summer Struve and Roach, who came back to McDonald to work with him during his vacation from teaching at Tucson, had measured the spectrogram, analyzed it, and had a paper ready for publication on 17 Leporis and P Cygni, another star with an extended atmosphere that Struve had observed at McDonald in May.[45] Just before this paper, Struve had published one himself on the ultraviolet spectra of several normal A and B (hot) stars, also obtained with the same spectrograph.[46]

To mark the completion of the eighty-two-inch reflector, Struve arranged a research symposium, always his favorite type of meeting. It was held at the end of the first week in May at McDonald Observatory, in conjunction with a meeting of the Southwest Division of the American Association for the Advancement of Science (AAAS) at Alpine, Texas, the home of Sul Ross State Teachers' College. The college's president was eager to cooperate with the mighty Universities of Chicago and Texas.[47] The Warner and Swasey Company provided

$5,000 in traveling expenses to Mount Locke for some twenty lead-ing research astronomers, selected by Struve, who rubbed shoulders on the first day of the dedication ceremonies with the members of the Texas Archeological and Paleontological Society, the Panhandle Plains Historical Society, the Texas Folk Lore Society, and the Texas Academy of Science, all constituent societies of the AAAS. They were invited to the mountain for one day.[48]

Benedict, who as dean and then president of the University of Texas had done so much to make the observatory a reality, had died in 1937 before it was completed. His longtime associate, John W. Calhoun, like Benedict a southerner who had taken his undergraduate degree at the University of Texas, followed by graduate work in mathematics at Harvard and then a long teaching career at Austin, had been the leader in urging the building of a student observatory on campus, even before William J. McDonald's death. In 1925 Calhoun became comptroller, in charge of the university's budget and finances, and in 1937 he was named Benedict's interim successor and served as presi-dent for two years. Just two months before the dedication the Texas Board of Regents chose Homer P. Rainey to head the institution; he attended the dedication as president-elect at Struve's personal invita-tion.[49] Rainey was a native Texan who had done his undergraduate work at little Austin College, earned master's and doctor's degrees in education at the University of Chicago, then taught education at Ore-gon, and at age thirty-one had become president of Franklin College in Indiana. From there he moved on to the presidency of Bucknell, then went to head an educational foundation in New York. When he returned to Texas for the dedication and then inauguration as presi-dent in 1939, he was destined to last only five years before a political fight, fired by his own high ambitions, drove him from office. Bene-dict, Calhoun, and Rainey represented types of university president very different from Hutchins, but like him they supported Struve and his plans for building and operating McDonald Observatory all the way. The combination of public interest in astronomy—the study of the universe—with a romantic, impressive mountaintop observatory, a top-flight research department respected everywhere, and a director who flattered them constantly but came up with hard-headed plans that always succeeded proved irresistible.

Struve had tried very hard to have his own president there too, inviting Hutchins months before the event. The president had agreed to come and make a speech, but as the reality of spending days away from his busy office in Chicago came closer, he put off making

Figure 36   Group at McDonald Observatory dedication (1939). From left, Henry Norris Russell, John S. Plaskett, Otto Struve, Harlow Shapley, and University of Texas president-elect Homer T. Rainey (1939). Courtesy of the Center for American History, the University of Texas at Austin.

definite travel plans, and in the end he sent word to Texas that his doctors considered him too weakened from a (previously unannounced) "attack of influenza" to travel. Instead Hutchins sent an inspiring message of congratulations, which he had composed himself, lauding the spirit of cooperation between the two universities and "the distinguished director of the observatory," whose "unusual comb[ination] of energy, tact and scientific knowledge" had made possible the building of the observatory[50] Gale read the president's words at the dedication, while physicist Arthur H. Compton, who was to succeed him in 1940 as dean of the physical sciences at Chicago, gave an uplifting speech, much less succinct than Hutchins's message, titled "The First of the Sciences." With further speeches by Shapley, Plaskett, Charles J. Stillwell of Warner and Swasey, Joaquin Gallo, director of the Mexican National Observatory at Tacubaya, Struve, and Rainey, all presided over by Henry Norris Russell, the ceremonial part of the dedication on Mount Locke ended. After supper on the mountain and a chance to look through the telescope as soon as the skies became dark, the paleontologists, historians, and folklorists were bused back to Alpine, leaving the field to the astronomers.[51]

Struve, a fervent believer in not wasting weekends, had scheduled the research symposium for Saturday and Sunday, May 5 and 6, with morning, afternoon, and evening (after-dinner) sessions. Although he and all the other Yerkes astronomers were doing research on stars and gaseous nebulae, Struve had chosen "galactic and extragalactic structure" as the topic, looking into the future with very prescient eyes. He had assembled an outstanding group of research astronomers and astrophysicists; among the sixty-odd present were at least sixteen current or future observatory directors. Robert J. Trumpler (who years before had urged Benedict to locate McDonald just as far south in Texas as he could, for studying the galactic center) described his important ongoing research on galactic star clusters; Baade spoke on his work on measuring accurate magnitudes of stars and galaxies; Joel Stebbins reported on his photoelectric measurements of the reddening and extinction of starlight by interstellar dust in the galaxy; and Hubble told of his studies of the forms and classification of "extragalactic nebulae" (galaxies). Adams and Wright, who did not give papers themselves, each presided at one session. As members of the home team, Struve, Elvey, William W. Morgan, Kuiper, and Chandrasekhar gave papers on their own work. The Yerkes director had hoped to have Arthur S. Eddington from England, Bertil Lindblad from Sweden, and Jan H. Oort from Holland, but the time was unpropitious, in the uneasy truce between the Munich agreement of September 1938 and the outbreak of World War II in September 1939, and the first two begged off. Only Oort, Edward A. Milne of Oxford, with whom Chandrasekhar had worked closely, and Albrecht Unsöld of Kiel were on hand from abroad. Cecilia Payne-Gaposchkin was the one woman astronomer at the dedication; Struve had always admired her astrophysical work. Nearly all the talks were very forward looking, and several provided glimpses into problems that would still be live ones fifty years later.[52]

With the telescope in operation, Struve was making his final additions to the McDonald staff, consciously going outside the group of former Yerkes students who had previously been the main source. Instead he hired two recent Lick Ph.D.'s. One was Daniel M. Popper, highly recommended to him by Wright and Trumpler for his skills in stellar spectroscopy. Popper had finished his graduate work at the University of California in 1938 and then had spent a year at the Eastman Kodak Laboratory in Rochester, New York, becoming an expert in photographic emulsions and processes for spectroscopy. The other was Horace W. Babcock, who had just completed his thesis on the rotational velocity field of M 31, the Andromeda galaxy, which he measured spectroscopically from the radial velocities of gaseous nebulae in its spiral arms. Babcock had obtained most of these spectrograms with the thirty-six-inch Crossley reflector at Lick Observatory, but he had also used the same spectrograph with a small, five-inch reflecting telescope to take a few exposures with a long slit crossing the whole galaxy. He had come to the McDonald Observatory dedication, where his work had drawn wide interest and admiring comments. Struve decided to add him to the Yerkes staff in a temporary position. In those old days before affirmative action, one of the points Wright stressed in his recommendations of both of them was that their fathers were productive, valued research workers: William Popper was a professor of Semitic languages at Berkeley, and Harold D. Babcock was a solar astrophysicist on the Mount Wilson staff. Both young men were to go far in astronomy themselves, though only for their first few years at McDonald and Yerkes Observatories. Popper was stationed at McDonald and Babcock at Yerkes, but both were expected to do most of their observing with the eighty-two-inch reflector in Texas.[53]

Popper was to start work just after Struve had left Texas at the end of May to receive his first honorary doctor's degree, from Case Institute of Technology in Cleveland. Its administration and its astronomy department, headed by Jason J. Nassau, were very close to the Warner and Swasey high command, who no doubt had initiated this recognition of the Yerkes and McDonald director to mark the completion of the telescope project. Since Struve would not be able to see Popper before he started work, he sent instructions to Kuiper, who would be in Texas, to be passed on to the young man. Popper should understand, he wrote, that Elvey was in charge on the scene while Struve was away, but all the scientific direction would always come from the director himself. Popper would have to do a certain amount of routine work, but he would get plenty of observing time for himself and would be expected to make good use of it. He should coordinate his observing program with Kuiper's, who should assure the young man "that his work will not be used by you or others without proper credit, and we expect that all independent contributions will be published under his own name." Finally, above all, Kuiper should stress to Popper "that the only important thing is the scientific output." That was Struve's credo.[54] Both Babcock and Popper soon were working hard, getting results and publishing papers. Struve was pleased.[55]

Finally the last piece fell into place when Pol Swings arrived back at Yerkes from Belgium at the end of August. After his 1931 working visit to Yerkes, Swings had founded an Institute of Astrophysics at the University of Liège and had made himself a leading authority on stellar and molecular spectroscopy. He was especially close to the Swedish laboratory spectroscopist Bengt Edlén, with whom he worked on ions that could be observed in hot stars. Struve kept Swings informed of the progress of the work on the eighty-two-inch reflector and encouraged him to plan to come back to work with him as soon as it was completed. In contrast to many other European astrophysicists of the time, the young Belgian was determined to get real observational data with large telescopes in a good climate and use it to analyze interesting, uncommon types of stars. At Liège he worked assiduously on the ultraviolet spectra of hot stars, using spectrograms Struve had obtained at Perkins Observatory and sent him.[56] In 1935, long before the eighty-two-inch was completed, Swings came back to Yerkes for four months' research with Struve, followed by visits to Mount Wilson—where this time Adams welcomed him and gladly let him measure all the plates he wanted—to Lick, to Berkeley, to McDonald, where he saw the peripatetic director again, and to Princeton to discuss molecules in stars with Russell. Then he stopped off briefly at Harvard over Christmas 1935, seeing Kuiper, who was then still on the staff there, and Chandrasekhar, on the visit that led to his Yerkes appointment. Swings embarked for Cherbourg just before the new year, reminding Struve in his last letter from Washington that he intended to come back again soon.[57]

As soon as the Yerkes director was certain the eighty-two-inch reflector would be in operation by the spring of 1939, he began arranging to bring Swings over from Europe. Struve recommended him for a visiting professorship for six months, beginning September 1, at a salary of $2,500. Swings himself had applied for a fellowship from the same Belgian Research Council that had supported his first visit; if he received it he would be able to stay longer and do less teaching. It did come through, and the Belgian astrophysicist planned a stay of a year and a half in the United States, the first and last six-month periods at Yerkes and McDonald, split by six months at Mount Wilson and Lick Observatories.[58] He arrived at Williams Bay on August 28, 1939, and within a few days was on his way to McDonald Observatory with Struve. The weather was perfect, and working together they obtained a large number of spectrograms with the Cassegrain spectrograph, raw material for many joint papers. Their collaboration was to be an exceptionally fruitful one. Both were demon workers, Struve brimming with ideas, some of them quite wrong but all worth trying and some amazingly successful, while Swings had the detailed, up-to-the-minute familiarity with physical theory and laboratory spectroscopic results necessary to interpret them.

But on September 1, three days after Swings had arrived at Yerkes Observatory and even before he and Struve set off for Texas, Adolf Hitler's German armies invaded Poland, launching World War II. Struve, with his strong ties to Russia and Germany, experienced a long period "of a feeling of utter hopelessness." He knew a "bloody struggle" for "everything good and decent" lay ahead.[59] He was right, and as it turned out in personal terms, Swings could not return to Belgium until 1945, after it had been liberated, Liège was a smoldering ruin, and Hitler had died in his bunker in Berlin. The whole world had changed. Struve and Swings were to write many joint papers based on McDonald spectrograms, but by the time the war was over the Yerkes director, like many other Americans, was to find that almost everything he had lived for was different.

# CHAPTER NINE

## An Extraordinarily Fine Group, 1936–1942

Otto Struve's plans had begun to bear fruit at Yerkes Observatory years before the eighty-two-inch reflector went into operation at McDonald. With William W. Morgan already on the scene in Williams Bay, the three young "foreigners" all joined the University of Chicago faculty in 1936. The first to arrive was Gerard P. Kuiper, after his year at Harvard. In the spring he had met a "dear girl," Sarah P. Fuller, a Mount Holyoke graduate whose family had long been connected with Harvard. Kuiper swept her off her feet, bought a car, learned to drive, married Sarah in June, took her on a honeymoon trip to Maine, Quebec, and around Georgian Bay, and arrived at Williams Bay at the beginning of August.[1]

They moved into the house Frank E. Ross had occupied; he was spending half of each year in Pasadena and was separated from his wife. Struve had assigned him a small, run-down cottage that was satisfactory for the Wisconsin summer in place of his former two-story house with a view of the lake. The director notified Elizabeth Ross, who was living in Madison with their son, a student at the University of Wisconsin, to get her possessions out of the observatory house well before the Kuipers arrived and vetoed her hope to spend the entire summer there.[2] The newlyweds moved in without incident, and Kuiper plunged into observing double stars visually with the forty-inch refractor on the few nights of good seeing and using a small, low-dispersion spectrograph he had quickly thrown together to get spectrograms for classification of the faint, low-luminosity stars that were his main field of research. In January 1937 he went to McDonald to observe for the first time, taking Sarah with him and staying for three months. There he observed double stars with the twelve-inch refractor that Struve had moved to Texas. With the very clear sky, high altitude, and favorable latitude of Mount Locke, Kuiper was able to measure close pairs to declinations far south of the equator.[3]

The next to come was Bengt Strömgren, who arrived in Chicago near the end of September 1936 with his wife, two infant daughters, and a young Danish nursemaid. Struve had found an apartment for them to rent near the campus, and Strömgren started teaching in the fall quarter. At the same time the Yerkes director put a graduate student, Gordon W. Wares, to work making tracings of stellar spectra for Strömgren to use in his research. He was analyzing them to measure the relative abundance of sodium (and from it of the other heavy elements as well) with respect to hydrogen in stellar atmospheres. Struve visited Strömgren in Chicago as soon as he could to talk over his results, set aside a house for him in Williams Bay, and sent him a train schedule, so that the Danish theorist could come with his family for the weekends and continue research there with Struve, Kuiper, Morgan, and Wares. Working seven days a week plus a few nights at the telescope seemed perfectly normal to Struve. The personable Strömgren, who had frequented Niels Bohr's institute in Copenhagen, was soon on close terms with the physicists and mathematicians in Chicago, giving talks on his recent work on stellar interiors, which, like his new results on their atmospheres, revealed a much lower abundance of heavy elements than earlier believed.[4] Just before Christmas the brilliant Louis G. Henyey completed his Ph.D. thesis on interstellar dust, and Strömgren came to Williams Bay with Walter Bartky, his colleague on the campus, for the final examination. Henyey passed with flying colors and immediately joined the faculty as an instructor.[5]

Subrahmanyan Chandrasekhar, Struve's other new theorist, arrived at Williams Bay just before Christmas 1936. Dean Henry G. Gale, a confirmed racist, had absolutely refused to have him on the campus in Chicago, although the younger physicists, especially Arthur H. Compton, would have welcomed him. After returning to Cambridge in the spring from his trip to America, Chandrasekhar had continued his theoretical work on stellar interiors, now directed more toward comparisons with observational data on the masses and luminosities of stars, and the quantitative conclusions that could be drawn from them. Kuiper, who had become his good friend during his months at Harvard, had encouraged him to follow this path.[6]

In June Chandrasekhar went to the American consulate in London to ask about getting a visa. He soon found that to the Department of

State, citizens of India were not particularly welcome. He notified Struve, who quickly mobilized the resources of the University of Chicago behind the young astrophysicist's case. The university counsel provided the Yerkes director with a sheaf of precedents and a brief of the precise wording to use in preparing Chandrasekhar's "credentials," to be forwarded to him "in duplicate" for presentation to the consul. If that failed, the lawyer indicated, the university administration was prepared to "[take] up the matter with various Government officials," right up to the commissioner general of immigration. That proved unnecessary; Struve's formal letters "cleared up the difficulties," in Chandrasekhar's words, and the consulate officials approved his application for a visa.[7]

At the end of July he sailed for India to see his family and his fiancée, Lalitha Doraiswamy, who had been a graduate student in physics, earned a master's degree, and become the headmistress of a middle school. They married in September and a month later sailed back to England together.[8] In Cambridge Chandrasekhar worked hard, trying to finish two papers on his latest stellar-interiors research before leaving again for the New World. Struve wrote asking him to teach three graduate courses on the subject, in the first reorganization of the astronomy and astrophysics program since early in Frost's directorship. The old program had been "entirely inadequate and [did] not give the student a satisfactory preparation for research," the director wrote (in no small part from his own experience). Under his new plan, there would be a two-year cycle of eighteen courses, each one quarter long, so the students would take three courses a quarter, except in summer, and learn all of astrophysics in two years no matter when they started. Always sensitive to Chandrasekhar's unexpressed needs, Struve arranged for him and his wife to stay as guests of the Kuipers until their own house had been made ready for them.[9] It was the home that Hans Rosenberg, the middle-aged refugee from Hitler's Germany, had been occupying with his family. Struve had been glad to have him at Yerkes as a temporary faculty member, his salary paid with funds received from national foundations and a Chicago Jewish organization, but he had never been willing to give up one of the precious permanent faculty positions to Rosenberg. Now the pioneer photoelectric astronomer had had to vacate the university house that Struve had let him use until a younger, permanent faculty member needed it. Research always came first with the Yerkes director. The Rosenbergs stayed briefly as guests of George Van Biesbroeck and his wife, then left on their odyssey that ended in Constantinople.[10]

The Chandrasekhars sailed from Liverpool in December and landed at Boston in the middle of the month. After a few days visiting his friends in Cambridge, they went on to Chicago by train and, after a weekend there, arrived at Williams Bay three days before Christmas. Struve had been at McDonald Observatory, working on nebulae with the small Schmidt camera, but had made it a point to be back at Yerkes Observatory before Chandrasekhar got there, with a day to spare. All the travel arrangements went smoothly, and after a few weeks Chandrasekhar and Lalitha were in their own home. The director provided his newest faculty member with a general letter of introduction, explaining that "Dr. Chandrasekhar of Madras, India and Cambridge University, England" was now "a valued member of our scientific staff," to help smooth the way for him and his wife in the little Wisconsin community, strangers in a strange land. He had gone to work on his research as soon as they had arrived.[11]

Struve, however, was already worried about losing Strömgren. The Danish astrophysicist had accepted his faculty appointment on condition that he might give it up and return to Copenhagen after fifteen or eighteen months. Struve, eager to get him, had been confident he could break down Strömgren's resolve after he arrived, but he was finding that hard to do. The Danes wanted Strömgren back. His father, Elis, would be seventy years old in 1940 and due to retire as director of Copenhagen Observatory. His fondest hope had been that Bengt would succeed him, and especially now that the pleasant younger Strömgren had been "called" to the Chicago faculty, other Danish scientists, including Bohr, realized with heightened respect how much they needed him at home. Thus Bengt Strömgren, who was on leave from Copenhagen University, received an official offer from its administration in late December, conditional on his returning by April 1, 1938. He would be promoted to full professor, making it certain that he would be the next director. But he would have to promise to stay in Denmark for at least two years. Struve hastened to see President Robert M. Hutchins in Chicago between Christmas and New Year's Day. Very probably he had already conceived the plan he was later to reveal, to make Strömgren director of Yerkes, so that he himself could (paradoxically) have more time for research and at the same time rule over Yerkes, McDonald, and the Department of Astronomy on the campus as a sort of superdirector. Strömgren was too valuable a researcher to be kept in Chicago, Struve told Hutchins. Bengt and his family should be transferred to Williams Bay, which Struve knew Sigrid would like better than Chicago, and where her

Figure 37  Yerkes Observatory staff (1937). Seated in front from left, Alice Johnson, Subrahmanyan Chandrasekhar, Mary R. Calvert, Bengt Strömgren, Marguerite Van Biesbroeck, Gerard P. Kuiper, Lalitha Chandrasekhar, Jesse L. Greenstein. In second row, George Van Biesbroeck at far left, Louis G. Henyey fifth from left, Edith Kellman sixth, William W. Morgan seventh; in rear row, Otto Struve third from left, Frank R. Sullivan fourth, E. Lloyd McCarthy (very tall) fifth. Courtesy of Yerkes Observatory.

husband could spend full time on research. Struve recommended that he be promoted to associate professor, with a salary raise of $1,000 to $5,000 a year, effective immediately. Hutchins agreed, and Strömgren got the money and the promotion, retroactive to January 1, 1937, but made no promise to stay beyond April 1, 1938. Struve wrote to Bohr and to Elis Strömgren, emphasizing the scientific advantages of Yerkes, the telescope soon to be in operation in Texas, and "the exceptionally brilliant and enthusiastic group of astronomers that we have assembled in Williams Bay." Then he hoped for the best. He could do little else.[12]

Strömgren worked very actively on research in Chicago through the winter quarter, making occasional weekend trips to Williams Bay with his wife to arrange for their move in the spring. He was in close touch with the latest theoretical developments in Europe and briefed Struve on Carl F. von Weizsäcker's early ideas on nuclear-energy generation in stars and the consequent transformation of hydrogen into helium. Strömgren realized that simply assuming that helium was the next most abundant element after hydrogen made it possible for him to improve his determination of the hydrogen abundance from stellar masses and luminosities, and the new value he derived was quite close to the one still accepted today. He and his family moved to a rented house in Williams Bay at the end of the winter quarter, and he became even more closely involved in theoretical work on interpreting observational data to derive the physical properties of stars and nebulae.[13]

Struve constantly sought to strengthen the research atmosphere at Yerkes Observatory. In 1936 he brought Bart J. Bok from Harvard for the month of June, to lecture on his work on star counts, stellar statistics, and his attempts to derive the structure of the Milky Way from them. Struve hoped to make this pale imitation of Harlow Shapley's "summer schools" a tradition, and the following year he imported Cecilia Payne-Gaposchkin for six weeks in May and June, for another series of lectures, on photometry and stellar astrophysics. He regarded her as "one of the most outstanding astronomers in America" but still paid her only at a rate equivalent to $3,000 a year.[14]

One of the first fruits of Struve's new astrophysical team at Yerkes was a joint paper he wrote with Kuiper and Strömgren on the interpretation of ε Aurigae, the bright supergiant that had the light "curve" (periodic variation) of an eclipsing binary with a period of twenty-seven years, radial velocity variation to match, but no sign of the spectrum of a second star in the system. Struve and Christian T.

Elvey had published a preliminary discussion of some of the peculiarities in the profiles of the spectral lines in 1930, based on the Yerkes
radial velocities that Frost and Walter S. Adams had begun measuring
in 1899. Struve and Elvey's paper had added little to the understanding of the system, but now with Strömgren supplying his expertise on
low-density astrophysics, applicable to the very tenuous atmosphere
of the supergiant, and Kuiper his double-star and dynamical skills
and experience, Struve and his new collaborators developed a consistent physical picture. The companion was an even larger, cooler "infrared star" that revealed its presence (in the optical spectral region)
only by adding to some of the characteristic low-temperature absorption lines as its atmosphere moved across the face of the brighter
F-type supergiant behind it.[15]

Small wonder that Henry Norris Russell, in 1937, evaluated the
staff Struve had assembled at Yerkes and McDonald as "an extraordinarily fine group of young men."[16] Chandrasekhar, of whom Russell
had been skeptical, was one of the most productive of the group. He
had finished most of his work on stellar interiors and was beginning
to shift his research to stellar dynamics. As he began to get acquainted with this topic he started writing a monograph on the theory of stellar structure, summarizing his work in that field. His book
was to be the second in a series of astrophysical monographs published by the University of Chicago Press, another project the tireless
Struve had organized. Each book, written by an outstanding expert,
would summarize the state of a particular field. Written at the research level, they were intended for graduate students and active
research workers. One new monograph was to be published each
year. Bok had written the first, *The Distribution of the Stars in Space*,
based on his lectures at Yerkes in the summer of 1936, and Chandrasekhar had made a good start on his manuscript the following
spring.[17]

As always, publication costs were a major consideration. Struve
had persuaded Hutchins to authorize a publishing subsidy of $1,000
for each monograph, to go to the University of Chicago Press. Bok's
monograph, a little over 100 pages long, fitted neatly into this
scheme. But Chandrasekhar's book promised to be longer, perhaps
170 pages, and his manuscript would contain many mathematical
equations, making it expensive to typeset by hand. The manager of
the Press suggested that Struve could make up the difference from
Yerkes funds. He exploded. Always intensely loyal to staff members
who produced research results, he fought the Press, getting competing

bids from other publishers, proposing wild marketing schemes, and finally going straight to Hutchins, who he knew was especially supportive of Chandrasekhar. His book would be "one of the most important astronomical publications ever issued," the Yerkes director insisted. The president told the Press to go ahead with the book. He promised to allot the needed funds.[18] Chandrasekhar's manuscript grew far beyond his earlier estimates. The editor of the Press, Dean Gordon J. Laing, trying to hold down the publication costs, suggested that Chandrasekhar could leave out some of the equations that were already well known to the astronomers who would read the book. He added that he would be glad to send the book to an outside reader, Donald H. Menzel (the Harvard theorist who had gotten his job after Struve had turned it down in 1932) or Frederick H. Seares (the Mount Wilson editor) for an independent opinion. Struve replied with another tirade. He was the editor of the astrophysical monographs. He knew what was best for them. Chandrasekhar wanted every one of those equations to stay in, and they would stay in. Menzel was incompetent, and Seares was not at all familiar with the subject. The Press should forge ahead instead of putting obstacles in Chandrasekhar's path. Laing weakly backed down. He knew that Hutchins was behind Struve and Chandrasekhar and that the Press would have to publish the book as written. Ultimately, no doubt at Hutchins's or Gale's request, Struve handed over to the Press $2,000 from a windfall received when the Chicago World's Fair closed its books with a slight profit and gave it to the university.[19] Morgan, the onetime aspiring English literature student, read and corrected the entire manuscript, then the proofs of the book.[20]

When *An Introduction to the Study of Stellar Structure* appeared in January 1939 it was over 500 pages long, three times the original estimate. It bristled with equations. And it was a very important book, the only up-to-date one on the subject, destined to be widely used as a graduate textbook wherever astrophysics was taught for the next quarter of a century.[21]

Gale, whatever he felt about Chandrasekhar's outstanding research, was adamantly opposed to having him on the campus. In the spring quarter of 1938 the Astronomy Department was to offer an elementary survey course at the downtown University Center in Chicago, on Michigan Avenue in the heart of the city. Courses there were designed to attract students who already had jobs and wanted to learn more about a subject, not earn a degree. The astronomy course was to meet one day a week, on Saturday, for two hours.

Struve followed his preferred policy of having the course taught by different faculty members from Yerkes each week, who would drive to Chicago to speak on the subjects they knew best. It was an ideal arrangement for such a course. Struve intended to send seven faculty members, including himself and Chandrasekhar. Knowing Gale's strong views against anyone with a dark skin, the director took the precaution of writing Hutchins in advance, explaining the situation. Struve reported Gale's "[feeling] . . . that it would not be advisable for Dr. Chandrasekhar to participate in any of the work on the campus" but went on to say that since this course was to be given in the University College, headed by a different dean, he was "not certain" this restriction should apply. Struve wrote that he himself would much prefer to have Chandrasekhar lecture in the course; the students would want to see him, and he could explain his own work much better than anyone else. The president's reply was succinct: "By all means have Mr. Chandrasekhar lecture." [22]

When he learned the news, Gale was extremely displeased. He claimed that Struve had breached an "understanding" they had reached when Chandrasekhar was appointed to the faculty. Gale wrote that he had been "quite upset" on the one occasion when Chandrasekhar had come to Chicago to attend a physics colloquium on the campus. Ominously, he added that he wanted to discuss "some matters concerning Chandrasekhar's monograph" (then still being written) with the Yerkes director. Struve replied immediately. Carl F. Huth, dean of the University College, had gladly accepted Chandrasekhar as a lecturer. He had written Struve, "There certainly will be no objection to Professor Chandrasekhar. We have had another Hindu gentleman teaching in University College for two years now and there has not been the slightest indication of an adverse reaction." Struve himself believed that all "our practical astronomers" should become acquainted with the students, and that hearing their lectures would "encourage many good students to take up astronomy." (He probably believed this himself, but it seems a will-o'-the-wisp idea given the type of students who took the downtown courses.) Furthermore, Struve wrote, President Hutchins was "very enthusiastic about the plan and he thought that it would not be desirable to exclude Chandrasekhar." Therefore, without agreeing that there had been any "understanding," Struve asked Gale to permit Chandrasekhar to give a lecture in the course. The Yerkes director was "really sorry" that Gale felt "as you do about Chandrasekhar." Struve himself found him "an extremely likable person, . . . not at all the dangerous type of political propagandist that we had feared." Most important to Struve, Chandrasekhar was "without any doubt the most brilliant theoretical astronomer now living; and since we are all deriving an immense amount of information from him, it would be almost fatal if we should be forced to give him up."

Gale bleakly pointed out that Chandrasekhar was scheduled to give two lectures, not one (he was right; Struve was working from an outdated preliminary schedule), but since it was downtown instead of on the campus and since his director felt so strongly about it, he would agree this time. He made it clear, however, that he would still not approve Chandrasekhar's lecturing on campus. The Indian astrophysicist's promotion from research associate to assistant professor had been made while Gale was hospitalized the previous year, but since the acting dean had approved it he would not object now. He was clearly implying that he would not recommend another promotion, to associate professor, but Struve knew that Gale was due to retire in a year and a half. The director hurried to Chicago to see the dean, who had written that he would be very sorry if this matter had clouded "the perfectly harmonious relations that have always existed between us." In his office Gale consented to Chandrasekhar's giving either one or two lectures in the course and, according to Struve's written record of their discussion, "said that it would make no difference whether the course was to be on the campus or downtown." The dean said he had approved the salary raise Struve had recommended for Chandrasekhar for the coming year and promised that money would be made available to publish his monograph. Evidently Gale still considered Chandrasekhar black and therefore undesirable but recognized that he had to support his observatory director as the president was doing.[23]

Most of the faculty members on campus in Chicago did not share the dean of physical science's bigoted ideas. This was made clear when the Quadrangle Club, the university faculty club, tried to add a few more members from Yerkes Observatory. As nonresident members, they would be able to lunch at the club, where many of the faculty members from all departments met daily. Struve was already a member and always lunched at the Quadrangle Club with Gale, Bartky, some of the physicists, or his friends from the Innominates, a group of senior faculty members from all the departments in the Physical Sciences Division.[24] Struve wanted to invite Chandrasekhar, along with others from Yerkes, to apply for nonresident membership in the Quadrangle Club, but he first made sure he would not be refused. The Yerkes director wrote Carl R. Moore, a professor of zoology and the club membership chairman, explaining the situation. Some years ago, he said, Dean Gale had "rather violently . . . objected to Chandrasekhar's presence in the Quadrangle Club," and he did not want to put his star theoretician in a position where he would be subjected to that treatment again. Chandrasekhar, Struve wrote, was "one of the most brilliant physical scientists in the world" and "beyond doubt the most capable member of our astronomy department." He was also a nephew of C. V. Raman, the Nobel Prize winner in physics for 1930 and a member of the "Brahm[a]n class." "Personally, I think that it is rather ridiculous and not very flattering to our white civilization that a man of his type should be excluded," Struve concluded, but he wanted to be certain Chandrasekhar would not be rejected. Moore agreed with him and passed the letter on to the Quadrangle Club Council. At their next meeting, later that month, they unanimously voted that their Quadrangle Club "would be honored by membership of such a man." In contrast to the dean, they welcomed Chandrasekhar to the campus.[25]

Meanwhile Strömgren was completing his work on determining the hydrogen and helium content of stars from their observed temperatures, masses, and luminosities. His result, that hydrogen was the most abundant element, helium next, and all the heavy elements together (in the "Russell mixture," the proportions found on earth) much less abundant was controversial at the time. Most astronomers then thought that the heavy elements were the most abundant, but Strömgren's paper was one of the most convincing in changing their minds.[26] He contributed a good deal of the material on "practical" applications of the theory of stellar structure for Chandrasekhar's monograph. Strömgren had begun working on the theory of stellar atmospheres to interpret the strengths of absorption lines quantitatively in terms of the abundances and thus to check his conclusion based on stellar interiors data. It was to become his main work in a year or two, but at Yerkes Strömgren quickly became interested in the new results that Struve, Elvey, Henyey, and Jesse L. Greenstein were getting with the McDonald Observatory nebular spectrograph. They were finding nebulae, some bright but most large and faint, wherever there were hot O stars in the Milky Way. The Danish astrophysicist realized they were probing the same interstellar medium that Struve and others had observed in the guise of absorption lines in the spectra of distant stars. Strömgren launched into a theoretical interpretation of the new emission-line data, using and extending the same low-density physics concepts that had been so important in the ε Aurigae paper. The paper he began at Yerkes, completed in 1939, became a benchmark in understanding interstellar matter.[27] For many years it was the classic, most widely quoted paper in the field.[28] William W. Morgan gradually switched his research away from peculiar A stars, the subject of his thesis and early independent work, to spectral classification. Struve had seen him as a junior edition of himself, a Yerkes Ph.D. who needed to get out in the world and learn more about astrophysical theory. Struve urged him to go to Harvard or Lick and work there for a time, absorbing new ideas from their directors and staff members, as he himself had done. Morgan, married and with two small children, resisted the idea, saying he could not afford it. Instead, reading, studying, and observing on his own, he learned more and more about spectral classification. The Henry Draper catalog, produced at Harvard by Annie J. Cannon and her assistants, was the standard catalog of spectral types. Based on objective-prism spectra, it was a one-dimensional scheme, in which a star with a spectrum like the sun's was labeled G2, a star like Sirius A0, a star like Betelgeuse M2. It did not differentiate between dwarfs, giants, and supergiants, although they had been recognized long before. At Mount Wilson Adams pioneered in two-dimensional spectral classification. He and the astronomers who worked with him took slit spectrograms of stars primarily to measure their radial velocities, classified the stars by spectral type "on the Draper system," then, from the ratios of strengths of particular lines, estimated the absolute magnitude of the stars, which they gave in numerical form. Morgan, with his nonphysical, unconventional way of thinking, came to realize this was not a good method. Every time a better distance determination was made for one of the stars used to calibrate the absolute magnitude scale, the system changed, and all the absolute magnitudes of the classified stars changed as well. It was more logical to give a spectral "luminosity class" along with the spectral type. Morgan invented the symbols ranging from Ia and Ib for supergiants through III for giants to V for main sequence stars, so that the sun's full spectral classification became G2 V. Thus the absolute magnitude calibration was separated from the classification process.

Morgan also came to understand that for accurate spectral classification it is essential to use homogeneous data with (in modern terms) a very good signal-to-noise ratio. This he achieved by using the same spectrograph for all stars, exposing them all to the same level, using a special low-contrast, fine-grain developer, and widening the

Figure 38  William
W. Morgan (1939).
Courtesy of Yerkes
Observatory.

spectrograms to get the maximum possible signal and store it effectively on a photographic plate. Finally, Morgan realized that the "system" of classification should be defined by a relatively few "standard
stars" of each spectral type and luminosity class rather than by supposed values or equalities of specified line ratios, or the appearance or
disappearance of specified line rations, as in the older Draper and
Mount Wilson systems. In his mind spectral type and luminosity class
were discrete coordinates, much like quantum numbers, rather than
continuous variables.[29] Morgan's paper announcing these results was
necessarily critical of the earlier Mount Wilson work in these fields,
though he did his best to mute that aspect. Struve apologized to
Adams, insisting that Morgan, "an exceptionally able young man,"
had meant no criticism and realized that he was simply building on
the Mount Wilson results. The Yerkes director wished he could send
Morgan out to work with the Mount Wilson astronomers and get to
know them better, he added. Adams accepted the apology but issued

no invitation to the young Yerkes spectroscopist to come and use the big telescopes in California.[30]

In his second long paper on spectral classification Morgan demonstrated the great advantages of his own system. The accurate types and luminosity classes he obtained by his careful methods could be calibrated accurately in terms of absolute magnitude and color index (the "color" of the star, expressed in numerical terms). There was much less scatter than with the older classifications, indicating that Morgan's method was more accurate and physically correct. Previously published results, which seemed to show that fainter stars of a given spectral type were redder than nearby ones, turned out in fact to result from systematic errors in the spectral classification of the fainter stars. On the other hand, with Morgan's accurate types the intrinsic color index of a distant star could be predicted and, compared with an accurately measured color index, could be used to determine the "color excess" or reddening by interstellar dust. The reddening, in turn, would give the total extinction of light by dust along the path to the star. Correcting for this lost light, the difference between apparent and absolute magnitudes of the star tells how much it has been dimmed by distance and hence measures the distance. This is the well-known spectroscopic distance method; Morgan simply showed that his accurate spectral types and luminosity classes, together with some of the first photoelectrically measured accurate color indices and magnitudes of stars from Germany, gave much better results for the distances of stars than any earlier work.[31] Morgan soon realized that combining his types with the photoelectric measurements of high-luminosity O and B stars that Joel Stebbins and his collaborators were making just seventy miles away in Madison would make it possible to begin mapping the Galaxy.[32]

That would come later; for the present Morgan was working on finding the best spectroscopic criteria to use for each range of spectral types and luminosity classes. His ability to recognize patterns, and his freedom from preconceived ideas based on theoretical ideas of how lines "should" change with temperature or luminosity (many of which were wrong in those primitive days), perfectly suited him for this task. As a local product who resisted leaving Williams Bay and who was not sought after by Harvard or Mount Wilson, Morgan did not get the salary raises and promotions that came so easily to Strömgren, Kuiper, and Chandrasekhar. Struve knew his former thesis student would not take a job elsewhere, although Bok had praised Morgan's second paper as "one of the most fundamental contributions in recent years" and kept hoping he would come to Harvard.[33]

Morgan was still acting as assistant director, but without the official title. He was left in charge of day-to-day activities while Struve was gone to McDonald, with instructions to report to him frequently, never less often than once a week, on what was happening at the observatory and on any important letters that had arrived, especially from Hutchins or Gale. But Struve never revealed his plans to Morgan, who consequently operated in an information vacuum. The director's secretary, Lillian Ness, who typed all his letters and opened his mail, knew his mind better than Morgan did. A workaholic like Struve, she shared his strict, demanding attitude, and they got along well together; she served in many ways as his executive assistant. Ness resented Morgan, whom she regarded as an irresponsible youngster, and she sometimes undercut him if he did not attend to his assistant director's duties as quickly as she thought he should. Struve cautioned her not to push Morgan too hard, but he himself played them off against one another. He could not afford to have just one set of eyes reporting to him during his month-long absences from the observatory.[34]

Philip C. Keenan came back to Yerkes in 1936 as an instructor after one year at Perkins Observatory. He was the faculty member Struve sent to teach on the campus after Strömgren moved from Chicago to Williams Bay. Keenan, a bachelor, could move back and forth more easily than a married man, coming to the little village whenever he did not have to teach on the campus. Struve made sure that his appointment was for twelve months, with a one-month vacation, like everyone else at Yerkes, rather than for nine months, with the whole summer off, like the faculty members on campus. "I do not believe that it will make any difference to him, and of course it involves no financial readjustment," was the director's reasoning.[35]

In that same year the Astronomy Department finally got a telescope on the campus, so the students could familiarize themselves with observing. Frost had had grandiose plans for an astronomy building with a library, faculty offices, graduate student workrooms, and an impressive dome for a large telescope. His dream never became a reality. Struve settled for a five-inch refractor in a small dome, erected on the roof of Ryerson Laboratory, the physics building where Gale had his office.[36] The person who did most of the instruction with this telescope was not Keenan, however, but Thornton L. Page, who had been a Rhodes scholar in England, where he studied astrophysics. Struve, far more enthusiastic about research productivity than teaching experience or skill, had tried a second time to bring Marcel Minnaert from Holland to Chicago or, after he declined, Theodore Dunham from Mount Wilson Observatory. Luckily for the students on campus, neither accepted.[37]

Gale wanted a teacher on campus, not a researcher who would disappear to Yerkes or McDonald at unpredictable intervals. After Minnaert declined, the dean had mildly asked Struve if they should not bring Bartky, who was committed to teaching on the campus, into the discussion too. Struve did not want Bartky's opinion and ignored the suggestion. After Dunham had refused the proffered appointment, Gale insisted. He had met Page, then twenty-four years old, in the summer of 1937. The young astrophysicist was home from Oxford for the summer and passed through Chicago with his father, Leigh Page, a well-known teacher of mathematical physics at Yale. They were driving west to visit the California observatories and national parks. Thornton Page was to return to England for one more year, as chief assistant at the Oxford University Observatory, get his Ph.D., and then come back to the United States. He was personable, well trained in modern astrophysics, and clearly would be an excellent teacher. Gale decided to offer him an instructorship to begin in the fall of 1938, and Struve, who had failed to land any of his top candidates, had to agree. However, he arranged that Page would spend only two quarters of his first year on the campus, then come to Yerkes for the spring and summer quarters, with an observing trip to McDonald Observatory. Keenan would temporarily replace Page on the campus in the spring.[38]

Although Struve had hoped to persuade Strömgren to stay at Yerkes rather than return to his new professorship at Copenhagen, the call of the young Dane's homeland won out. He was attached to his family; his father was old and in poor health. Adolf Hitler was threatening Europe, and trouble clearly was coming. Strömgren was determined to return to Denmark. Struve accepted the inevitable but hoped to lure him back soon. He brought Strömgren to McDonald to observe with him just before his departure for Europe; the Dane's wife and young children accompanied him to Texas and enjoyed their first glimpse of the mountains and plains of the West.[39] Just before boarding the ship at New York, Strömgren dropped down to Washington where, with Chandrasekhar, he took part in a conference of physicists and astrophysicists, organized by George Gamow and

Merle Tuve, to discuss possible nuclear sources of stellar energy generation. All four of them, like von Weizsäcker and most theorists back to Arthur S. Eddington, believed these reactions would prove the answer to the long-standing question of where the energy radiated by stars ultimately came from. It was at this conference that Hans Bethe realized he could solve the puzzle. Soon afterward he put forward the basic ideas of the proton-proton cycle and the carbon cycle, the main energy-producing reactions in stars.[40] Struve sent Strömgren a bon voyage telegram on behalf of "all your friends at Yerkes Observatory."[41]

Settled back in Copenhagen, the Danish astrophysicist continued working very productively on stellar atmospheres and interstellar matter. He could see that the new idea of Rupert Wildt—that the negative ion of hydrogen, $H^-$, must be an important source of opacity in stellar atmospheres, including the sun's—promised to remove long-standing discrepancies between theory and observational data. Incorporating $H^-$ into the model stellar atmospheres Strömgren was beginning to calculate became his main priority. He wrote Struve, who hoped he would return to Yerkes in 1939, that although he wished to retain his status as a University of Chicago professor on leave, he could not come back soon, nor could he predict if he ever would. Writing less than a month after the Munich crisis, in which Neville Chamberlain, the prime minister of England, bought "peace in our time" by handing Czechoslovakia over to the tender mercies of Hitler, Strömgren could clearly see that a major European war was inevitable. He intended to remain in Denmark with his family and his father and share whatever happened to his countrymen.[42] Pol Swings, Struve's Belgian collaborator, advised him that Strömgren really meant it; he, and even more so his wife, was determined to stay in Copenhagen.[43]

The Yerkes director, however, with his single-minded devotion to astronomical research, was far less perceptive about human beings than his sympathetic European friend. With McDonald Observatory's big telescope about to go into operation, Struve already felt overworked in his administrative duties. He had tried to shed some of them by delegating them to junior staff members, but no doubt recognizing that he would demand they act exactly as he would have done (without clearly disclosing his basic policies), they had avoided accepting them. The only exception was Morgan, acting director in Struve's absences, though not in name, who did as he was asked but no more.[44] Struve, in spite of Strömgren's and Swings's warnings, went ahead with the plan he had long been preparing. Under it, he would move to the campus in Chicago as head of the Astronomy Department. Strömgren would return to Williams Bay as director of Yerkes Observatory, responsible directly to Struve. An assistant director, "if we can find one," also responsible to Struve, would supervise McDonald Observatory. (Struve evidently had little confidence in Elvey, whom he described in less than glowing terms as "a reliable administrator and a satisfactory [research worker].") As head of the department Struve would live in Chicago and continue to edit the *Astrophysical Journal* but would have permanently assigned houses on Mount Locke, for his observing visits there, and at Williams Bay. On campus, Struve visualized that his role as "a person of sufficient experience and training and of a sufficiently aggressive personality" would be "to reorganize the department along modern lines, making it essentially independent of the department of mathematics and aligning it very closely with the department of physics, although retaining complete independence in all matters of research and instruction." He would supervise Page, "whose experience is limited" but who might "turn out to be a good teacher and a capable administrator," but Bartky was conspicuously absent from Struve's plan. Presumably he would be handed back to the Mathematics Department, or perhaps he would be kept to teach celestial mechanics. Struve diplomatically did not mention his fate; privately he considered Bartky "essentially a mathematician" with little interest in astronomy who had "really never done anything in astronomical research." His teaching experience and abilities, like Page's, counted for little in Struve's eyes. The Yerkes director thought this proposed reorganization would give him plenty of time for research, but apparently he allowed no time at all for communicating with his subordinates.[45]

Dean Gale, although "not entirely happy with respect to all [the] points" of Struve's plan, backed him on it in general terms. They forwarded it to President Hutchins, who approved it.[46] Just as Swings, Kuiper, and Strömgren himself had predicted, however, the Danish astrophysicist declined the directorship of Yerkes Observatory. He considered it for several weeks after receiving Struve's letter but then, in late November, cabled his regrets, "though deeply impressed by excellence of plan." In a long follow-up letter Strömgren explained that although the director's reorganization scheme was excellent, the opportunities for research at Chicago, Yerkes, and McDonald were wonderful, and the "friendly cooperation" he knew he would enjoy with Struve was very attractive, he nevertheless intended to stay at

Copenhagen. There, close to the Bohr Institute, his opportunities for theoretical work would be great; he hoped to build a new observatory outside the city, and he would live in "peace and harmony." He and his wife both preferred to stay in Denmark. Chicago would be wonderful scientifically, but their personal considerations were more important. If his plans for the new observatory in Denmark fell through, or if there was a European catastrophe, he might reconsider his decision, but if he did the initiative would come from him. It was an extremely polite, thoughtful letter, and in fact it forecast much of the rest of Strömgren's life.[47]

To Struve, for whom research was the only value, Strömgren's decision was incomprehensible. The Yerkes director was deeply disappointed. He felt sure the time would come when the young Copenhagen astrophysicist would regret it. To Swings, Struve was more direct. Strömgren was "a great fool." Not only had he rejected "an exceptional opportunity for himself and for the advancement of science," but his decision had erased "several advantages, financial and otherwise, which I could have secured for the Observatory if he had accepted."[48] No one but Strömgren could take over the Yerkes directorship in Struve's opinion, as he had already told Hutchins and Gale, so he soldiered on himself.

In Copenhagen, Strömgren went on working even more productively. Stimulated by the results Struve, Elvey, and others had obtained with the McDonald nebular spectrograph, he investigated theoretically the photoionization of interstellar gas and discovered the great importance of the radiation from the relatively rare, hot O stars in fixing the physical conditions in space. He inspired atomic physicist Harrie S. W. Massey, in London, to calculate with David R. Bates, his student, the first reasonably accurate values of the opacity of the $H^-$ ion, so important for Strömgren's own research on model stellar atmospheres. He finished his own paper on the interstellar gas and sent it to Struve for publication in early 1939, then sent Massey and Bates's paper to him in the fall of that year. Both papers became classics. Struve continued to hope that Strömgren would come back to Yerkes, but the young Dane politely but firmly declined each of his overtures.[49]

Strömgren saw Kuiper and Chandrasekhar again in Europe, at the international Conference on Novae, Supernovae, and White Dwarfs, held in Paris on the eve of World War II. This meeting was the second in a series of annual astrophysics conferences organized in France, subsidized by a wealthy publisher, and intended to rival the extremely

Figure 39    Conference on Novae, Supernovae, and White Dwarfs, Paris (1939). In front row Cecilia Payne-Gaposchkin is second from left, Henry Norris Russell third, Arthur S. Eddington fifth; in second row Pol Swings is third from left, then Gerard P. Kuiper, Bengt Strömgren, Subrahmanyan Chandrasekhar, and Walter Baade. Knut Lundmark is standing in front of Chandrasekhar. Courtesy of Yerkes Observatory.

prestigious Solvay conferences in physics. Some fifteen outstanding research workers were invited, headed by Russell and Eddington, and including Payne-Gaposchkin, Walter Baade, Bengt Edlén, Knut Lundmark, the three Yerkes representatives, and Swings, who was soon to return to America for a longer stay than he expected.[50] At this conference Eddington, who was becoming increasingly a mystic, continued to support his own private version of the equation of state of a degenerate gas at very high density, the basis of the theory of white dwarfs. Chandrasekhar, using the correct Fermi-Dirac equation of state approved by all the physicists, had derived an upper limit to the mass a white dwarf could have, approximately 1.4 solar masses, as we understand today. Larger mass white dwarfs cannot exist. Eddington could not accept this result, and in spite of having been one of the pioneers in explaining relativity theory, he tried to argue against applying it correctly in the equation of state. Chandrasekhar had debated him politely on this subject from his first year in England. All the physicists, and the younger astronomers, knew that Chandrasekhar was right, but Eddington's immense prestige kept the argument alive. At the Paris meeting it was clear to all that Chandrasekhar's version of the theory was correct and that his "Chandrasekhar limit" was a reality, but no one, not even Russell, was willing to press this point on Eddington. Chandrasekhar, though triumphant, was deeply hurt.[51]

When Strömgren left America in March 1938, Struve arranged to use the money budgeted for his salary to bring Karl Wurm, a young German theoretical astrophysicist, to Yerkes for a year. Wurm worked largely on the physical interpretation of comets and nebulae, neither of which was a prime research subject to Struve.[52] Struve made no attempt to keep Wurm on the faculty; he still hoped Strömgren would return. Instead, the Yerkes director brought Albrecht Unsöld, the rising young astrophysicist from Kiel, Germany, as another temporary visitor. Unsöld was pioneering in applying the latest quantum-mechanical results to the interpretation of stellar spectra, to study the structure of the stars' atmospheres and the abundances of the elements. Now he needed observational data, and he was eager to come to America to get it. He hoped to spend three months each at Mount Wilson and Yerkes Observatories in 1939. However, with tension rising daily between Germany and the United States in the aftermath of the Munich agreement and Hitler's takeover of the Sudetenland from Czechoslovakia, Adams was not at all eager to have Unsöld visit Mount Wilson Observatory. Its most famous staff member, former Rhodes scholar Edwin Hubble, the Carnegie Institution of Washington of which it was a part, and Adams himself were all strongly pro-English and hostile to Germany. For Struve, only scientific ability and results counted. He tried to interest Adams in helping bring "one of the world's leading specialists in the field of astronomical physics" to America. The Mount Wilson director was clearly unimpressed. He made no effort to provide support, nor did he invite the Kiel astrophysicist to Pasadena.[53]

Unsöld came to America anyhow, with only three months' salary as a visiting professor, $1,250, from the University of Chicago. He wanted to use the new McDonald eighty-two-inch reflector to obtain high-dispersion coudé spectrograms of stars for analysis back in Germany. He brought his wife with him; like many German tourists they crossed the Atlantic on a North German Lloyd ship, interrupted their train trip to Chicago to visit Niagara Falls, and stayed in the Hotel Bismarck on the way to Williams Bay. By then Struve, at McDonald, had tested the coudé spectrograph and found that it would produce satisfactory spectrograms for the bright stars Unsöld planned to observe, although it was slow. By the end of March the director was back in the Midwest and drove the Unsölds to Williams Bay.[54] Liselotte, Unsöld's wife, turned out to be an outspoken Nazi; her husband, although he was a strong supporter of the "Ein Volk, Ein Reich" pan-German expansionist policy of the Nazis and openly expressed anti-Semitic ideas, was probably not actually a party member himself, a knowledgeable observer later concluded.[55] In May Liselotte returned to Germany, while her husband went on to Texas for the dedication of McDonald Observatory and stayed to observe with Struve. They had good weather, and though there were some problems, they obtained numerous spectrograms for their joint program. By mid-June Unsöld was back in Williams Bay, and on August 11 he sailed for Hamburg on the *Bremen*, less than a month before Hitler's invasion of Poland and the British declaration of war.[56]

At Yerkes Struve measured the wavelengths from their spectrograms, and then his assistant, graduate student Frances Sherman, traced them with a microphotometer. Struve sent the paper tracings, from which Unsöld could measure the intensities and profiles of the spectral lines, to him in Germany. The precious photographic spectrograms he would not entrust to the high seas "until later." He had arranged with Unsöld to number their letters to one another so they would be able to recognize if any had not gotten through. The Kiel astrophysicist had arrived back in his office in late August and waded through the pile of papers on his desk in eager anticipation of getting to work on $\tau$ Scorpii, the first and most important star on his program. He was not to publish the results until 1942, when no communications, numbered or otherwise, would pass back and forth between Kiel and Williams Bay.[57]

Struve was always interested in physics and in improving interactions with physicists, perhaps because he realized how weak his own background in it was. Thus when Gale suggested that the Physics and Astronomy Departments might organize a joint symposium for the summer of 1938, he jumped at the opportunity. Struve proposed molecular spectra, in which the Chicago Physics Department was strong, as a topic. Astronomers were investigating molecules in stars, planets, and interstellar matter, and since the dean had not named any specific level of support, Struve recommended bringing in Russell, Swings, and Arthur Adel of Lowell Observatory, for a total of $800 in travel funds, to complement Wurm, who would be available under his visiting appointment. Gale quickly made it clear that he could not provide that kind of money; Hutchins would allocate $250 for astronomy, the physicists could add $150 to $200, and they could get speakers from closer to home than Belgium or even Arizona.[58]

Robert S. Mulliken, Struve's closest friend in the Physics Department, took the lead in organizing their end of the conference. Within a few months the two of them had the program outlined. Russell was going abroad for the summer, and they could not afford to bring Swings, but the others they wanted would come. Adel; Rupert Wildt, who had identified methane and ammonia molecules in the spectra of the major planets; Nicholas T. Bobrovnikoff, the Yerkes product who was now doing molecular spectroscopy as director of Perkins Observatory; George H. Shortley, the Ohio State theorist of quantum mechanics; laboratory spectroscopist Gerhard Herzberg, another refugee then in Canada; and Wurm were the main outside speakers. Mulliken and Hans Beutler complemented them from the campus Physics Department.[59] The meeting was held at the George Williams College Camp, a YMCA-operated resort and conference center on the shore of Lake Geneva directly below the observatory. For the four full days of the conference, room and board at the camp cost $24.75 per person in a double room, with a $1.70 supplement for those who wanted a single room. Under Struve's stern eye all the Yerkes staff members attended, as well as many of the physicists from the campus. Among the attendees from outside were John T. Tate of the University of Minnesota, a molecular spectroscopist who was editor of the *Physical Review*; Julian E. Mack, Ragnar Rollefson, and John G. Winans of the nearby University of Wisconsin faculty in Madison; Edward Teller from George Washington University; and Dorothy N. Davis of Vassar College, who a few years later was to do an outstanding thesis on molecules in stellar atmospheres at the University of California.[60]

At the meeting in late June, Morgan, as a representative of the Yerkes faculty, gave a paper on the use of molecular bands for classifying stellar spectra, and Struve and Kuiper gave "general interest" evening lectures on contours and intensities of spectral lines in stellar atmospheres and on the mass-luminosity relation and on white dwarf stars, respectively. A boat ride on Lake Geneva and a visit to the University of Chicago gardens at Wychwood, the former estate and nature sanctuary of trustee Charles L. Hutchinson, completed the program. For those who wanted more, the Physics Department held two successive symposia on campus the following week, on nuclear physics and on cosmic rays.[61]

This scientific meeting, the first formal one Struve organized as director, was successful and established the pattern for the McDonald dedication the next year and for the symposia he organized for later American Astronomical Society (AAS) meetings he hosted at Yerkes. Astrophysics was always at the fore, participation by physicists was essential, recognized experts, always including some from his own faculty, gave invited papers on specific, linked topics, and though they were expected to start with a brief review, recent, important research results were to make up the main body of their talks.[62]

On the Yerkes Observatory staff, Louis G. Henyey flowered in astrophysics under Strömgren's tutelage, and in 1939, encouraged by Struve, he applied for a Guggenheim Fellowship to spend a year in Copenhagen, working on research with his former teacher. Henyey got the fellowship, a signal mark of respect for his abilities, but with the beginning of World War II in September, he and other Guggenheim fellows were advised not to go abroad. He postponed the fellowship a year, in the vain hope that the war would end by September 1940. Struve thought Henyey might use the fellowship to go to Mount Wilson then, but instead he chose to go to Cornell, to work with Bethe in quantum mechanics and nuclear physics, the bases of stellar opacity and energy-production rate calculations.[63] Greenstein, who had been a National Research Council postdoctoral fellow at Yerkes for two years, earned appointment as an instructor in the fall of 1939. A Harvard Ph.D., he was "a man of unusual ability and fairly good training," in Struve's estimation. At first the director was skeptical of Greenstein's observing skills with the eighty-two-inch reflector, but he was soon won over by his energy and productivity. The new instructor was sent to the campus one quarter a year (he commuted by car a few days each week) in place of Keenan, who was switched to a purely research appointment. Struve helped Keenan try to find a position at some other university, thinking of him as a theorist, but in fact he was moving more and more into spectral classification, working with Morgan on this program.[64]

Just as Struve was closest to Mulliken in the Chicago Physics Department, he had the most contact with Gregory Breit at the University of Wisconsin. Breit was a theorist who was working in nuclear reactions, like Bethe, and gave a colloquium on the subject at Yerkes in November 1938. He found much more astrophysics going on at Yerkes than he had realized and invited Struve to speak at the physics colloquium at Madison "about interstellar hydrogen or any other subject that you think suitable."[65] Stebbins, who had been the leading outside candidate for the Yerkes directorship, and Leonard R. Ingersoll, the chairman of the Wisconsin Physics Department, enthusiastically seconded the invitation, and a month later Struve drove up to Madison and gave a "very stimulating talk" on interstellar absorption and emission lines. He received $4 traveling expenses for the 150-mile round trip between Williams Bay and Madison. (Chicago was more generous; Breit had gotten $4.50.) At the dinner following his colloquium Struve mentioned the problem of the decay rate of the metastable energy level of the hydrogen atom; this further aroused Breit's interest. Struve briefed him on it, and within a few months Breit and Teller worked out the quantum mechanics of the two-photon emission process and calculated the decay rate. Their highly accurate result confirmed Struve's observational result and provided further physical data for investigating nebulae and stars with extended atmospheres.[66]

A great triumph for Struve about this time was the publication of a series of papers by himself, Kuiper, Greenstein, Page, and graduate student Jocelyn R. Gill, on β Lyrae. It was one of the brightest variable stars in the sky, and its regularly repeating light curve had made it also one of the first eclipsing binaries to be recognized. But its orbit determined spectroscopically from the radial velocities of what appeared to be two B stars did not agree with the orbit derived from the light curve. Struve had assigned graduate student Helen M. Pillans to work on this problem for her master's thesis in 1934. She had concluded, from spectra taken with the forty-inch refractor, that one of the stars was not normal, and that a "nebula" or gas cloud was involved in the system. Struve, analyzing her data, had seen that one of the stars was a supergiant, losing mass from its outer layers into space. The combination of that flow with the velocities of the stars in their orbits produced blended spectral lines that, measured as single lines, gave erroneous measured radial velocities and hence an incorrect orbit.[67] Many problems remained, however, and Struve, his colleagues, and his assistants continued obtaining spectrograms of β Lyrae.

In 1941 he published his long paper showing clearly from all the observational data he had by then, including Greenstein and Page's, and Gill's, that there is a common gaseous envelope about both stars but only one of them is a B star, which is shedding matter into the disklike envelope. This low-density gas mimicked a supergiant's atmosphere. From the observations, Struve could pinpoint the regions in the envelope where the gas was densest and the directions it was flowing. Then Kuiper, using the same dynamical techniques related to the gravitational forces in a rotating double-star system he had earlier applied to $\epsilon$ Aurigae in the joint paper with Struve and Strömgren, supplied the theoretical interpretation of the flow and of the form and nature of the disk. Furthermore, the same methods were applicable to understanding many other close binary systems, with different mass ratios and separations of the two stars, as Kuiper showed.[68] These papers led to a very important step forward in analyzing close double stars and the nature of their evolution. Struve was to return to $\beta$ Lyrae and to the whole subject time after time, culminating in his Henry Norris Russell lecture in 1957, years after he had left Yerkes Observatory.[69]

Struve constantly supported a wide definition of astrophysics. In 1940, when Vladimir Rojansky, a physicist at the University of Pittsburgh, submitted a paper to the *Astrophysical Journal* suggesting that the mysterious meteorite explosions recorded in Tunguska, Siberia, three decades earlier had been a massive example of annihilation of "contraterrene matter" (antimatter), Struve welcomed it. First Rojansky had to shorten the paper, but that was simply par for the course, as everyone who published in the *Journal* knew.[70] Likewise, when Grote Reber, a radio engineer who had erected his own early "dish" antenna at his home in nearby Wheaton, Illinois, discovered "cosmic static" coming from the Milky Way, Struve invited him to give a colloquium on it at Yerkes. After the talk Struve published Reber's paper in the *Astrophysical Journal,* also reduced in length, even though he had previously published much the same material in the *Proceedings of the Institution of Radio Engineers.*[71] The part Struve cut out was Reber's attempt to calculate the rate of what astrophysicists call free-free emission by ionized hydrogen gas; the radio engineer had not gotten it quite right. The Yerkes director had assigned Henyey and Keenan to do the calculation correctly and published their paper just after Reber's.[72]

Ross, the elderly wide-field photography expert, did not fit into Struve's plans at all. Although pleasant and peppery, Ross had long ago made it clear that he preferred California to Wisconsin and was spending half of each year in Pasadena working on the two-hundred-inch project. He would be sixty-five in 1939 and due to retire, but he hoped to be kept on past the regular retirement age, for he needed the half-time full professor's salary he was still being paid at Chicago. He had not become a faculty member until he was fifty, and his pension would be small. But he had never been a real member of the team at

Yerkes. Struve carefully advised Gale in formal terms that although Ross was a very distinguished astronomer who had received many honors in America and abroad, he would not recommend that his appointment be extended but would leave that decision to the dean. Gale knew exactly what that meant and notified his half-time professor that he would be retired on October 1, 1939.[73] To his friends' amazed delight, just before that date Ross married his third wife, Anna Olivia Lee, the daughter of his former Yerkes colleague Oliver J. Lee, and set off with her on a nineteen-day automobile trip to Pasadena.[74]

In the Ross matter, as in everything Struve wanted to do, Hutchins supported him completely, as long as no additional money was involved. Mostly the president did so because Struve had only one priority, research, and succeeded so well in it. Partly it was because Struve was still middle-aged, like the president himself, and had assembled a brilliant young staff, including Chandrasekhar, to whom Hutchins was especially attracted. And partly it was because Struve gave the president his complete, unquestioning support. In contrast, many of the older heads of other departments, and senior professors, clashed frequently with Hutchins. He was a brilliant leader, with strong ideas of his own and little reverence for any received opinions, and a witty speaker who could easily wound a serious, plodding specialist. Like Struve, Hutchins preferred a dictatorship (his own) to true faculty democracy in running his organization. Many senior professors were fighting to change the rules under which the university operated, basically to give the faculty more control. Struve indignantly rejected their campaign. He believed all the faculty members should recognize and support the president's efforts to get the best research workers in every field, to pay them "adequate salaries," to reduce their teaching loads, and to weed out the less successful and the elderly, unproductive placeholders. Struve said he could never have built up the Yerkes staff if he had had to get votes of approval from the faculty members who were there when he took over. A majority of its members had then "consisted of incompetent or unproductive persons. . . . [T]he [astronomy] department required a fairly thorough process of cleaning, and I do not see how any elimination of incompetent persons can be successfully carried out under the voting system." Hutchins encouraged Struve to convey his views forcefully to his senior colleagues in other departments, and the Yerkes director did so. The president appreciated it. There was never a question in Struve's mind of trading his support for a quid pro quo; the two men simply thought alike. But the result was that Hutchins started off with a receptive attitude toward Struve's proposals, quite different from the way he could look at requests from some of his older faculty critics.[75]

In the fall of 1939 Hutchins summoned Struve to Chicago to get his opinion of who should succeed Gale as dean of physical sciences when he retired the following summer. Under the normal procedures of Chicago (and most universities) at that time, only current faculty members of the division were considered. Struve listed the three main qualifications for the job as scientific eminence and promise, administrative ability, and youth. He considered scientific ability an absolute necessity, while Hutchins probably gave more weight to administrative experience and to youth. The Yerkes director named physicists Arthur H. Compton, James Franck, and Robert S. Mulliken as the best possibilities; Compton and Franck then had Nobel Prizes, and Mulliken was to receive one years later. Franck, at fifty-seven, was clearly too old; Struve considered Compton (who was forty-seven) too old also. On the other hand Mulliken, then forty-three, was young enough. (Struve himself was forty-two at the time, and no doubt this had something to do with just where he drew the line between youth and age.) Mulliken had no administrative experience or skills; Struve had found him "rather indecisive." Yet in the end he supported Mulliken as the best candidate, "everything considered." Struve's unstated reservation about Compton no doubt was that he had much wider horizons than a typical scientist, was involved in public affairs, gave speeches, and made pronouncements, while Mulliken, like the Yerkes director, was dedicated wholly and completely to scientific research.[76]

After the New Year Hutchins collected formal opinions from all the chairmen in the Physical Sciences Division. Compton was the favored candidate, but he declined the post.[77] Like Struve, he wanted more time for research, not less. Vice President Emery T. Filbey, Hutchins's right-hand man, then tried to persuade Struve that Bartky would be an acceptable dean. A teacher and administrator with little in the way of a research record, he would have been much more in the mold of Gale than Compton. Struve parted company with Hutchins on this idea. He was strongly opposed to Bartky, four years younger than himself, whom he claimed he liked "personally." But Struve had absolutely no respect for the campus astronomer, who was becoming increasingly a mathematical statistician, as a scientist. He had never done any real research in spite of much prodding by the director. Struve had objected vehemently to the salary raises Gale had given to

Figure 40  Walter Bartky, campus astronomer and later dean of physical sciences. Courtesy of the Department of Special Collections, University of Chicago Library.

Bartky and Page the previous year. The Yerkes director had compared Bartky very unfavorably with Henyey. Struve took absolutely no account of the very important teaching contributions the two campus astronomers were making. He was "humiliated" that Gale had "disregarded" his recommendation that they not be given raises and felt that "an injustice [had] been done to some of the most brilliant and loyal employees of the University (such as Henyey), whose only mistake has been to stick to their jobs instead of parading their social graces in Chicago." Gale had told Struve he was overreacting and had gone ahead with the raises.[78] But in 1940 Hutchins did not go ahead and appoint Bartky dean of the Physical Sciences Division.

Instead, Hutchins persuaded Compton to accept the deanship, with Bartky under him as dean of students in the division, in effect the assistant dean. Struve, under pressure from Filbey, felt he could not fight this appointment, but he emphasized to the vice president that it was "satisfactory" to him only "provided it is not in anybody's mind a temporary arrangement intended to pave the way for Bartky's eventually becoming dean." Filbey assured him the administration had no such intention, but in fact it was exactly what was to happen five years later.[79] Just before his retirement on June 30, 1940, Gale was honored by the faculty of the division at a luncheon on the campus; among those who attended from Yerkes were Struve, George Van Biesbroeck, Morgan, and Chandrasekhar.[80] Gale's reaction is not on record. Shortly after his retirement, Struve managed to push through raises for Kuiper and Morgan, to bring their salaries a little closer to Bartky's, and a promotion to associate professor for Chandrasekhar. (Kuiper was already at that level, but Morgan was not.) Filbey had advised the director he would hold up the promotion until just after Gale's retirement, to avoid any trouble with him. Chandrasekhar had just completed the manuscript for a new book on dynamical astronomy, which Struve described as "one of the most outstanding contributions to dynamical astronomy ever made," and he called its author an astrophysicist "who has surpassed all living astronomers in theoretical skill." He had also just received a very attractive offer to return to India as a full professor at Allahabad, and Struve was afraid he might take it.[81]

In 1940 Hutchins did go along with another of Struve's plans, probably against his better judgment. The Yerkes director, still feeling overworked administratively, followed Compton's lead and asked for an assistant director to take some of the load off his shoulders. Morgan was not really interested enough to hold the job on an official basis, Struve thought, and Kuiper was "not temperamentally suited to [an] administrative position." In both judgments he was quite right, although if he had given either of them more of his confidence, more real responsibility, and more money at an earlier stage the story might have been different. Instead he suggested hiring Bok, from Harvard, as assistant director. The Germans had invaded and occupied Denmark, and it was certain that Strömgren would not return for a long time, if ever. Hutchins agreed to Struve's proposal.[82]

Bok was then Struve's closest confidant in astronomy. Born in Europe and grown to manhood there, both of them had become United States citizens as soon as they could, but they were highly conscious that they were still considered "foreigners" by the old-line Americans who dominated astronomy in this country. Passionately devoted to science, they itched to organize research along what they considered rational lines and get rid of the superannuated "old boys" who ran the AAS as a social organization rather than as the powerful instrument for modern ideas they wanted to make it. They tirelessly discussed every possible young candidate for positions at Yerkes, and though Struve discounted Bok's fulsome praise of recent Harvard Ph.D.'s, especially those who had done their thesis research with him, he always solicited the younger man's opinion.[83] Bok was approximately the same age as Morgan, Kuiper, and Chandrasekhar, who were far less positive about him than Struve was. Bok, though a great enthusiast and a hard worker who had tremendous drive, was relatively uncreative. In his research he simply repeated the old statistical star-count methods he had learned while doing his Ph.D. thesis under P. J. van Rhijn at Groningen.[84] But Struve wanted him, and at a departmental meeting in November he told them Bok was "my preference—my reward."[85]

By then the offer was completely different from what Struve had originally intended. He had stopped in Cambridge in late October, after a meeting in Philadelphia, to talk confidentially with Bok. The Dutch-born astronomer had immediately declined to become assistant director at Yerkes under Struve; he said he could not take over the administration there when the director was away. No doubt he realized that doing so would quickly lead to friction between them. Instead Bok proposed that he be appointed as head of the astronomy group on the campus in Chicago. There he could build up the teaching side of the department, if not in astrophysics, at least in stellar astronomy rather than celestial mechanics. Bok was a forceful, outgoing person and an excellent teacher, especially in elementary courses. Struve no doubt recognized that his friend could be his agent in pushing Bartky out of the Astronomy Department. Hutchins and Compton were enthusiastic about getting a real astronomy teacher whom Struve wanted on the campus. But Bok's demands were high. He wanted essentially a separate department on campus, with a separate budget under his own control, not Struve's. He demanded his own group, doing his kind of star-count research under his direction. In it he wanted Greenstein, redirected from the astrophysics he was beginning to enjoy, as well as James Cuffey, another recent Harvard Ph.D., to be added to the Chicago contingent. Bok wanted a lighter teaching load, time to obtain data with the eighty-two-inch reflector (or to send someone from his group to get it for him), and of course a higher salary than his current $6,000 at Harvard.[86]

All this Struve was willing to give Bok. The director desperately wanted an ally and confidant, even if he was seventy-five miles away in Chicago. Struve notified Shapley that he was trying to lure Bok away from Cambridge, informed Bartky that "the best man in the world . . . who created the strength of the Harvard organization" after Shapley had failed to do so might soon be taking over the department in Chicago, and asked Compton to support the offer as strongly as he could. The Yerkes director let Bok know that Hutchins was prepared to go as high as $7,000 a year to get him. It would be simple, Struve told Bok, to "shift" Greenstein from Yerkes to Chicago, and if necessary he would get rid of Keenan to create an opening for Cuffey.[87]

Bok came to Chicago the week after Thanksgiving, gave a colloquium in the Physics Department on the campus, met Compton and Bartky, went on to Yerkes with Struve and gave another talk there, and returned to Cambridge to await the formal offer. It came in mid-December, specifying a $6,700 salary, with Greenstein, Page, and a "roving assistant" (presumably Cuffey) under his direction on the campus. But Shapley, who wanted to keep Bok, persuaded Harvard president James B. Conant to match the offer. The deal was arranged just before Christmas. Shapley took Bok to see Conant, who promised to reduce his teaching load, to relieve him of the elementary and celestial mechanics courses and let him teach a graduate course instead, and to give him all the assistants he needed for the "Milky Way bureau" he had long dreamed of directing at Cambridge. Bok stayed. He tried to explain to Struve that Russell, and Frank Schlesinger at Yale, had advised him that "the Chicago job [was] a close second . . . to the best opportunities for milky way research in the world."[88]

Struve was disappointed, but he soon came up with another reorganization plan, while Bok, though apologetic, was happily hiring a new assistant at Harvard College Observatory and planning how best to displace Annie J. Cannon without hurting her feelings as he took over the spectral classification work she had so long headed.[89]

As 1941 unfolded, the war in Europe increased in intensity. All the Yerkes faculty members and visitors, now from England and its allies only, continued their highly productive research, concentrating on it as well as they could as the world thundered. Struve was arranging a symposium, to be held in conjunction with a meeting of the AAS at Yerkes, marking the fiftieth anniversary of the founding of the University of Chicago. The university celebration was called "New Frontiers of Learning," and the administration encouraged each department to organize a symposium as part of it during the 1941–42 academic year. Early in 1940 Struve had decided on "The Atmospheres of the Stars and the Interpretation of Stellar Spectra" as a topic. He planned that he, Morgan, Kuiper, Chandrasekhar, and

Swings would all give papers in it, as well as Strömgren, if the war should be over by September 1941 and he could get out of Copenhagen. There was little chance of that, but they were all recognized leaders in their fields. Yerkes under Struve had arrived as a powerful research force, and he knew it. He picked the week just after Labor Day, when College Camp would be available for the visitors, and arranged with Stebbins, then president of the AAS, to have the meeting there.[90]

Struve divided up the $500 he was allotted by the university for the symposium well; with amounts like $50 or $85 for traveling expenses he was able to bring some of the foremost astrophysicists in America as invited speakers.[91] The AAS meeting was the first held at Yerkes Observatory since 1922 and the best attended in the history of the society up to that time, with about 140 members present and some fifty guests (spouses and graduate students who were not yet members). Sunday, September 7, was devoted to a conference of teachers of astronomy, the council meeting in the afternoon, and an evening reception at the observatory. In the next two days forty-seven papers were presented, forty-one orally and the rest "by title only" (if the author had not come to the meeting nor arranged for a substitute to read it). There were no parallel sessions, and this was the first meeting at which the presiding officer (Stebbins) used a timer with a bell to cut off lengthy presentations. Five inches of rain drenched the meeting, forcing cancellation of a "chuck wagon" (outdoor picnic) supper, a boat ride on the lake, and an observing session with the forty-inch refractor. The society dinner, marking the end of the AAS part of the meeting, was held in the College Camp dining room. Mary H. Frost, the former director's widow, was present and gave a little talk, as did two of the visiting astronomers. Struve had made a special effort to invite all the former Yerkes graduate students and alumni to come to this meeting as a reunion; many of them did, and they all knew her well.[92]

The next three days, Wednesday through Friday, were given over to the symposium. Everyone was invited to remain for it, and many did, particularly the astrophysicists and spectroscopists. Struve had drawn up a taxing schedule consisting entirely of invited papers, two or three each morning, another two each afternoon, and two more each evening. Plenty of time was left for full discussion after each paper. The Yerkes director had paid careful attention to wide geographical representation; among the invited speakers were Russell and Wildt from Princeton; Martin Schwarzschild from Columbia; Donald H. Menzel from Harvard; Robley C. Williams and Leo Goldberg from

Michigan; Paul W. Merrill and Rudolph Minkowski from Mount Wilson; and Arthur B. Wyse from Lick; in addition to Struve, Morgan, Kuiper, and Swings from his own staff. Van Biesbroeck, Chandrasekhar, Greenstein, and Page had all given contributed papers in the earlier AAS sessions.[93]

At the special University of Chicago semicentennial convocation held on the campus in September after the meeting, Russell received an honorary doctor's degree as the outstanding astrophysicist of the time. Each department had been allowed to recommend one candidate; Struve had named him and had lobbied hard to have Adams, the former Yerkes graduate student who had succeeded so well, receive another.[94] That was not to be, however, for another four years, at the end of World War II. In December 1941, three months to the day after the AAS meeting had begun at Williams Bay, Japanese bombers launched their surprise attack on Pearl Harbor. The United States was in the war. Everything changed at Yerkes Observatory, as it did in America and in the world.

CHAPTER TEN

# World War II, 1939–1945

World War II changed the lives of all Americans. Those at Yerkes Observatory were no exception. The war put its scientific research activities on hold, but after it ended they came back stronger and more productively than ever. Otto Struve, the Yerkes director, emerged shaken and depressed, however, never again to feel the same deep joy and excitement he had felt in building a research organization in the 1930s, while the team of brilliant young scientists he had assembled achieved recognition on their own after the war and soon took over the leadership from him.

Struve, with his roots deep in Europe—in Germany as well as Russia—felt "utterly hopeless" when the two countries signed their "nonaggression pact," with its secret agreement to divide Poland, which German troops invaded on September 1, 1939. He considered Adolf Hitler and Joseph Stalin both "lying dictators," and unlike most Americans, he never changed his mind about either one of them. Their "pernicious ideas" threatened "everything good and decent," and he knew it would take "a bloody struggle," in which astronomy would play little part, to defeat them.[1]

The eighty-two-inch reflector had gone into operation early in 1939, and Struve and Pol Swings were observing with it a few days after the war broke out in Europe. Their "run" had been scheduled long in advance, and for Struve using the telescope productively to obtain data on stars was the highest priority there was.[2] Almost equally important was the problem of obtaining the funds necessary to keep two large research observatories in operation. The McDonald bequest had paid to build the observatory on Mount Locke and had provided the money for Struve's trips to Texas while it was being constructed, but once it was "substantially completed" the income from the endowment that could be used for operating expenses was limited to $10,000 a year. This was enough to pay Christian T. Elvey's salary as assistant director at McDonald and to hire one assistant and a single maintenance employee, together with publication costs for the scientific papers and a little travel, but it was not nearly enough to operate the second largest telescope in the world effectively at a distant site. The University of Chicago was hard pressed for money for all its many departments. It was already putting more money into McDonald Observatory than the University of Texas was. President Robert M. Hutchins, like many administrators after him, advised Struve to look to the big foundations for funds rather than to his own institution. Struve knew that solution would not work; the Rockefeller Foundation was committed to building the two-hundred-inch reflector on Palomar Mountain, and the Carnegie Institution of Washington to operating Mount Wilson Observatory, but he tried loyally. He never succeeded. Neither did Vice President Frederic Woodward's attempts to raise money from wealthy Lake Geneva residents. McDonald Observatory continued to operate on a shoestring.[3]

One possibility Struve considered very seriously was closing Yerkes Observatory and moving the forty-inch refractor to McDonald Observatory, with its clear skies and good seeing, where the telescope would be much more effective. In various versions of this plan, Struve visualized transferring part or all of the staff to Chicago, where they could live, teach, do their research, and commute to Texas to observe with the telescope. Those not in Chicago (all, in one version of the plan) would move to McDonald Observatory. Few of the Yerkes faculty members except Gerard P. Kuiper were enthusiastic about leaving Williams Bay when Struve revealed his plans to them at a staff meeting the day before Hitler's tanks began rolling into Poland. As the director himself realized, the "human element" militated against the move. So did the cost; each version of his plan, including the alternative of modernizing Yerkes Observatory and staying there, had an estimated cost between $250,000 and $300,000. Of this amount $50,000 to $100,000 represented an endowment, which was to provide the income to attract a "[man] of outstanding ability" to replace Bengt Strömgren, whose return to Denmark had "weakened us considerably." Another large sum was to build a large Schmidt telescope, the recently invented wide-field camera system that would be extremely useful for surveying the sky. Struve's plans were quite unre-

Figure 41   Yerkes Observatory (ca. 1942). Courtesy of Yerkes Observatory.

alistic; to solve the problem of raising funds to operate two observatories, he envisioned spending more money to enlarge one or the other of them.[4]

Like all astronomers, Struve was convinced of the great importance of his subject. He thought nonastronomers shared it, and on one occasion he advised Vice President Emery T. Filbey that perhaps the rich Lake Geneva residents would get serious about raising endowment funds for the university if they knew that the university administration was thinking of closing Yerkes Observatory and moving the telescope to Texas. Whether as a result of this advice or inadvertently, Hutchins let slip the news at a public appearance in April 1940. It created a brief sensation in the Chicago and southern Wisconsin newspapers, consternation in the staff members, employees, and their families, and not the slightest flicker of interest in the lakeshore property owners. They blandly continued to advise Struve and senior astronomer George Van Biesbroeck that John D. Rockefeller Jr. should provide the money and that they would try to introduce the astronomers to him.[5]

Struve had another plan, however, which did not succeed then either, but that much later became the basis of the national observatory system of today. His idea was to build up McDonald Observatory, with the large Schmidt telescope and possibly others, and then to operate it as a "cooperative" observatory in collaboration with several other universities. In this way the investment and fixed costs at the site, as well as the astronomers' traveling expenses, could be shared. Only by some such plan, Struve believed, could astronomers from midwestern and eastern universities have the chance to do research with a large telescope at a good, clear-weather site. Otherwise the Pasadena astronomers with their one-hundred-inch telescope on Mount Wilson, and the two-hundred-inch on Palomar, then expected to go into operation very soon, would monopolize observational astronomy. Struve did not want to see that happen. Instead his proposal called for shared observing facilities at McDonald Observatory, with an additional telescope paid for by a large foundation, and operating expenses paid by the universities who used it, in proportion to the time their astronomers spent observing with the telescopes.[6] He traveled to New York and tried to sell his plan to Warren Weaver at the Rockefeller Foundation and to Frederic Keppel at the Carnegie Corporation; both expressed polite interest but provided no funds. They were supporting Palomar and Mount Wilson, respectively, and saw no need for another big telescope elsewhere.[7] Struve tried to interest Harlow Shapley in joining the plan, but the Harvard director felt his institution's needs for instrumental equipment were "well taken care of," a proposition that every one of his staff members would undoubtedly have disputed.[8] The University of Texas was unwilling to provide any more funds, and Hutchins made it clear that Chicago would not do so either.[9]

The Yerkes director published an article in *Scientific Monthly* to spread his idea and drew up individualized plans for Yale, Michigan, and Virginia. They all had observatories of their own with small telescopes in poor climates, but they still had dreams of better ones. Not one of them would give them up to shift its funds into an existing large telescope at a good site that it would not control.[10] By the spring of 1940 it was obvious that Struve's cooperative observatory plan was dead. Only Frank K. Edmondson at Indiana and Willem J. Luyten at Minnesota were interested enough to try to get money from their institutions. Edmondson succeeded, and in 1941 Indiana University allotted $1,275 to pay for fifteen nights' observing with the eighty-two-inch reflector, to be used for his radial-velocity program.[11] That was the only fruit of all Struve's efforts then, but a decade after World War II the national observatories inspired by this early idea—Kitt Peak and Cerro Tololo—came into being, funded almost entirely by the new National Science Foundation.

In 1941 Struve's next plan was to set up a theoretical astrophysics institute on the campus, built around Subrahmanyan Chandrasekhar, who would move there. With Dean Henry G. Gale by now retired, the main impediment was gone from the administration, and the Indian theorist was initially enthusiastic about the plan. Bart J. Bok had just declined the position on the campus Struve had worked so hard for, and the director knew that Hutchins would be glad to bring Chandrasekhar there in his place. Paul Ledoux, a young Belgian theorist who had escaped from Nazi-dominated Europe via Sweden, Russia, Siberia, and a long sea voyage to America, would work with him on the campus.[12] However, later that spring Chandrasekhar decided that he would prefer to remain in Williams Bay rather than chance Chicago. He had just received a formal job offer from Allahabad University for the same professorship he had informally declined "for the time being" the previous year. This time he considered it very seriously; it was "the best chance I could ever get (now or in the future) in India." But he decided on Yerkes Observatory, "one of the most active centres of astronomical research in the world." Struve, who thought Chandrasekhar was "the only person" in America who could build up astrophysics, with Europe now essentially out of the picture because of the war, could hardly press him to move to Chicago.[13] Instead Struve had Gunnar Randers, a young Norwegian theoretical astrophysicist who had also escaped the Germans, appointed to a temporary teaching position on the campus, and Ledoux joined Chandrasekhar at Yerkes.[14]

Struve, always determined to have a strong research presence on the Chicago campus, next envisioned moving Kuiper there. The Dutch observer (an American citizen since 1937) blew hot and cold on this idea. He considered it almost entirely in terms of his own research and proposed making the move in a year and a half, when he hoped to have completed his observing program on nearby stars, before starting another one. In one version of the plan Chandrasekhar would be in Chicago with him; Kuiper favored this idea. Struve tried to prod him into making a decision, but on December 7, 1941, the Japanese attacked Pearl Harbor, America entered the war, and all plans for the future, including this one, suddenly changed.[15]

Struve's deepest thoughts and feelings could not change, however. Most Americans considered the Soviet Union a welcome ally in the struggle against Hitlerism, but to the former imperial Russian and White army officer, whose brother had died fighting the Bolsheviks and who had himself been driven out of his native land, it was not so simple. He sympathized with the Russian people and with individual Russian astronomers who stayed out of Communist anti-intellectual and antiforeign campaigns, but he hated their leaders, especially Stalin, whom he considered a criminal. Aid to Russia, if it was controlled by the leaders of the Soviet Union or their American sympathizers, was "political[ly] bias[ed]" in his mind. A loyal American citizen since 1927, Struve did his best to suppress these feelings but could not; the effort kept him tense and nervous all during the war. In July 1941, just days after the Soviets switched to the Allied side, Struve stated for the first time that he was tired and ready to give up the Yerkes directorship, a refrain he was to echo more and more frequently in the coming years. His first thought was to press America's new allies for information on the fate of astrophysicist Boris P. Gerasimovich, who had disappeared in 1936, denounced as an enemy of the people. He had been shot in 1937, as Struve learned for certain only years later. On the other hand, the Yerkes director was sympathetic and worried about the fate of Pulkovo Observatory, almost completely destroyed by the Germans in the siege of Leningrad.[16]

Immediately after Pearl Harbor, Struve's orders to Elvey were that only absolutely essential tasks should be continued at McDonald Observatory. Observing with the telescope came first. Other "essential" tasks were to complete an aluminizing chamber for the eighty-two-inch mirror, to finish the twenty-inch Schmidt telescope under construction at McDonald (this was never done, initially because John Mellish, the optician who had contracted to make the delicate glass corrector plate, fled Illinois and was eventually jailed in California for nonsupport of his ex-wife), to continue and extend Elvey's own observing program on the light of the night sky, and to safeguard the telescope, the dome, and its heavy electrical motors, all irreplaceable in wartime. The scientific work should not be disrupted, but in case of an air attack (McDonald was one of the most conspicuous landmarks in the mountains of west Texas) the eighty-two-inch mirror should be removed from the telescope and stored in the steel aluminizing chamber, which would protect it against bomb fragments though not direct hits. Walter S. Adams had taken very similar steps at Mount Wilson; both directors agreed that science would continue up to the moment of attack, which they did not expect but for which they believed they must prepare.[17]

A brief, welcome respite from tension came in the dedication of a new Mexican observatory at Tonantzintla, near Puebla, some seventy miles south of Mexico City. Its main instrument was a twenty-four-inch Schmidt telescope for wide-field photography and objective-prism spectral classification. Shapley had played a major role in advising the Mexican scientists and government officials on the project, and Bok was also heavily involved in it. The dedication, planned before Pearl Harbor, brought many top American astronomers to Tonantzintla for a symposium, with all expenses paid by their hosts. Struve, Elvey, and William W. Morgan made up the Yerkes contingent; Henry Norris Russell was also there from Princeton, Adams from Mount Wilson Observatory, and young Nicholas U. Mayall from Lick. Struve's paper was based on his spectroscopic measurements of stellar rotational velocities and his speculations on how those rotations might be linked to the formation of planetary systems and double stars, a subject he was to return to time after time for the rest of his life. Morgan's paper on spectral classification was especially stimulating to the Tonantzintla astronomers, who were soon to begin using their new telescope in this field.[18] At the dedication Struve met young Guido Münch. a brilliant student of mathematics and astronomy at the National Autonomous University in Mexico City. As a direct result, Struve later hired Münch as an assistant at McDonald Observatory in the spring of 1943; he went on to Yerkes that fall and earned his Ph.D. in 1946.[19]

In the summer of 1942 Struve and Robert S. Mulliken, of the Chicago Physics Department, arranged another joint symposium on spectroscopy on the campus. To Struve such meetings were especially important in getting physicists interested in astrophysical problems and keeping the astronomers up to date on the latest laboratory and theoretical results in atomic and molecular spectroscopy. The speakers included Swings and Elvey, Hans Beutler of the campus Physics Department, and several other laboratory spectroscopists, including once again Gerhard Herzberg, a refugee from Hitlerism, who was then working at the University of Saskatchewan. He had recently identified the $CH^+$ molecule in interstellar matter, a perfect example of Struve's ideal.[20] In an after-dinner speech at the close of this symposium, the Yerkes director gave his somber views on astronomy in wartime. Physics would make an important contribution to the war effort, he said; astronomy would not. Undoubtedly he was aware that Enrico Fermi and a group of young physicists he headed had moved to Chicago in April to work, with their local colleagues and graduate students, on a hush-hush project in Ryerson Physical Laboratory and under the west stands of Stagg Field, formerly the home of the

Maroons football team. Struve probably did not then know that the project was the "atomic" (fission) bomb, and that less than six months after the joint spectroscopy conference Fermi and his team would complete enough of their uranium-graphite "pile" to produce the first artificial, self-sustaining nuclear chain reaction. Many astronomers were "fairly practical physicists," Struve said (Bok was later to call them "the best damn second-rate physicists you can find"), and most of them were doing their part in the war effort. Then Struve called for more effective use of large telescopes, national planning of astronomical research (giving Frost's radial velocity program, which he had worked on, as a giant example of the failure of unplanned research), and extolling the advantages of his hoped-for cooperative observatory.[21]

Ledoux, Randers, and Swings were marooned by the war. Many of the younger Yerkes and Chicago faculty members, beginning with Elvey, Horace W. Babcock, and Thornton L. Page, were soon leaving for wartime technical development projects. Struve knew that they were patriotically serving their country in essential tasks, but his single-minded devotion to astronomical research made him grumble and question their motives.[22]

One of Struve's first responses to the war was to organize special courses in navigation at Williams Bay and in downtown Chicago, aimed at prospective aviators and sailors, would-be merchant marine or naval officers, and anyone else who hoped to serve somewhere. He also stepped up offering a navigation course for regular students on the campus. Van Biesbroeck, Kuiper, Louis G. Henyey, Jesse L. Greenstein, and John Titus, another young Yerkes instructor, all participated in teaching these courses. However, they were more a remnant of World War I thinking than a vital need for World War II, in which sophisticated electronic navigation systems were coming into increased use. The astronomers' courses withered after the first few years of the war, as the navy and air force developed their own highly specialized navigators' training programs.

To preserve his organization, Struve knew he needed a wartime weapons project of his own at Yerkes Observatory. Research scientists, graduate students, and machinists in the instrument shop wanted to serve their country, to get and keep draft deferments, and to earn more money. They were all being besieged by offers from defense contractors and from other research centers that had already converted to development work; Harvard College Observatory had an optical research laboratory hard at work designing and producing

Figure 42    Otto Struve at his desk in Yerkes Observatory director's office
(1946). Courtesy of Yerkes Observatory.

aerial photography lenses for the air force, and Mount Wilson had
many ordnance and air force development contracts. In the summer
of 1940, with the Battle of Britain raging and America beginning to
tool up for war, Hutchins had appointed a faculty committee on
"university work in national defense," with Arthur H. Compton, the
new dean of physical sciences, as one of its members. Compton had
canvassed his department heads for suggestions, and Struve had
replied that Yerkes faculty members, with their skills and experience
in optics, could design Schmidt cameras for aerial photography, applying the newest technology of the time to warfare.[23] Nothing came
of his suggestion. A year and a half later, just a few days after Pearl
Harbor, Struve offered the services of Yerkes Observatory to Vannevar Bush, president of the Carnegie Institution of Washington and
since 1940 the head of American science mobilized for war. In his letter Struve barely mentioned optical design, and Bush's Office of Scientific Research and Development (OSRD) minions merely filed the
letter without replying to it. Struve pushed ahead through the university, independent of Washington, and by that summer the Yerkes Optical Bureau was in existence, supplying designs and test data for new

optical systems directly to the Ajax Manufacturing Corporation, a
military supplier in Chicago. For the first time in its history, part of
the Yerkes Observatory building was marked off as a security area,
and Struve warned the staff that work "of a highly secret character"
would be going on. Everyone working on the project was to lock up
all letters and computations at night, and everyone on the staff should
be watchful for "unauthorized persons" who might seek to "gain . . .
access to this material." Henyey and Greenstein became the main optical designers, aided at first by Van Biesbroeck and Daniel M. Popper. In 1943 Struve succeeded in getting the OSRD to take over direct
support of the Optical Bureau from Ajax.[24]

In 1943 Popper left Yerkes on leave to work "for the duration" on
the Manhattan Project in Berkeley, while Van Biesbroeck spent the
major part of his time in astronomy, helping Struve and Morgan keep
Yerkes and McDonald active in research throughout the war. Henyey,
a brilliant applied mathematician, and Greenstein, a hardworking
practical theorist and a highly effective expediter, continued as full-
time optical designers. Fred Pearson, who had been Albert A. Michelson's optical technician on the campus, moved to Yerkes to make the
prototype lenses and mirrors for the gunsights, periscopes, tank telescopes, and Schmidt cameras they designed. He was a valuable asset;
when Struve had asked Shapley for help in getting support for the
wartime optical development work at Yerkes, the Harvard director
had responded that his own organization could take care of it all
without help from Yerkes and suggested that Richard S. Perkin would
be glad to hire Pearson for his rapidly growing Perkin-Elmer optical
firm. Then Compton had taken Pearson back to the campus for some
months and released him to Yerkes again only in a trade for part of
Henyey's time, so that he could come to Chicago periodically to supervise the Manhattan Project calculators (humans working with
electromechanical calculating machines) there. Earlier Shapley had
recommended Greenstein for another wartime project in New York
City. Struve had to fight hard to keep his team together and was being
worn down by the struggle.[25]

Chandrasekhar had been on leave at the Institute for Advanced
Study in Princeton for the fall of 1941, working on highly mathematical problems of the statistics of the fluctuating force fields in situations where many "particles" or "bodies" contributed to them. One
application was to the electrical force on an atom in a stellar atmosphere, another was to the gravitational force on a star in a cluster or
galaxy. He collaborated with the brilliant John von Neumann, a pro-

fessor at the Institute, on part of this work. Chandrasekhar's research was attracting wide attention, and Struve knew that his star theorist would certainly receive many attractive offers once the war ended. The Yerkes director carefully wrote long, newsy letters to him, holding out a vision of the future of astrophysical research at the University of Chicago. When Chandrasekhar finished the manuscript of his book on stellar dynamics, published as a monograph in the astrophysical series, Struve objected violently to the idea that the Press should have it reviewed by any outside astronomer or physicist. Chandrasekhar did not believe that more than one or two scientists "outside our own group" would be competent to do it, and the Yerkes director simply stated, very forcefully, that since he himself was the editor of the series and he knew the manuscript was excellent, the Press should go ahead and publish it. It did.[26]

Chandrasekhar learned the news of the Pearl Harbor attack at Princeton, shortly before his scheduled return to Williams Bay. Before long, through Von Neumann, he received an offer to become a civilian consultant at the Ballistics Research Laboratory of the Army Ordnance Department at Aberdeen, Maryland. Many American astronomers and astrophysicists were there, some in uniform and others as civilian scientists. Edwin Hubble was head of the exterior ballistics division, and young astrophysicist Martin Schwarzschild, then an enlisted man, was at Aberdeen for a time before going on to Europe as an officer. Chandrasekhar, who worked on theoretical problems connected with shock waves in the explosion of a shell or bomb and in the atmosphere, had a half-time appointment, alternating three weeks at Aberdeen with three weeks in Williams Bay. As a native-born Indian he was not then eligible to become a United States citizen, but in the struggle against Nazi Germany his heart belonged to America. During the war Chandrasekhar, by now a recognized world leader in his field, was promoted to associate professor in 1942 and full professor in 1943, and he was elected a member of the Royal Society of London in 1944.[27]

Morgan, with two young children and little training in physics or in the highly precise astrometric measurements that were so applicable to tracing the trajectories of artillery shells or rockets, remained at Yerkes doing astronomical research throughout the war. His work in spectral classification came to fruition in the publication of his monograph *An Atlas of Stellar Spectra, with an Outline of Spectral Classification,* with Philip C. Keenan and Edith Kellman, Morgan's assistant, as his coauthors. All the main ideas and most of the work were his own, but he was loyal to his collaborators. He wrote the text for the forty-page booklet that accompanied the fifty-five sheets of illustrations of spectra, which he carefully labeled and marked to show the key identification criteria for each spectral subtype and luminosity class. Alfred H. Joy, longtime Mount Wilson stellar spectroscopist, reviewed the monograph very favorably in the *Astrophysical Journal*, and Morgan always believed this review had earned him his promotion to associate professor in 1943. He was the local product who stayed on the Yerkes faculty and had not received any prestigious job offers from outside, the one sure path to a glowing recommendation for a salary raise or promotion from Struve. Keenan was gone from Yerkes by the time the MKK spectral atlas appeared in print, on leave to work with the Naval Bureau of Ordnance for the duration, and Kellman left the observatory soon after it came out, to become a science teacher at the Williams Bay High School, the job she had trained for.[28]

Struve himself, in the time he had left over from running two observatories, pushing his cooperative observatory idea, organizing and defending his optical design center, editing the *Astrophysical Journal*, and struggling to keep it alive as the number of overseas subscriptions (and payments for them) dropped to zero, devoted himself to research. He was amazingly productive with the new McDonald Observatory reflector, particularly in collaboration with Swings. They obtained spectra of peculiar stars, stars with extended atmospheres, binary stars, shell stars, planetary nebulae, and even a comet. Struve did most of the observing, although his collaborator did his share too; they both measured the spectrograms, the director's computers reduced the measurements to wavelength or velocity units, and Swings did most of the astrophysical analysis that went into their papers. Not content with all the low- and medium-dispersion spectrograms he obtained at McDonald Observatory, and disappointed by the relative slowness of its coudé spectrograph, Struve borrowed a series of spectrograms of $\alpha^2$ Canum Venaticorum that had been taken by Olin C. Wilson at Mount Wilson Observatory. It is a peculiar A star, familiar to Struve from his early work on the Yerkes radial-velocity program. This star's peculiarity is that is has a very rich spectrum of rare-earth ions, and he and Swings had a field day measuring and identifying thousands of lines in these high-resolution spectrograms. They published the results in a 139-page article in the *Astrophysical Journal* in 1943.[29]

Swings was a tower of strength, Struve's closest confidant, collaborator, and constant aide. The director wanted desperately to keep the eighty-two-inch reflector in constant use, and Swings spent months at McDonald obtaining research data. In one period he observed for thirty nights in a row, a few interrupted for one or two hours by clouds, but not a single one bad enough that he could get a full night's sleep. During the days he would develop the spectrograms, inspect them carefully, measure some of them, and write papers based on his measurements. In his spare time he acted as Struve's eyes and ears on Mount Locke. Swings gently poked fun at a young graduate student who came with him to help observe and learn by doing: "Evidently the boy has to sleep in day-time and is not able to take part in the work of measurement, reductions and discussions." Yet the Belgian visitor never became exhausted and lashed out at others as Struve himself was increasingly prone to do, particularly at Elvey. Struve would not give the McDonald Observatory assistant director the independence he needed but continued to micromanage and second-guess his subordinate's decisions.[30]

Between observing, writing papers, and peacemaking, Swings traveled around wartime America by car in 1942, when gasoline was still available, giving lectures at a free Franco-Belgian university in New York City, visiting Harvard, Yale, and Princeton, and taking part in an important spectroscopy conference at Ohio State University. He was widely regarded as one of the outstanding astrophysicists working in America during the war.[31] Swings had come to Yerkes as a visiting scientist, supported by a Belgian fellowship and a stipend from Chicago. Struve had no difficulty having him appointed as a visiting associate professor at a salary of $4,500 for the 1940–41 academic year, after Belgium had been overrun by the Germans. Swings was horrified by hearing the invasion news nightly on the radio as he observed at McDonald in the spring of 1940, but he kept steadily at work, producing paper after paper.[32] Then in the fall of 1942 he was appointed a research fellow for three months at Lick Observatory, where he lectured, wrote more papers, and strengthened his contacts with the California astronomers. He had visited Pasadena and Mount Wilson Observatory on several previous occasions, before or after sessions at McDonald. Swings was at Mount Hamilton when he heard the news of the Japanese bombing of Pearl Harbor.[33] Struve wangled a promotion to visiting professor for him for 1942–43, but no increase in salary came with it, because of "the terrible strain of the war upon the university's finances," according to the administration.[34] Swings was not to hold the position long. With his homeland occupied by the Germans, he wanted to get into a wartime job, and by 1943 it was possible for foreign nationals to get clearance to work as civilians on such projects. Swings found a position at the Ray Control Company in Pasadena, designing and testing the analytic spectrographs it produced for synthetic rubber and gasoline plants. There he was making his contribution to the war effort in a balmy climate that he and his wife enjoyed much more than wintry Wisconsin or arid west Texas, and close enough to the Mount Wilson Observatory offices for him to work on his favorite stars and spectra on occasional evenings, Sundays, and holidays. Struve was deeply depressed by his collaborator's departure, although intellectually he knew that Swings had done the right thing. Probably the Belgian astrophysicist would never return to Yerkes and McDonald, probably it was better not to continue any collaboration too long, the director rationalized. He would soldier on at the observatory as best he could until everyone else had left for war work, he moodily wrote, then turn it over to a few secretaries and find a new job himself.[35]

Kuiper also left Williams Bay for war service in 1943. Until then he had taught a wartime navigation class, but he could see that it was rapidly becoming redundant as radio and radar position-finding systems came into operation. That summer he went on leave to the Harvard Radio Research Laboratory, headed by Stanford electrical engineer Frederick E. Terman. Its main task was developing radar countermeasures. At Cambridge Kuiper, like Swings in Pasadena, could keep in touch with astronomy, if only by attending colloquia and informal discussions at Harvard College Observatory. In addition, he arranged for one or two months' leave each year from the wartime laboratory, to return to McDonald Observatory, paid by the University of Chicago, and work with the eighty-two-inch reflector.[36]

In fact he did so once only, over Christmas 1943 and New Year's 1944, when he took every clear night with the telescope for more than a month. Like Struve, Kuiper was a compulsive observer. Although he spent most of this observing time finishing up his program of spectroscopic investigation of the dwarf stars near the sun, he also made his first foray into the planetary astronomy that was to become his lifework after World War II. During the long winter nights at Mount Locke, Kuiper obtained spectra of the largest satellites in the solar system (ten in all), specifically looking for signs of possible atmospheres on any of them. On Titan, the giant moon of Saturn, he did detect the presence of methane gas, revealed by an absorption band in the red spectral region. Methane was already known to be a component of the atmospheres of Jupiter, Saturn, Uranus, and Neptune, but this was the first discovery of an atmosphere on any satellite. Although Kuiper stated to an older friend that he had taken the spectra only because all the planets with large satellites happened to be in the nighttime sky that winter and he had extra observing time at his disposal, certainly his mind had been prepared to seize this opportunity by the spectroscopic symposium at Chicago in June 1942, at which Rupert Wildt had given his survey paper on the atmospheres of the planets. Earlier Kuiper had twice taken similar spectra of Pluto when it had "happened to be in the sky" without success, as well as long-slit spectra of Jupiter and Saturn, to see if there were differences in their atmospheres between pole and equator. Fortune favors the hardworking who are well prepared and persistent.[37]

In 1944 he went overseas as a civilian scientist attached to the Eighth Air Force, based in England and bombing Germany and the German-occupied parts of the Continent. With his fluent Dutch, German, French, and English, Kuiper was deeply involved in intelligence activities, probably including contacts with the underground resistance in Holland. Most of his work was reading and analyzing captured or intercepted reports and documents and debriefing French and Belgian scientists who were liberated as the Allied armies swept across their countries. He met his two brothers, who had made their way to Eindhoven, near the Belgian border. One of them had been "underground" for more than a year, hiding from the Germans. In 1945 Kuiper joined the ALSOS scientific mission, headed by Samuel Goudsmit, traveling close behind the lines, reading and digesting captured technical reports and interrogating German scientists. As the war in Europe ended in the late spring and early summer of 1945, Kuiper rescued the elderly physicist Max Planck, who was in the eastern zone of Germany, to be occupied by Soviet forces, and brought him back to the western zone and on to Göttingen. A few months later Kuiper was able to have astronomers Jan H. Oort and Marcel Minnaert (whom he had recommended in 1935 for a faculty position at Chicago), as well as one of his own brothers, sent from liberated but devastated Holland to England for a few months of recuperation from wartime privations.[38]

Page also played an important role in the war. After a short period as a civilian scientist at the Naval Ordnance Laboratory, he had returned briefly to the campus and completed a paper on the spectra of planetary nebulae, based on his observational data from McDonald. Then he received a commission in the navy and served at sea in the Pacific. He became one of the early practitioners of operations research, attempting to apply scientific methods and quantitative reasoning to making real-time tactical decisions.[39]

Compton, the new dean of physical scientists, was one of the top leaders of the atomic bomb project. In early 1942 he had organized the Metallurgical Laboratory, the cover name for the fission research center he headed in Ryerson Laboratory on the campus. In addition to directing it, Compton made frequent trips to Project X, the huge gaseous-diffusion plant for separating uranium isotopes at Oak Ridge, Tennessee. He had no time at all for the ongoing research and teaching in the physical sciences, including astronomy. Struve was severely critical of Compton, believing he was incapable of organizing his time effectively. In this the astronomer was no doubt wrong; the dean had far more to do than any single person could handle, and winning the war had to come first. The obvious solution, although only a partial one, was to name an associate dean who could act for Compton on all regular university duties. The natural candidate was Walter Bartky, who as assistant dean in charge of students was already effectively Compton's deputy. Yet this was just the step Struve had foreseen, feared, and warned against. Now he objected strongly to it. The Yerkes director did not respect Bartky as a research scientist; hence he discounted all the campus astronomer's very real teaching and administrative talents. Instead of Bartky, Struve once again recommended his friend Mulliken for the post. An outstanding research physicist, Mulliken had been Struve's candidate for dean in 1940, and Hutchins had not chosen him then; there was little chance he would name him associate dean now, and he did not. To Struve the dean had to be one of the top research scientists in the university; he had accepted Compton on this basis but he could not accept Bartky. Hutchins, from previous experience with Mulliken, believed he would not be an effective administrator, but was sure that Bartky would. His decision was clear. Struve had no choice but to work under his former subordinate in the dean's office, but he did not like it.[40]

One new faculty member joined the Yerkes staff during the war, W. Albert Hiltner, who came on a postdoctoral fellowship in 1942 and stayed as an instructor in 1944. Hiltner had been a graduate student in astronomy at the University of Michigan, where he did his Ph.D. thesis on quantitative intensity measurements of stellar spectrograms. At Yerkes Hiltner quickly demonstrated his skills in experimental physics techniques, particularly the new vacuum-aluminization process for coating mirrors, far superior to the older chemical silvering method. They made him useful in the optical bureau, and Struve counted on him for postwar instrumental help.[41]

The Yerkes Observatory graduate teaching program that Struve had been so eager to set up, with a formal course sequence including theoretical astrophysics, began to bear fruit. Edwin G. Ebbinghausen and Gordon W. Wares earned their Ph.D. degrees in 1940, followed by John A. O'Keefe in 1941, Louis R. Henrich and Wasley S. Krogdahl in 1942, William P. Bidelman and Ralph S. Williamson in 1943, and Margaret Kiess Krogdahl in 1944. Only Ebbinghausen had done his thesis under Struve's direct supervision; Wares, Henrich, the two Krogdahls, and Williamson had all worked with Chandrasekhar, and O'Keefe and Bidelman with Morgan. Nearly all of them went straight to wartime weapons development posts or teaching jobs, but more than one became important figures in American astronomy in their subsequent careers. The official United States "good neighbor policy" brought several young Latin American astronomers to Yerkes, who learned while helping Struve in his highest-priority aim, to keep the observatory producing research results. Jorge Sahade and Carlos U. Cesco from Argentina observed with him at McDonald and published a joint paper with him on the spectroscopic binary BD Virginis, while Mario Schönberg, from Brazil, worked with Chandrasekhar on the theory of stars with isothermal cores, a very early step in the quantitative study of stellar evolution.[42]

Finally, in the spring of 1945, with the Germans at last defeated and victory over Japan in sight, America began to relax. Adams, the outstanding Yerkes product who had left for Mount Wilson with George Ellery Hale in 1903, wrote that he would be able to come to Chicago to accept the honorary doctor's degree for which Struve had recommended him, and which Hutchins had promised to award as soon as he could attend a convocation. In June the Mount Wilson director, who had stayed on the job past the normal retirement age because of the war, made the trip east and received his Sc.D. diploma. Struve pumped him for information on large plane gratings, for a new coudé spectrograph he hoped to have built for the McDonald eighty-two-inch telescope, to replace the unsuccessful prism instrument from which he had earlier hoped for so much but gotten so little. Adams met with the physicists on the campus and also accompanied Struve

Figure 43   Walter S.
Adams. Courtesy
of the National
Academy of Sciences.

to Williams Bay, where he gave a colloquium and met the astronomers' wives at a tea. He was now highly complimentary about everything at Yerkes.[43]

Two months later, just days after American B-29 bombers had
dropped the two atomic bombs that brought about the final defeat
and surrender of Japan, Russell assured Struve that his Yerkes faculty
was "by far the best group of astronomers anywhere in the world."
And as the director proudly added in reporting these words to President Hutchins, the eminent Princeton astrophysicist "did not made
an exception in the case of Harvard." Yet at the very moment Struve
had reached his goal of making the University of Chicago the preemi-
nent astronomy power on the globe—better even than Shapley's Harvard College Observatory, which once had seemed so great and so
self-proud to the young Russian recently arrived in the New World—
he was deeply depressed and could not savor his triumph.[44] In April
Dean Compton had announced that he had accepted the position of

chancellor at Washington University in St. Louis and would soon be leaving Chicago. Bartky succeeded him immediately as acting dean and was the natural candidate for permanent appointment to the post. All Struve's fears had come true. He did not for the moment accept the idea that his former subordinate in the Astronomy Department on the campus should now become his superior in the academic chain of command, between him and Hutchins. Struve dashed off a letter to the president expressing his sorrow at Compton's departure, together with a strong plea for an outstanding scientist and administrator as the next dean, which to him would eliminate Bartky from consideration.[45]

Hutchins had surely decided long before that Bartky would succeed Compton whenever the ambitious, distinguished Nobel Prize winner decided to accept one of the offers he was constantly receiving. There was no hope of changing the president's mind, but Struve nevertheless launched an all-out campaign to do so. He met with the Policy Committee of the Division of Physical Sciences, made up of the chairmen of all its departments, and learned that he was the only one who was opposed to Bartky. The others thought the acting dean had been doing the administrative work well, and they did not really want a great scientist dictating to them. They were willing to recommend Struve for the post as dean, but he would not consider accepting it himself. He could not persuade any of them to take it, and he proposed the names of such distinguished outside scientists as Gaylord P. Harnwell, Lee A. DuBridge, and J. Robert Oppenheimer as possibilities. Everyone but Struve realized they would not accept; they were destined for greater things, and in fact Harnwell soon became president of the University of Pennsylvania and DuBridge of the California Institute of Technology, and Oppenheimer was named director of the Institute for Advanced Study. The Yerkes director next recommended several other equally distinguished outside candidates, none of whom wanted to be a dean, and finally in desperation he persuaded his fellow chairmen to recommend that Hutchins appoint Bartky for only a limited term as acting dean while they continued the search for a satisfactory permanent dean. This solution the president rejected; such lame-duck appointments "[had] not been very satisfactory in our experience in the past," he stated. However, the president was willing to give the committee a few more months, keeping Bartky on the job as associate dean until they agreed on a candidate.[46]

In the end the committee recommended Bartky for the appointment, with only Struve dissenting. There was no other choice. Struve accepted their verdict but wrote a formal letter to the president force-fully stating his reasons for opposing the appointment. Bartky had done no significant research, he had never supervised any graduate student's Ph.D. thesis, and he had, as associate dean, made "unwar-ranted changes in my recommendations for appointments, promo-tions and salary increases in the astronomy department, sometimes without consultation with me—probably based upon a misunder-standing of his qualifications to judge the details of the administration of astronomy." In conclusion, Struve asked Hutchins to relieve him of his duties as director and chairman and allow him to go back to full-time research. This threat of resignation brought a phone call from Hutchins's secretary, summoning the director to Chicago for a con-ference. There, in an hour-long discussion, Struve reiterated all these arguments to Hutchins, giving as examples of Bartky's reversals of his recommendations cases in which the associate dean had valued teach-ing qualifications more than Struve had and had opposed large raises for Henyey, Greenstein, and others based on their research alone. Hutchins told Struve he had made every effort to find a dean "with higher standing in research," in the Yerkes director's words, but could not do so. Bartky had the support of "the majority of the division" (actually, of almost everyone but Struve). Hutchins "urged" him to remain director, promised him a distinguished service professorship, and said that if he insisted on resigning the only choice for his succes-sor would be Kuiper, whom Struve had previously stated could not handle the job. Finally Hutchins promised that "there would be no interference from the dean's office"—that is, that Bartky would not be allowed to overturn Struve's recommendations. On that basis he withdrew his threatened resignation of the directorship but began casting about for more ways to carve out time for his own research.[47]

Losing this struggle and ending up with a new dean he had bitterly opposed was bad enough, but in addition Struve now had another unwanted administrator between himself and Hutchins. The latter had put his own reorganization plan into effect on July 1, 1945. He became chancellor, at the apex of the hierarchy, with Ernest C. Col-well below him as president, in charge of the "educational admin-istration" of the university. Thus Struve would have to go to Col-well, through Bartky, with his recommendations for promotions and requests for more money to operate Yerkes and McDonald Observa-tories, while Hutchins would become a remote court of last appeal, presumably spending most of his time thinking deeply, planning for the future, and dealing with the trustees. Colwell, two years younger than Hutchins, was a longtime theology professor at Chicago, a writer on the Bible, and dean of the Divinity School, who had been named vice president the previous year. He was much more of a plodding academic bureaucrat than a brilliant idea man like Hutchins, and Struve undoubtedly feared the worst from him. Colwell's first official action, confirming Bartky as dean of the physical sciences, did nothing to change the director's mind.[48]

Back in April, with the allied armies still engaged in Germany and the atomic bomb still a deeply secret, untested design-in-progress, Struve had warned Hutchins against "an impending drive on the part of members of the Physical Sciences Div[ision] whose work had been of great practical value [the physicists] to usurp the official resources of the University and such facilities as buildings, equipment, etc." This drive, he feared, "would impair the work of departments whose function had been of only secondary importance in connection with the war effort" (such as astronomy). Now in early August, with the war suddenly brought to an end by the fruits of those early experiments in Ryerson and under the stands at Stagg Field, and the University of Chicago's previously secret plans unveiled for new Institutes of Nuclear Physics and for the Study of Metals, to be built around Fermi, Harold C. Urey, and their teams, Struve's worst fears were realized. The Institute for Spectroscopy that he and Mulliken had dreamed of, and another large telescope on Mount Locke for a Chicago-led cooperative observatory, would be far down the scale of priorities as ordered by Colwell and Bartky. To Shapley and Russell, now his major confidants, Struve poured out his deep concern over "the entire trend of events at the University of Chicago." The general situation had been going from bad to worse since Compton's resignation, and Struve had "very little hope for anything approaching a reasonable post-war policy on the part of the central administration of the university." He still believed in Hutchins, but in little else. Struve was ready to give up the "misery" of being director and wanted to lose himself in "some quiet scientific job." Thus did the director of one of the outstanding astronomical research institutions in the world face the moment of his country's victory in 1945.[49]

# CHAPTER ELEVEN

## Golden Years, 1945–1950

As World War II came to an end, most of the Yerkes astronomers who had gone on leave to weapons development projects came back to their peacetime research positions just as quickly as they could. Only a few left for greener pastures. Horace W. Babcock took a coveted job at Mount Wilson Observatory, the first staff member hired by its new director, Ira S. Bowen, who replaced Walter S. Adams, well past the normal retirement age by the war's end. Philip C. Keenan accepted a position at Perkins Observatory of Ohio Wesleyan University, where he could devote himself to spectral classification with its sixty-nine-inch reflector. Working on naval ordnance during the war, he had kept in touch with astronomical research and had come back to Yerkes during one brief vacation in 1943 to measure a few spectrograms and write up a paper. Otto Struve admired Keenan's astrophysical research and his work ethic and tried to keep him at Yerkes with a promotion to assistant professor and a raise, but he had decided to move on.[1]

George Van Biesbroeck reached the mandatory retirement age of sixty-five in 1945, and although Struve tried to get his appointment extended, Chancellor Robert M. Hutchins would not permit any exceptions to this policy. Van Biesbroeck stayed at Yerkes as an emeritus professor, observing double stars, comets, and asteroids with the old twenty-four-inch reflector George Willis Ritchey had built. Within a few years Van Biesbroeck and his wife converted their home, where their children had grown up, into a boardinghouse for graduate students, young instructors, research associates, and visiting astronomers. Informal discussions on astronomical research raged at every meal, moderated by the kindly old double-star observer, his wife Julia, and his sister Marguerite, for many years the part-time Yerkes Observatory librarian.[2]

Another Yerkes faculty member who did not return after World War II was Christian T. Elvey, Struve's first Ph.D. student. He had been excellent as the resident astronomer at McDonald Observatory, in charge during Struve's absences (the director did not like to give anyone the formal title of assistant director), but in the years after 1940 the two became increasingly estranged. Largely this resulted from Struve's long-distance micromanagement, compounded in 1941 by Elvey's moving from the mountaintop observatory to the nearby small town of Fort Davis, seventeen miles by winding road from the summit of Mount Locke. With growing children he and his wife wanted to be part of a community, and Fort Davis was a much more convenient location for an observatory office from which to order supplies, receive them, and send them on to the summit. Gasoline and tire rationing were to make frequent, unplanned trips difficult. But to Struve it meant that Elvey was not supervising the maintenance of the observatory, or the employees there, closely enough, and no doubt this was true also.

In the summer of 1942 Elvey returned briefly to Chicago, to take part in the spectroscopy symposium Struve and Robert S. Mulliken had organized (he gave an excellent paper on his pioneering night-sky research) and to look into his possible temporary assignment to Dean Arthur H. Compton's secret wartime Metallurgical Laboratory (atomic bomb) project. Apparently the job Compton had in mind for him was to develop remote-control systems for operating the nuclear "pile" then being assembled under the Stagg Field grandstand, similar in some ways to the remote-control system for accurately and safely pointing the eighty-two-inch reflector. Elvey discussed the job with physicist Robert R. Wilson, decided he was not qualified for it, and returned to Fort Davis. Struve, who wanted to present to the higher administration at all times a united front of a happy, dedicated Yerkes staff whose ideas were at one with his own, was disappointed and angry.[3]

Very soon after his return to McDonald Observatory, Elvey got into a shouting match with Arch Garner, a Texan who was the practical mechanic and foreman of the work crew, the one man who could fix almost every piece of machinery necessary to operate the telescope. Elvey was fed up with Garner's negative, critical attitude, which in the astronomer's opinion had reached the point of insubordination. He recommended firing Garner at once. Struve refused point blank and chastised Elvey severely for stirring up trouble and trying to get rid of an employee who was practically irreplaceable. Elvey immediately replied that if Struve would not back him up, he wanted to be relieved of all his supervisory responsibilities at once. He would resign that position, stay at Fort Davis only long enough to finish writing his papers on his night-sky research, and then return to Williams Bay to continue observing there. Even before Struve could get to Texas, Elvey had received an informal offer from the California Institute of Technology to join its secret military rocket project. Very soon after that Elvey accepted a formal offer from Dean Earnest C. Watson at Caltech; he left Texas for Pasadena in October just a few days after Struve's arrival at Mount Locke.[4]

At Caltech, Elvey, along with Franklin E. Roach, his predecessor at McDonald Observatory, and Nicholas U. Mayall were three key young astronomers on the rocket project. Their work involved large amounts of field testing in the Mojave Desert northeast of the Los Angeles basin. Elvey found time to write several papers on his night-sky observational results, and late in the war he was allowed to continue this research part time, since it was closely related to the properties of the upper atmosphere, by then so important for jet aircraft and rockets. He kept in touch with Struve, who was initially cold and formal but gradually softened his attitude. However, in late 1944, as the war appeared to be reaching its end, Struve formally notified Elvey that there would be no job for him in Texas and that he was not recommending that his University of Chicago faculty appointment and leave be extended. Others at Yerkes, including Struve's secretary Lillian Ness, were much friendlier to Elvey and were sorry to see him go.[5] After the war ended he and Roach built up upper-atmosphere research at the Naval Ordnance Test Station at Inyokern, in the Mojave Desert, the successor of the Caltech project, and then in 1952 he joined the University of Alaska faculty as professor and director of its Geophysics Institute. He and Roach, two of Struve's early Yerkes Ph.D.'s, are often considered leading pioneers in night-sky and upper-atmosphere research.

After Elvey's precipitous departure from Texas in 1942, Struve, with Compton's help, succeeded in having Elmer Dershem, an elderly, unproductive physicist from Chicago, appointed astronomer in charge of maintenance at McDonald Observatory. He was a long-winded, tiresome talker, extremely conservative and all too likely to blame President Franklin D. Roosevelt personally for any shortages of necessary supplies, but he and his wife were willing to live on Mount

Locke, where he could supervise the maintenance and repair anything himself, if necessary, given enough time. Garner saw that his days were numbered and left a few months after Dershem's arrival, but the observatory continued to operate just as well without him.[6]

In 1946, with many more scientists available again after the wartime shortage, Struve decided to replace Dershem, who was only one year from retirement. The Yerkes and McDonald director decided on Arthur Adel, an outstanding spectroscopist who had specialized in infrared measurements of the molecular absorption bands in the spectrum of the earth's atmosphere. He had been on the Lowell Observatory staff in Flagstaff and was now a junior faculty member in the University of Michigan physics department. Adel was anxious to get back into astronomy and volunteered for the McDonald position, which he had heard about from a friend, in those days before job advertisements and open searches. At McDonald he could continue his infrared research, which required only frequent sunlight, an infrared spectrometer, and a detector. Struve recommended Adel's research for outside support, causing a minor blowup as Gerard P. Kuiper, who was also involved in the proposal, modified it to increase his own role and decrease the prospective new staff member's, without consulting him.[7]

Struve smoothed this over, and on September 27 Adel and his wife arrived on Mount Locke, moved into the observatory house that had been prepared for them (their furniture was still on the way), and had dinner with some of the astronomers who were there. The next day Adel announced he was resigning before he had ever begun work, and they drove down the mountain and back to Michigan. He gave as his reason his wife's health; she had never been to McDonald before, although Struve had suggested they visit before deciding to take the job. No doubt she had envisioned it as something like Lowell Observatory, at the edge of a thriving western city, with schools, museums, and a college, rather than as an isolated observing station with a shifting population of fifteen or twenty souls.[8]

Struve was sorry to lose Adel and gave him the chance to reconsider, but he did not press him to come back. The director had submitted Adel's infrared research proposal to the Research Corporation, a private foundation that supported scientific research then as now. It granted $10,000 for Adel's research even before he had arrived at McDonald, and Struve arranged for it to go wherever the molecular spectroscopist relocated. Adel took the funds to McMath-Hulbert Solar Observatory in Michigan, where he had found a temporary position for two years, and then used the money to get started at Arizona State College (now Northern Arizona University) back in Flagstaff. Struve persuaded Dershem to return for one more year on Mount Locke, and then in 1947 he hired Paul D. Jose to succeed him as "astronomer in charge of maintenance" at the Texas observatory. Jose was a Michigan Ph.D. who had been at the McDonald dedication in 1939 and had been the caretaker director of Steward Observatory in Tucson during World War II. He was a capable mountain superintendent and cooperated effectively in observing with several astronomers, including Struve and Pol Swings, though he did not do much research on his own as Elvey had.[9]

One new staff member appeared on the scene in 1945, just as World War II ended. He was Gerhard Herzberg, a laboratory spectroscopist who had written well-known textbooks on atomic and molecular spectroscopy. When Swings had left Yerkes in 1943 for his wartime job in Pasadena, Struve, realizing how valuable he had been, wanted to replace him with another physicist well versed in spectroscopy. His attention soon centered on Herzberg, a refugee from Hitlerism who had settled at the University of Saskatchewan in Canada. He had recently identified a previously unknown interstellar absorption line with the molecule $CH^+$, and an emission band observed in comets with $CH_2$. Struve had met him at conferences on spectroscopy at Yerkes in 1938 and 1941 and in Chicago in 1942 and was impressed by his interest in astrophysics. Mulliken encouraged Struve's desire to bring Herzberg to Yerkes, saying that the physicists would welcome him and would want him to be associated with their department as well as astronomy. The German spectroscopist was eager to come but made it clear that he wanted a full professorship and a laboratory of his own. All the initiative came from Struve, and after Herzberg had visited Chicago, given a colloquium, and met President Hutchins, the Physics Department drew back. With many top nuclear physicists now on the campus, including Enrico Fermi, several of the senior professors wanted to keep as many postwar, permanent positions open for them as they could, rather than using one to hire a spectroscopist who was interested in astronomy. In the end Compton authorized offering Herzberg only an associate professorship, although at a higher salary than he had expected. Herzberg was highly appreciative of Struve's "definiteness and [the] straightforwardness of [his] proposals" but deeply disappointed in Compton, Mulliken, and the other physicists.[10] Struve apologized to him and promised that if Herzberg came, he would work as hard as he could to have him promoted to full professor quickly. The Yerkes director diplomatically explained the spectroscopist's feeling of rejection to Compton and Mulliken, and at his urging they wrote Herzberg, assuring him that they really wanted him on the Chicago faculty. Observing at McDonald, Struve himself typed a very persuasive letter to the German physicist, emphasizing the opportunities for research on the Yerkes faculty while Carlos U. Cesco and Jorge Sahade operated the telescope for him. Herzberg could not resist, especially since Struve had promised to outfit a spectroscopic laboratory for him at Yerkes, in one of the rooms George Ellery Hale had designed fifty years earlier for just that purpose. Herzberg accepted.[11]

However, he could not leave Canada and move to Williams Bay to begin his new job. Because of the war, the Canadian government barred any university faculty members from leaving their posts to accept positions in the United States (which in general paid considerably higher salaries than comparable institutions in the Dominion). The regulation was justified on the basis of the need for teachers, but it applied to Herzberg even though the University of Saskatchewan had a surplus of physics professors and though his reason for accepting the Yerkes position was to get the superior research opportunities of the McDonald eighty-two-inch reflector and a well-equipped spectroscopy laboratory of his own. So in spite of frequent wire pulling by Struve, Herzberg had to remain in Saskatchewan throughout the entire 1944–45 academic year and was granted his exit permit only in July 1945, after the war in Europe had ended and, coincidentally, he had received his Canadian citizenship.[12] Struve had kept the position open for him, and Herzberg was soon traveling to McDonald to observe infrared spectra of planets, stars, and nebulae and building up his laboratory at Yerkes. He installed a large-grating spectrograph and a long absorption tube with which he could mimic the path lengths of air, $CO_2$, $NH_3$, $CH_4$, and ultimately even $H_2$ traversed by radiation in planetary atmospheres. Struve had worked hard to bring the laboratory spectroscopist to the Yerkes faculty, and Herzberg more than justified his efforts by the research results he produced.[13]

By the end of the war in 1945 Struve was exhausted. He was tired of all the letters, phone calls, conferences, and other administrative duties connected with every appointment (like Adel's and Herzberg's), every new instrument (like the infrared photographic spectrograph), every promotion, and every plea for funds. When Hutchins appointed Walter Bartky as the new dean of physical sciences, an appointment only Struve had opposed on the "policy committee" that advised the chancellor, the Yerkes director had threatened to resign. Subrahman-yan Chandrasekhar hurried to Chicago to express his concerns to the chancellor. Then Hutchins had summoned Struve to his office and had rhetorically asked him to propose the name of someone who should and could succeed him as director. He had named Kuiper, Jan H. Oort of Leiden Observatory, Bart J. Bok, and William W. Morgan as possibilities. Hutchins had rejected them all and persuaded Struve to stay on the job, promising him a coveted distinguished service professorship and saying there would be "no interference" from Bartky's office. Struve had then agreed but had "insisted" that he had to reduce his administrative load and suggested a plan he had evidently long been considering: that he would become "chairman" of the department, retaining "final responsibility" for it, but would make Kuiper director at Yerkes and W. Albert Hiltner director at McDonald, both under him. Hutchins agreed to the concept, providing all departmental recommendations would go through Struve and "that these men would not start coming to this office." [14]

Hence Struve soon submitted a written reorganization plan, modeled in some ways on Hutchins's own recent elevation of himself from president to chancellor, to whom the new president and other high administrators reported. In Struve's plan he was to be chairman of the department, a sort of superdirector analogous to chancellor in the new university hierarchy. As chairman, he would be in charge of "outlining" all the research in astronomy, would continue as managing editor of the *Astrophysical Journal,* would himself "engage primarily in personal research," and would write a book on astronomical spectroscopy. He would remain in residence at Yerkes and would have a house at McDonald; he would assign the observing time on the eighty-two-inch reflector. He would remain in charge of the cooperative program at McDonald Observatory (then limited to Indiana and Minnesota) and of any expansion of it in the form of new telescopes, which he still hoped for.

Under his plan there would be a director of Yerkes Observatory ("perhaps . . . Kuiper"), who would be in residence there, and an astronomer-in-charge at McDonald ("perhaps Hiltner"), who would divide his time between Mount Locke and Williams Bay with houses at each. They would be in charge of "routine administration" at their respective sites and "preparation of budget requests," but the chairman would be the sole liaison between them and the administration. All their actions would be subject to his approval; they would "consult" with him, and he would have the ultimate responsibility for all recommendations to the administration. Finally, there would be a division of "astronomy on the University of Chicago campus," where Struve wanted to build up the teaching of advanced astrophysics. He planned to transfer one faculty member there, "perhaps L[ouis G.] Henyey," and put Thornton L. Page in charge of the general astronomy courses and of keeping the students busy in the observatory on the roof of Ryerson Laboratory.[15]

Struve had wanted to clear the whole idea with Kuiper before submitting it, but the prospective new Yerkes "director" did not get back from writing his final wartime reports at the Harvard Radio Research Laboratory until November. However, he had made it clear to Struve that he had a lot of research he wanted to do as soon as he could, and both Hutchins and Struve doubted that Kuiper would be able to put the problems of the observatory above his own projects.[16] Whatever the reason, Struve modified his reorganization plan, left Kuiper out of it, and agreed to retain the directorship himself if Hutchins wished. Still, Struve said he wished to withdraw "as soon as possible" from matters of routine administration and that Hiltner would become assistant director of both observatories. Hutchins accepted this plan, but Struve then confessed he had not asked Hiltner if he would accept the post.[17]

It was a strange appointment; Hiltner was only three years past his Ph.D. and had joined the faculty as an instructor just the previous year. Struve had recommended him for promotion to assistant professor effective January 1, 1946, only a few months before proposing him for assistant director. The director had described him to Bartky as "unquestionably the best observational astrophysicist that I have come in contact with in the past nineteen years," surely a vast overstatement.[18] But with Van Biesbroeck retired Struve desperately needed someone with the mechanical and instrumental skills he himself lacked, and that Hiltner had demonstrated he possessed in abundance. Morgan, who had been acting as assistant director of Yerkes without the title or a formal appointment, had been competent but had done only the minimum amount of administrative work, confirming that he was far more interested in research. He almost never went to Texas to observe, preferring to confine his large program of spectral classification of reasonably bright stars entirely to the forty-inch refractor. Clearly the assistant director had to be at McDonald frequently, and again Hiltner fitted the bill. Struve apparently never notified Morgan that he had been supplanted; he learned it only when Hiltner, who accepted the post, took over his functions after January 1,

1946.[19] By that time the new assistant director was well into the design of a new coudé spectrograph for the eighty-two-inch reflector, for which Struve had secured $12,000 from the University of Texas; the Yerkes instrument makers Carl Ridell and John Vosatka would devote much of their time and effort to building it under Hiltner's supervision. Yet within six months Struve was fuming about his new assistant director's "unsatisfactory performance" and had begun chastising him. The director was hard to please. In all these negotiations he dealt directly with Hutchins, ignoring President Ernest C. Colwell and Bartky as much as he could and notifying them of the chancellor's decisions only after they were made. Bartky, on the other hand, was unfailingly polite, considerate, and helpful in his responses.[20]

In March 1946 Struve made his first postwar trip to Europe, flying across the Atlantic for the first time via London to Copenhagen. He, Harlow Shapley of Harvard, and Joel Stebbins of the University of Wisconsin were the three American representatives to a week-long conference, called by the officers of the International Astronomical Union (IAU) to get it on its feet again. It had nearly died as communications between enemy countries became impossible during World War II. Their appointment symbolized that the three of them, along with Adams and Henry Norris Russell, both now too old and feeble for the strenuous trip, were the top leaders of American astronomy. Struve stayed with Bengt Strömgren and his family in the director's residence next to the Copenhagen Observatory throughout the conference. Struve had gotten in touch with the Danish astrophysicist soon after the Allied triumph in Europe and had learned of his brilliant work on model stellar atmospheres while German soldiers patrolled the streets of Copenhagen. It was a large project, based on the latest quantum mechanical ideas and on reams of computations carried out by Strömgren and his assistants, which showed that hydrogen is abundant in the atmospheres of stars just as he had previously shown it to be in their interiors.[21] During the IAU conference in Copenhagen, Struve renewed his contacts with several leaders of European astronomy, including Grigory A. Shajn of the Soviet Union, with whom he had published a joint paper years before. Struve was apprehensive that Shajn and A. A. Mikhailov, the other Soviet delegate, might be opposed to him because of his anti-Bolshevik stance, but he found them both "friendly and sincere." He enjoyed talking with them and invited them to come to visit Yerkes and McDonald Observatories.[22]

After the IAU conference, Strömgren arranged for Struve to be elected a foreign member of the Royal Danish Academy of Sciences and then for him to receive an honorary doctor's degree at Copenhagen University. It was to be awarded at the four hundredth anniversary of the birth of Tycho Brahe, in December 1946. In honor of this first observational astronomer Strömgren, as dean of the faculty of sciences, had brought about the awarding of honorary degrees to ten outstanding astronomers: Shapley, Oort, Shajn, Mikhailov, and representatives of France, Great Britain, Sweden, and Holland, in addition to Struve and Hutchins, the one university administrator on the list. (In the end the chancellor did not attend.) As director of Copenhagen Observatory Strömgren invited Struve to come and stay for three months before the ceremony, to work undisturbed on research. Struve did so, bringing his wife with him. On the way he attended the Zeeman conference in Amsterdam in September, the first big international spectroscopy and astrophysics meeting after the war. In Copenhagen, Otto and Mary Struve stayed in a special visitors' apartment in the observatory, and though there were still some privations in postwar Europe (Strömgren advised them to bring enough coffee, tea, and soap for their entire visit, as well as sheets for their beds), there was also pageantry (he suggested they bring "full evening dress if possible" as well).[23]

As Struve prepared for his departure for Europe in the summer of 1946, a new crisis, which he had foreseen, erupted. Russell was due to retire, and after Svein Rosseland, who had escaped from Norway and spent the war years at Princeton, announced his decision to return to Oslo, the university offered its research professorship to Chandrasekhar. It was a tremendous honor for the thirty-five-year-old Indian theorist to be considered the successor of Russell, the greatest American astrophysicist of his time. Chandrasekhar had earned it by the outstanding research he had done, which had already brought him election to the Royal Society of London in 1944 and the national prize for mathematics awarded by Andhra University in India in 1945. Princeton University offered Chandrasekhar a salary of $10,000, over 50 percent more than he was earning at Yerkes, and Russell and Shapley pressed him to accept. Struve thought that if he were a theorist he would accept the offer himself, though the blow to Yerkes would be "terrible" if Chandrasekhar left, and he hoped he would stay. Recognizing the theorist's great "esteem and admiration" for the chancellor, Struve urged Hutchins, who was on vacation, to call Chandrasekhar in for a personal conference just as soon as he re-

Figure 44    Subrahmanyan Chandrasekhar in his office (1957). Courtesy of the late Prof. Chandrasekhar.

turned. The director and Warren C. Johnson, the acting dean (Bartky was also on vacation) recommended a raise to $8,500, along with more money in the budget for his publications and assistants, and Vice President Lawrence A. Kimpton immediately approved the recommendation. The $8,500 salary was what Struve was being paid at the time, and much more than Kuiper, Morgan, or Herzberg was making.[24] On August 26 Chandrasekhar wrote Princeton that he would definitely accept the research professorship, and Russell quickly replied that he was delighted and honored to have him as his successor.[25]

Hutchins, however, was determined to keep Chandrasekhar. He had been attracted to the brilliant Indian astrophysicist ever since he had hired him away from Harvard in 1936, and he did not intend to lose him now. From his vacation retreat the chancellor wrote Struve that he wanted to keep Chandrasekhar "at all costs" and would see them both as soon as he returned to Chicago. Struve showed the letter to Chandrasekhar as an earnest of the chancellor's intention and made an appointment for the two of them to come to Chicago on

September 7, the day Hutchins was to be back. After Chandrasekhar had received the letter from Russell confirming his appointment, however, he told Struve that he had accepted the Princeton offer and that it would be a waste of Hutchins's time to see him. The Indian theorist apologized for the "trouble" he had caused and said the deciding factor was not the higher salary, but the honor of succeeding Russell. Struve accepted these statements but asked his younger colleague and friend if a distinguished service professorship would change his mind. Chandrasekhar said it would not, nor would he be interested in any administrative post either at Yerkes or in Chicago. Nevertheless Hutchins insisted on going ahead with the conference, and on September 7 Struve and Chandrasekhar met him and Bartky in Chicago. The chancellor used all his considerable charm and personal magnetism to convince the young theorist to stay at the University of Chicago. Was not starting his own tradition at the midwestern university more important than continuing Russell's tradition at Princeton? Who now remembered Lord Kelvin's successor at Glasgow? Was not Chandrasekhar's professorship the question, not Russell's? On a more mundane level, Hutchins stated he would match the $10,000 salary and make Chandrasekhar a distinguished service professor. The brilliant theoretical astrophysicist was touched, but he did not commit himself.[26]

A few days later Struve left for Denmark. He hoped for the best, thanked Hutchins for his efforts, told him how deeply honored Chandrasekhar had been by the chancellor's "extraordinarily kind attitude" and "confidence in his work," but said he believed he would go through with his commitment to Princeton. The director was wrong, however; Hutchins had won out. His skillful, sensitive persuasion had been precisely right to win over Chandrasekhar. On October 7, a month after their previous conference, he came back to Chicago alone to see the chancellor at his downtown *Encyclopaedia Britannica* office. Evidently Hutchins confirmed that he had meant every word he said, and a few days later Chandrasekhar decided he would stay. In mid-October he wrote Russell to tell him he had changed his mind, and to Struve to say that his close contacts with his observational colleagues at Yerkes had won out over the attractions of pure theory at Princeton. Struve was overjoyed and saw a bright future for Yerkes; Hutchins was pleased and proud too and wrote Russell to ask, somewhat tongue-in-cheek, his "forgiveness" for the "terrific pressure" he had put on his new distinguished service professor to stay. The old Princeton professor was willing to laugh it off as "hoss tradin'" between gentlemen, which could always be done fairly, courteously, and yet strenuously. The following year the gifted younger theoretical astrophysicist Lyman Spitzer Jr., who had been Russell's own Ph.D. student at Princeton, succeeded him in the research professorship.[27]

Yet Struve must have felt a chill in his heart when he heard Hutchins offer Chandrasekhar the distinguished service professorship he had promised him a year earlier, but not given him, and a salary higher than he himself was then making as director. Struve was a strong believer in the hierarchical principle; he surely felt that he himself deserved a distinguished professorship before his brilliant junior colleague, and more money. In the event, Hutchins made Struve a distinguished service professor at the same time as Chandrasekhar and simultaneously raised his salary to $12,000, significantly above the younger man's. But Struve certainly realized that from then on he was no longer the chancellor's favorite; the former "boy director's" days were numbered.

While Struve was in Copenhagen in the fall, a delegation of Soviet astronomers led by Shajn finally visited Yerkes and McDonald Observatories, in response to the invitations from the Yerkes director and Shapley at the IAU conference earlier that year. It took time for both governments to arrange the trip, and there were delays with visas and airline reservations, but finally the group of nine Russians reached America. Five of them went to Yerkes and McDonald in November and December, where Chandrasekhar and Hiltner showed them everything. Although they completely missed Struve there, they saw him at the AAS meeting in Harvard at the end of December, where he stopped on his way back from Denmark. He still liked Shajn, by now the director of the Crimean Astrophysical Observatory, as the rebuilt Simeis Observatory had been named. The two of them published a second joint paper, on the spectra of carbon stars, based on spectrograms the Russian astronomer had obtained at McDonald Observatory during his brief visit.[28]

That same year Struve and other Yerkes astronomers took their first steps toward obtaining grants from military agencies for specific research projects, a very new concept at the time. The Office of Naval Research (ONR) announced its willingness to support research connected with its mission, and Struve, through Bartky, applied for money for the infrared work Kuiper was beginning with new solid-state detectors, for Herzberg's spectroscopic laboratory, and for a photoelectric photometry project to be set up once an expert on the subject could be chosen and added to the faculty. Initially none of these proposals was funded; the navy judged the first two "too astronomical" and the last too uncertain. However, in the following year the grant for Herzberg's laboratory did come through.[29]

Also, the army had brought the German V-2 rockets it had captured at the end of the war to America and was starting to let scientists use them for research projects. This work was centered at the Johns Hopkins Applied Science Research Laboratory, which was mounting an ambitious program for obtaining an ultraviolet spectrum of the sun from above the earth's atmosphere. Struve, with the help of his former student J. Allen Hynek, now at Ohio State University and a consultant to the Johns Hopkins group, obtained a contract for a cooperative Yerkes effort in this same direction. Under it Jesse L. Greenstein designed and had built an ultraviolet spectrograph, and though it was destroyed in a launch accident (a common fate of scientific instruments on rockets in those early days), he helped analyze the first ultraviolet solar spectrograms that were obtained.[30]

Yerkes Observatory was a beehive of research in its golden years immediately after World War II. A large staff and many visitors, particularly from abroad, were producing an exceptional scientific output on stars, interstellar matter, and planets. They did most of their observational work with the eighty-two-inch McDonald reflector. Struve himself was working on spectroscopic binaries and massive stars, trying to explore and understand the effects of rotation on the evolution of stars. Kuiper had shifted his attention increasingly to planetary astrophysics, applying infrared methods as well as optical. In 1948 he discovered the previously unknown fifth satellite of Uranus, eventually named Miranda, and in 1949 the second one of Neptune, Nereid. More important, he was investigating the atmospheres of all the planets spectroscopically.[31]

Chandrasekhar was working on the theory of radiative transfer in stellar atmospheres and the accurate quantum mechanical calculation of the opacity (or absorption coefficient) of $H^-$, the negative hydrogen ion, which Rupert Wildt, Strömgren, and Chandrasekhar himself, with Guido Münch, had shown to be all-important in cool stars like the sun. Münch, who had gone back to Mexico after earning his Ph.D. and holding a Guggenheim Fellowship, returned to Yerkes as a faculty member in 1947 and worked closely with Chandrasekhar on several of these problems. Among Chandrasekhar's Ph.D. students in this period, in addition to Münch, were Merle E. Tuberg, Marjorie Hall Harrison, Marshall H. Wrubel, Su-shu Huang, Arthur D. Code,

Henry G. Horak, and Frank N. Edmonds. He had his first physics Ph.D. in this period also, Esther M. Conwell, who came to Yerkes to work with him on the quantum mechanical problems of $O^-$ and $H^-$, the negative oxygen and hydrogen ions. Struve, always sensitive to the pitfalls that can arise in the academic jungle, made sure to get a letter from Frank C. Hoyt, executive secretary of the Chicago Physics Department, then in a state of rapid flux, approving her thesis topic and stating that Chandrasekhar, though not a member of the physics faculty, would be on her final examination committee. This protected them both, and she completed her thesis and received her degree in 1948 from a department by then dominated by nuclear physicists.[32]

Morgan was concentrating more and more on classifying O and B stars, the hottest, bluest stars, which because of their high luminosities and relatively narrow range of colors are ideally suited for probing the extinction of dust in interstellar matter. He was quick to grasp that the photoelectric measurements made by Stebbins and his collaborators at the nearby University of Wisconsin, and on trips to Mount Wilson, provided a new standard of accuracy. The combination of Morgan's accurate spectral types and luminosity classes with their colors opened a new epoch in mapping the galactic structure within a few kiloparsecs of the sun. Morgan's Ph.D. students in this period included Armin J. Deutsch, Nancy G. Roman, Douglas Duke, and Arne Slettebak. All of them and his former student William P. Bidelman, by now an assistant professor who had replaced Keenan on the faculty, worked closely with him on these problems. The only native-born American among the senior Yerkes faculty members, Morgan fitted in well in the small-town life of Williams Bay. He was elected to its Village Board for consecutive terms from 1943 to 1951 and was its chairman, unofficially the "mayor," for the last four years.[33]

Struve had hoped to add a photoelectric photometry expert to the Yerkes faculty, and he approached Albert E. Whitford, who had been Stebbins's right-hand man at Madison and then worked on radar at the Radiation Laboratory in Cambridge throughout World War II. However, Whitford preferred to return to the University of Wisconsin, where he would succeed Stebbins as director as soon as the latter retired. Struve then assigned Hiltner to switch from photographic spectroscopy to photoelectric photometry, a subject well matched to his experimental-physics talents and experience. Hiltner accepted the new responsibility, and before long he was measuring light curves of variable stars.[34]

At the war's end Swings had briefly accepted a permanent faculty job on the Berkeley campus of the University of California but then had resigned to return to Belgium as director of its Institute of Astrophysics, which he founded in Liège. But he returned often to Yerkes and McDonald (as well as to Mount Wilson Observatory) to do research and to lecture.[35] Struve also persuaded Strömgren to return as a visiting professor from July 1947 to March 1948. The Yerkes director had given up trying to get him to come back permanently, believing he would stay in Copenhagen as its director for the rest of his life. Besides Strömgren's research and teaching, Struve hoped to get his advice on younger European astrophysicists, among whom he might find "the brilliant young man" who would become the Chandrasekhar of the next decade![36] Strömgren did not succeed in divining who that paragon would be, but while he was at Yerkes he greatly extended his own theoretical work on the photoionization of interstellar matter by hot stars and the interpretation of the observations of interstellar absorption lines in terms of the abundances of the elements. In the summer of 1947 he gave a series of lectures to the faculty members and graduate students on this work, and in the fall, at Chandrasekhar's suggestion (he had been put in charge of the graduate teaching program), Strömgren taught the course on stellar atmospheres.[37]

Holland was a leading center of astrophysical research during and after World War II, and several Dutch visitors came to Yerkes. The brilliant young Hendrik C. van de Hulst, who as a student during the war had predicted the first radio-frequency emission line of interstellar matter, the twenty-one centimeter (wavelength) line emitted by atomic hydrogen, and had done a pioneering Ph.D. thesis on the physical properties of interstellar dust particles, was a postdoctoral fellow at Yerkes from 1946 to 1948. He worked on the nature of the solar corona, on radiative transfer, and on extensions of his research on dust. He was an outstanding theoretician, and Struve and Kuiper wanted to keep him on the faculty, but van de Hulst decided to return to a similar position at Leiden University instead.[38] Another young Dutch Ph.D., Adriaan Blaauw, came to Yerkes for six months at the end of 1947. An expert on analyzing the space motions of the stars and the kinematical groups they belonged to, Blaauw worked with Morgan, using his spectroscopic data on the O and B stars. An American, George H. Herbig, who had just completed his Ph.D. at Lick Observatory, was also working at Yerkes and McDonald for the 1948–49 academic year, under a National Research Council postdoctoral fellowship. He obtained spectra of stars involved in nebulosity and formed a lifelong admiration for Struve and Morgan as research workers.

At the more senior level Marcel Minnaert was a visiting professor for six months in 1946, and Oort for three months in 1947–48. Each of the two Dutch visitors gave a series of lectures on his specialty, respectively stellar atmospheres and galactic structure and dynamics, discussed research, and went to McDonald to obtain new data with the telescope in collaboration with Yerkes observers.[39] It is remarkable how many outstanding scientists came from abroad in those years, attracted by Struve's international reputation, and how much the students and staff members learned from them.

After Van Biesbroeck's retirement, Struve wished to start a regular parallax and proper-motion program at Yerkes. It would make good use of the forty-inch refractor, essentially a very long-focus camera outfitted with a guiding system and a photographic plateholder. No doubt Kuiper, who had needed just this type of data for his earlier research on the nearby stars, strongly supported the idea. Struve first attempted to hire Peter van de Kamp, a Dutch émigré who was carrying out a program of exactly this kind at Swarthmore College, with its eighteen-inch refractor. But he decided he preferred the attractions of a small, select, liberal arts college, where he was a well-known musician and orchestra leader as well as professor of astronomy, to joining the staff of a large research factory.[40]

Struve then turned to Kaj A. Strand, a young Danish astrometrist who had been a student of the great positional astronomer Ejnar Hertzsprung. Strand had escaped from Denmark and had served in the American Air Force in World War II. He joined the Yerkes faculty in 1946 as a visiting associate professor and soon had the astrometric program under way. Many of the graduate students were hired as part-time assistants to take the parallax and proper-motion plates, and one, R. Glenn Hall, did his Ph.D. thesis with Strand on this work.[41]

Many new faces came to Yerkes during these years, and others of course departed. Mary R. Calvert, who had been at the observatory since 1905, first as the assistant and computer for her uncle, Edward E. Barnard, retired in 1946. After Barnard's death in 1923, she had become curator of the Yerkes photographic plate collection and a high-level assistant, whose final salary, $1,800, nevertheless was just above a starting instructor's but well below a young assistant professor's. After her retirement Calvert moved back to Nashville, her birthplace, and worked part time in her sister's photography studio. Struve, who had come to Yerkes as a young student when she was already the senior computer on the staff, wrote her a heartfelt farewell, full of personal news of the observatory, but she received no pension from the university she had served so long and well.[42]

In 1947 Daniel M. Popper, a native of California who had grown up in Berkeley where his father was a professor, left to take a faculty position at UCLA, and that same year Henyey accepted a position at Berkeley, where he would become the number one (and only) theoretical astrophysicist in the department. Strand, after only one year at Yerkes, was appointed a full professor at Northwestern University and director of its Dearborn Observatory. But he kept an appointment as a research associate at Yerkes and remained in charge of the astrometric program. He returned every Friday for discussions with the students who worked with him and with his faculty colleagues, and to inspect the plates that had been taken and the measurements that had been reduced during the week. If it was clear, Strand usually observed the first half of the night with the forty-inch (if it was assigned to the astrometric program) before driving back to his home in nearby Evanston, Illinois. When Hall completed his Ph.D. in 1949, he went on the Yerkes faculty as an instructor, in charge of the program on the scene, but Strand continued as visiting overall head of it for some years.[43]

In 1948 Greenstein accepted an associate professorship at the California Institute of Technology, where he would be a member of the staff of the Mount Wilson and Palomar Observatories, with the two-hundred-inch telescope about to go into operation, and the one-hundred-inch still the largest in the world. Caltech was starting a graduate program in astronomy and astrophysics, and Greenstein would head it. That same year Herzberg resigned his professorship to return to Canada as director of the Division of Physics of the National Research Council in Ottawa. John G. Phillips, who had earned his Ph.D. under Herzberg at Yerkes in molecular spectroscopy and then been appointed an instructor, succeeded him in charge of the laboratory.

All these departing faculty members had gone to very good positions elsewhere, each of them exceptionally attractive for one reason or another. Yerkes was an extremely stimulating and productive research environment in those postwar years. In addition to a torrent of research papers, it turned out many leading astronomers and astrophysicists of the next generation. Four other Ph.D.'s of this period were Anne B. Underhill, who worked with Struve and Chandrasekhar; Daniel E. Harris, who worked with Kuiper; K. Narahari Rao,

Figure 45  Yerkes Observatory staff and students (1946). In the second row are two future foreign associates of the National Academy of Sciences, Pol Swings (left) and Gerhard Herzberg next to him, then, after Luise Herzberg, six future or current members, William W. Morgan, Otto Struve, Jesse L. Greenstein, Gerard P. Kuiper, Subrahmanyan Chandrasekhar, and Louis G. Henyey, with George Van Biesbroeck between Chandrasekhar and Henyey. Two other future members are Guido Münch (kneeling near middle of front row) and Arthur D. Code (standing in rear row, just to right of middle). Future observatory directors Arne Slettebak (at left end of back row) and Victor M. Blanco (third from right in back row, with open shirt) are also in this group, as are Anne B. Underhill (third from left in front) and Nancy G. Roman (second from right in same row). Courtesy of Yerkes Observatory.

who worked with Herzberg; and Robert H. Hardie, who worked with Struve. A photograph of the Yerkes faculty, staff, visitors and students taken in May 1946 shows among the thirty-nine faces eight current or future members of the National Academy of Sciences, generally considered the cream of American research scientists. They are Struve (the only one then a member), Kuiper, Chandrasekhar, Morgan, Greenstein, Münch, Henyey, and Code, in addition to two future foreign associates of the Academy, Swings and Herzberg. Two of these ten, Chandrasekhar and Herzberg, were future Nobel laureates, in physics and chemistry respectively. These men, and many of the other faculty members, scientific visitors, and Ph.D.'s of this era, won awards and honors in astronomy in this country and abroad. Yerkes Observatory was at the peak of its glory.

But behind the honors and glory, there were divisive fissures. Struve's relentless drive and his insistence on perfection had made Yerkes great. He had assembled a staff, supported and encouraged its members, and they had succeeded. But now, instead of one great scientist and many rising youngsters, there were several great scientists cooped up in one institution. That is not a stable situation. Kuiper, Chandrasekhar, and Morgan could no longer obediently follow Struve's directions. They had ideas of their own, and they now had the recognition that made it possible for them to go their own ways rather than his. That was why he had finally decided on Hiltner as assistant director, but even he had made a suggestion that drew Struve's wrath. The director was a benevolent dictator, but mature scientists do not accept dictatorships easily.

Struve was tired of administration—of all the letters, phone calls, conferences, and approval processes involved in hiring each new scientist, mountain superintendent, or machinist. He wanted more time to do research. Yet he could not conceive of stepping down as director; that was not an option in his time. In those days a department chairman or director was expected to be a top scientist who was paid more than anyone else and went on in the job until he retired or died. Struve still wanted to bring about an administrative reorganization along the lines he had put forward in 1946 but had failed to push through. He did not discuss it with his faculty in a group or in meetings; he made up his own mind on what he wanted to do, sold it to Chancellor Hutchins, and then tried to put it into effect.

Struve's new plan was a slight variant of his earlier one. Now it was to make Kuiper director of both Yerkes and McDonald, under himself as chairman in overall charge of the department. Chandrasekhar would be in charge of an institute of theoretical astrophysics (at Yerkes), Morgan would become managing editor of the *Astrophysical Journal*, freeing Struve from the day-to-day editorial judgemnts and management, and Page would be in charge on the campus. All of them would report to Struve, who would have his own separate office at Yerkes, with his own secretary, two computers, and an assistant, all of whom he might move to the campus if he decided to relocate there. He discussed this with Hutchins in Chicago in early January and got the chancellor's approval to go ahead with it.

Then Struve went to McDonald Observatory, where he was to observe, his stay overlapping Kuiper's by several days. Struve diplomatically revealed the new plan to him and soon was writing back to Lillian Ness, his loyal secretary and one real confidante at Yerkes, "It looks quite certain now that Kuiper will take the great assignment. He is already showing signs of increasing interest and activity. And I am showing signs of increased senility. So everything is as it should be." Before that letter Struve had written Chandrasekhar, telling him the general idea of the reorganization and asking him to "inform" Morgan and Hiltner. A few days later he wrote Herzberg along the same lines. He knew they would leak the news gradually to the rest of the faculty. That was Struve's way of reaching a decision and communicating it to his staff, and it was not greatly different from Shapley's at Harvard or Adams's at Mount Wilson.[44]

Struve wrote Russell, Shapley, Swings, and Strömgren to let them know in advance of the "fairly drastic" reorganization, which was to go into effect July 1, 1947. He explained that he would be chairman of the whole department and "honorary chairman," a title he liked (George Ellery Hale had held it at Mount Wilson after giving up its directorship to Adams) but that the University of Chicago did not in fact grant him. Struve stated he felt enthusiastic about the plan, although it would require "considerable determination" on his part to let go of the "administrative functions" he had carried for fifteen years, and added that he and Kuiper saw "eye to eye" on most questions. All of them replied that they applauded Struve's effort to get more time for research and hoped the reorganization would succeed, but they expressed polite skepticism about the other Yerkes astronomers' being willing to work cooperatively as administrators. Russell, in particular, was most direct in writing that Kuiper, whom he regarded as "an absolutely first-class research man" and "a fine executive," had a reputation for being "pretty positive in his views and might get into hot water." But, Russell concluded, Struve must know Kuiper better than he did. Adams, now retired, expressed the same thought in more generalized terms, regarding all the "extraordinarily able men" to whom Struve was giving new administrative responsibilities.[45]

In fact Struve had exactly these fears, but he desperately wanted out of the administration himself. He had cleared Morgan's appointment as managing editor of the *Astrophysical Journal* with the American Astronomical Society and its editorial board. In late March he formally recommended that Kuiper be appointed director of Yerkes and McDonald Observatories, adding that he was "willing" to be named honorary director himself. The University of Texas had already approved Kuiper's appointment, and University of Chicago President Colwell, Vice President Kimpton, and Dean Bartky came to Williams Bay to finalize the change. Kuiper was appointed director of both observatories, Struve was named chairman for a three-year term "on the recommendation of the department" (he had practically demanded an unlimited term), and Chandrasekhar was to be in charge of the theoretical section of the department and of the teaching activities at Yerkes Observatory. Struve was already complaining that he was working just as hard as before; now he attributed his problems to Colwell, Kimpton, and Bartky.[46]

Hutchins was having an extraordinarily difficult personal life in 1946–47, and his low profile and complete absence from the campus no doubt contributed greatly to Struve's bouts of depression. The chancellor's wife, Maude Phelps McVeigh Hutchins, who believed that her family was higher on the social scale than his, had been surprised by his early success in academe. Her own failure to attract recognition as a creative artist further soured her. Soon after they had come to Chicago in 1929, she refused to take part in any of the public roles of a university president's wife. She did not entertain trustees, faculty members, or students or accept invitations from them. She increasingly protested her husband's absences and became more and more demanding and confrontational in her relations with him. In October 1946 Hutchins took a nine-month leave of absence, during which he worked at the *Encyclopaedia Britannica* offices in the Loop. Struve was able to see him or telephone him there occasionally, but no more. In April 1947 Hutchins left Maude, moved into an apartment downtown, and never saw her again. It is understandable that he had little time in that year for the affairs of Yerkes Observatory.[47]

Yet when called on he would still help Struve. In 1947 the observatory was celebrating its fiftieth anniversary, and Struve, also born in 1897, was celebrating his fiftieth birthday. He persuaded the editor of

*Science,* the weekly magazine of the American Association for the Advancement of Science, to devote a large part of its September 5, 1947, issue to Yerkes Observatory. Adams wrote an article on the early history of the observatory, Struve and all the senior Yerkes professors wrote articles on the history of their brands of research, and he asked Hutchins to write an introductory article. The chancellor, despite his domestic problems, came through handsomely. His text began, "My association with astronomy is an association with Otto Struve. I have never studied the subject and never understood it. I believed in Mr. Struve. I believed in his judgement of men, in his standards of scholarship, and in his aims for his department." It went on in a paean of praise for Struve's scientific leadership, and although Hutchins was a little mixed up in his chronology and a few of his facts, his theme was correct. Yerkes Observatory in its 1947 manifestation was essentially Struve's creation.[48]

That same summer, very soon after officially giving up the directorship, Struve made his first recorded attempt to leave the University of Chicago. Congress had just passed a bill establishing the National Science Foundation (NSF), to support research and education in science. Vannevar Bush, president of the Carnegie Institution of Washington (CIW), the private foundation that operated Mount Wilson Observatory and several other research laboratories, had been deeply involved in having the bill introduced and in urging its passage. Struve approached Bush and expressed interest in an administrative job in the new foundation, "consistent with [his] experience." He wrote that he felt his work at Yerkes was finished and that he had "a strong desire to make a change." Struve invited Bush to obtain Hutchins's views on him if he wished to do so.

President Harry S. Truman vetoed the bill, however, although he supported its concept, because he felt the director of the NSF should be a presidential appointee, which it did not provide. Hence Bush could not follow up Struve's request. The CIW president indicated that he knew of no other scientific administration job commensurate with Struve's talents that was open.[49]

Struve published three historical articles on Yerkes Observatory, one of them in *Science* and the other two in *Popular Astronomy.* Like most directors writing about their institutions, he aimed not so much to make sure all the facts were absolutely correct as to give a positive picture of the current situation, and in that he succeeded admirably.[50]

Differences soon surfaced between Struve and Kuiper, who, it turned out, did not always see quite "eye-to-eye." The September 1947 meeting of the American Astronomical Society (AAS) was to be held at nearby Northwestern University, followed by an expedition to Yerkes Observatory to celebrate its semicentennial. Struve proposed his favorite type of symposium, one on stellar atmospheres. Kuiper, however, decided it should be on his field of interest, planetary atmospheres. In this case they both won. Kuiper might be director, but Struve had just been elected president of the AAS, so he scheduled the symposium on stellar atmospheres for the first day of the meeting in Evanston. Strömgren, Greenstein, Lawrence H. Aller, then at Indiana University, Alfred H. Joy, longtime Mount Wilson staff member, and Struve himself gave invited papers. Then, after the last day at Northwestern, everyone went on to Yerkes. Bartky appeared to welcome the Society, and Struve and Stebbins gave historical talks, ending the AAS meeting. But the planetary scientists stayed over Sunday and had their symposium, organized by Kuiper, in the first few days of the following week. Greenstein, Swings, Herzberg, van de Hulst, and Kuiper himself gave papers, along with meteorologist Carl-Gustav Rossby, chemists James Franck and Harrison Brown, and geologist Rollin T. Chamberlin, as well as several other astronomers, including Spitzer, now at Princeton, and Fred L. Whipple of Harvard. It was a truly interdisciplinary conference.[51] Kuiper arranged for the symposium papers to be published in a book, *The Atmospheres of the Earth and Planets.* Not to be outdone, the stellar and nebular astrophysicists got their book together too, containing some twenty-odd papers on nearly every topic in their field. The authors included all the senior Yerkes faculty members as well as several former ones, Yerkes Ph.D.'s, and frequent visitors. It even included a chapter on the origin of the solar system by Kuiper and one on comets by Nicholas T. Bobrovnikoff. Edited by Hynek, this volume, titled *Astrophysics: A Topical Symposium,* was the other tangible product of the Yerkes semicentennial. Both books were up-to-date reference sources on their subjects and were widely quoted and used in the following decade.[52]

To honor Struve, who had been on the Yerkes staff for twenty-six years and its director for fifteen, Kuiper, his successor, organized a faculty dinner. Kuiper planned the event for September and invited Hutchins to be there and make "a few remarks." The dates the chancellor was free to come conflicted with the planetary symposium, so Kuiper rescheduled the dinner for October 31, which Hutchins suggested as a "good date" for him. As it approached the chancellor bowed out, however, now saying he had to be in the East. Kuiper next invited Colwell, but he instead designated Vice President R. W. "Pat" Harrison as his representative. Harrison gave a fulsome speech of praise for Struve, and several of the senior professors said at least a few words also. After the dinner Struve thanked Harrison for coming, but the former director's comment that he hoped "that in the future my relations with the central administration will be as pleasant as they were in the past" sounded a foreboding note.[53]

In spite of the dinner and the speeches, Struve and Kuiper were soon at loggerheads. Both busy scientists, they apparently had no regular time set aside for conferences, but Struve nevertheless expected Kuiper to act just as he would have, had he still been director. If he did not, Struve did not hesitate to correct him, often quite brusquely. Their first disagreement came over Kuiper's desire to bring the German solar physicist Karl-O. Kiepenheuer to Yerkes on a temporary appointment. Struve was less enthusiastic and sent the director long, detailed "suggestions," actually closer to orders, which he was bound to resent.[54] When Struve submitted a long memorandum to Hutchins on the needs of the Astronomy Department, rather defeatist in tone, in the face of the fact that the two-hundred-inch reflector was soon to go into operation on Palomar and that the University of California had the money in hand to build a 120-inch reflector at Lick Observatory, he had not consulted Kuiper before writing it.[55] These were only a few of the many differences between the chairman and the director. Struve resented every action Kuiper took that he would not have taken himself; Kuiper resented Struve's meddling and obstructing his decisions. Meanwhile Hutchins, who had long been skeptical of Kuiper's ability to be an even-handed leader (largely based on Struve's previous reports), continued to insist that all recommendations to the central administration affecting the astronomy budget come from the chairman.[56] Thus Struve had the ultimate veto power but naturally thought he was using it wisely.[57]

Chandrasekhar had Hutchins's ear. Like everyone else at Yerkes, the Indian theorist hated the growing tension; they all wanted to concentrate on research, not engage in power struggles. Seeing Struve becoming more arbitrary and authoritarian, Chandrasekhar became increasingly supportive of Kuiper. He advised the director that Lillian Ness, still the observatory secretary, was loyal to Struve and was leaking information to him. Kuiper should get rid of her, and with her the "atmosphere of intrigue." The new director, who had been Chandrasekhar's close friend since their earliest days in Williams Bay, appreciated his counsel and advice but could not go that far.[58]

In April 1948 Bartky and Vice President Harrison came to Yerkes for a conference with Struve, Kuiper, Chandrasekhar, Morgan, and

Figure 46　Subrahmanyan Chandrasekhar, Gerard P. Kuiper, and Otto Struve, outside Yerkes Observatory (1949). Courtesy of Yerkes Observatory.

Hiltner on the astronomy budget, the teaching program, and the long-term research plans for the two observatories. It was a new step in participatory democracy for the Yerkes senior faculty.[59] Nevertheless the tensions continued to grow, and finally, at the end of December 1948, Kuiper reached his breaking point. Both he and Struve had become noticeably tense and irritable, and Kuiper feared for his own health. He wrote Bartky that he could not continue as director under Struve and must resign. Struve learned of it on January 4, 1949, when he was in Chicago to give the first lecture in a winter-quarter graduate course on the campus.[60] Bartky called him in and informed him of Kuiper's ultimatum, which the other senior faculty members had evidently backed. That evening Struve returned to Williams Bay; the next morning he was to leave for Washington by train for a meeting of the new ONR advisory committee he headed. He called Kuiper and Chandrasekhar to his office at the observatory for a conference; it continued from nine until one the next morning. A few hours later Struve left for Washington. Between trains in Chicago he telephoned Hutchins and offered his resignation; to his stunned surprise the chancellor accepted it. He had decided to go with the future—with Kuiper, Chandrasekhar, and the other senior faculty members.[61]

Struve had written a letter to Hutchins that morning giving his version of the conference the previous night. In it he said that Chandrasekhar had told him his own preference was to return to the former organizational setup with Struve as chairman and director. He replied that this was not an option. Then Kuiper and Chandrasekhar both told him that if he did not want to be director, he should really give it up. Struve did not agree, but as all his colleagues seemed to share their view, he would resign as chairman and devote full time to research, perhaps moving to the campus, an idea he had toyed with in the past.[62] This is the idea Hutchins seized on, telling Struve that no one at the University of Chicago had done as much as he had to build up a department, but that he should now step aside and move to the campus, where he would be supported. He should inform the members of the department that the chancellor had accepted his resignation as chairman "with the utmost reluctance," and that they should submit their suggestions as to his successor.[63] Privately Struve was very bitter against Kuiper and Chandrasekhar, who he believed had "squeezed [him] out of the organization" he himself had built.[64]

That was not the end, however. As soon as Struve returned from Washington, Bartky came to the observatory for a meeting with the faculty members. They elected Chandrasekhar acting secretary, or temporary administrator of the department, but he adamantly refused to become chairman or director. The Yerkes professors had been unhappy with Struve, but they realized they did not really want Kuiper as their leader. They discussed bringing in various possible new directors from outside, but no name aroused much enthusiasm except Strömgren's. Only the administration officials in Chicago and the faculty members and top secretaries at Yerkes were aware of Struve's ouster.[65]

For some time he had been considering switching his research from close binaries to spectroscopic measurements of the abundances of the elements in stars, which astronomers were just beginning to realize were closely connected with stellar evolution. Hutchins had put Struve in touch with David E. Lilienthal, the chancellor's close friend and head of the recently formed Atomic Energy Commission (AEC). Struve wrote Lilienthal proposing a program of AEC support of basic research on this topic, not for himself alone but for astronomy at large. Lilienthal was mildly encouraging, and soon, again no doubt at Hutchins's suggestion, Struve received an invitation from the new Institute of Nuclear Studies on the campus to accept a joint appointment in it and move to Chicago. It was the peaceful successor to the atomic bomb project, and Fermi, Edward Teller, and other nuclear physicists and chemists were eager to get into astrophysics. Struve considered the offer seriously but could not bring himself to leave Yerkes Observatory for a physics department, especially as there was a severe shortage of space on the campus. He even considered taking a year's leave of absence, to go "somewhere and live quietly" and work on his long-planned book on spectroscopy.[66]

Chandrasekhar and Hutchins, with his "paternal interest" in the Astronomy Department, solved the leadership problem, at least temporarily. Under their plan Struve's resignation was ignored but Kuiper's was accepted; the senior professors, including the two of them, Chandrasekhar, Morgan, and Hiltner, became the "council," which would meet regularly and decide on departmental policy and budgetary matters. Struve, as their elected chairman, would be responsible for transmitting their recommendations to the central administration and carrying out the policy. It was the solution he had wanted, but without the independence he regarded as a necessity for an effective director. The five met at Yerkes on March 9 and agreed to the council governance plan; Struve then accepted the chairmanship again. The directorships of Yerkes and McDonald were abolished, as was the assistant directorship, and he, as chairman with a three-year term, was responsible for all the administration. The council held its first meeting on May 17; at it they adopted a constitution restricting membership to all the associate professors and higher (the five of them) and naming Page as nonvoting secretary, with the proviso that they would decide whether or not to permit him to attend the meetings. In all the council met twelve times between then and February 2, 1950. Struve must have hated it, but he apparently carried out its decisions faithfully.[67] He was decidedly unenthusiastic about the council, replying to Hutchins, when asked for a report on its success, "Temporarily, the dove of peace is contentedly but somewhat uncertainly fluttering around the old domes of the Yerkes Observatory but there is no certainty that the new plan will necessarily work when the normal egos reassert themselves." To Vice President Harrison he merely said that the system was working out "fairly well" and that Kuiper was "cooperative." The reappointed chairman of the department was not happy.[68]

Scientifically, on the other hand, Struve was at the height of his success. In 1948 he was awarded the Catherine Bruce Wolfe Gold Medal, the highest honor of the Astronomical Society of the Pacific. He had been president of the American Astronomical Society from 1946 to 1949, and in 1948 he was elected a vice president of the International Astronomical Union. In 1949 he learned he had been named to receive the Henry Draper Medal of the National Academy of Sciences, the highest honor in American astrophysics, the following year. In March 1949 he had given the prestigious Vanuxem lectures at Princeton University, and in October he was to travel to London to accept the Herschel Medal of the Royal Astronomical Society (which he had been awarded in 1944). On both the latter occasions, as at the Bruce Medal ceremony, he lectured on his lifetime work on spectroscopic binaries, and in 1950 he published this material in his book *Stellar Evolution*.[69]

Thus, although it was hard for Struve to decide to leave Yerkes Observatory and the University of Chicago, his home and base since 1921, he knew it would not be difficult for him to find another position. On January 15, 1949, just a week after receiving the letter (later withdrawn) from Hutchins accepting his resignation, Struve wrote to Henyey telling him that he was in "a very receptive mood" for an offer from Berkeley and asking him to let the senior astronomers there know. Henyey was aware of the situation at Yerkes, but Struve told him it was "not a happy one" and that although the University of Chicago administration was showering him with praise, attention, and research funds, he knew his colleagues on the faculty "would feel better if we were not in the same place."[70] This was not a sudden whim on Struve's part; he had earlier told Swings and other close friends that the one thing he would leave Yerkes Observatory for was a job "under the famous live oaks of the Berkeley campus," an "ideal spot for a scientist to choose." He had always made the comment jokingly, but in retrospect it is clear that he had meant it.[71]

His timing was right, too. The Berkeley astronomy department, although small, had once been a power in American astronomy. But time had passed it by, and it now had one vacant faculty position, resulting from the death of William F. Meyer. Its chairman, Sturla Einarsson, was due to retire, and Henyey was the only astrophysicist on its small faculty. Einarsson considered it his duty to move the department toward astrophysics with the new appointment, and he had involved C. Donald Shane, who had been the one astrophysicist on the Berkeley faculty before becoming director of Lick Observatory at the end of World War II, in the search process. At a meeting on

December 29, 1948, the Lick faculty members had drawn up a "wish list" of candidates they would like to see appointed at Berkeley, not restricted by any considerations of availability, seniority, or age, and Struve had headed the list. Shane had already signed Struve up as the Alexander Morrison Research Associate for 1949–50, which would enable him to come to Lick for two months, at a time of his choosing, to do research there. He had visited Mount Hamilton frequently since Shane, a personable, far-seeing administrator, had become its director, as well as visiting Berkeley, where Jerzy Neyman, a mathematical statistician, was rapidly becoming interested in applying his knowledge and skills to astronomical data.[72]

Hence, as soon as Henyey received Struve's letter the University of California wheels started turning to bring him to Berkeley. The astronomers knew he was outstanding and that his presence could transform the campus department into a first-rate astrophysics center. Their only worry was why he wanted to leave Yerkes Observatory and whether he was the source of the "discord" there. Henyey reassured them that most of the "fault" was not Struve's but rather that of his senior colleagues at Yerkes. Satisfied with this explanation, they went after him. The Berkeley administration, headed by President Robert G. Sproul, was eager to make over its astronomy department into a first-class research group, to match its already very powerful physics, chemistry, and mathematics departments, and Shane and Einarsson assured him that Struve was just the man to do it.[73]

Sproul was receptive, and Einarsson arranged for Struve to come to Berkeley and meet with the astronomers and administrators. He gladly did so but warned them that his colleagues, especially Chandrasekhar, were urging him to reconsider and stay at Yerkes. This was certainly true, but it also had the effect of making the Californians want him all the more and raise their private estimates of what they would be prepared to give him in the way of salary and resources. Shane ultimately collected strong letters of support for Struve's appointment from Adams, Bowen, Stebbins, and Elvey. Longtime physics chairman Raymond T. Birge strongly supported it, as Neyman did also. The administration had approved an informal offer even before Struve and his wife arrived for their visit in early April. The Californians were even more eager to get him after he gave his colloquium and talked with everyone.[74]

From Berkeley Otto and Mary Struve went straight to Pasadena. In 1947 Struve had learned that Bowen, the recently appointed Mount Wilson Observatory director, would welcome guest observers at the sixty-inch and one-hundred-inch telescopes, particularly after the two-hundred-inch went into service and relieved the pressure from his own staff for observing time with the "smaller" telescopes. Struve had visited Bowen in Pasadena and formed solid ties with him; then as president of the AAS he had officiated at the 1948 summer meeting in Pasadena and atop Palomar Mountain at which the two-hundred-inch was dedicated as the Hale telescope. In early 1949 Struve began hinting to Bowen that he might soon be asking permission to come to Mount Wilson as a guest observer himself, without revealing that he was now quite serious about leaving Yerkes. He asked if he might come to Pasadena after his Berkeley visit and talk with Bowen about using the telescopes. The Mount Wilson and Palomar director welcomed him and, when they met, again assured him that he certainly would be able to use the sixty-inch and one-hundred-inch for his observing program.[75] Thus Struve knew that even if he lost his access to the eighty-two-inch McDonald telescope by leaving the University of Chicago, at Berkeley he would be able to rely on the Mount Wilson telescopes and spectrographs. As soon as he got back to Yerkes from California, he wrote Einarsson, thanking him for all the hospitality and spelling out his modest "demands" in the way of an assistant, measuring and computing "machines," secretarial help, a light teaching load, and authorization to leave the campus for observing trips and scientific meetings.[76]

But Struve was ambivalent about whether to leave Yerkes or stay. He knew it would be best if he left, but he had been there so long it was hard for him to imagine being gone. He really believed that he alone was competent to direct Yerkes and that it would fail with anyone else in charge. When he resumed the chairmanship (which in effect meant the directorship as well) in June 1949 he was clearly thinking of staying. He arranged with Shane to take his two-month Morrison Research Associateship at Lick in the spring of the following year, with two breaks when he could return briefly to Yerkes to catch up on the administrative duties. The Lick director agreed with all his wishes, reminding him gently that they all hoped he would take the Berkeley job too, but never applying any pressure.[77]

Struve and his wife planned a trip to Europe in September and October, partly as a vacation, but mostly for him to visit astronomers in Switzerland, Germany, Holland, Belgium, and France to discuss research as well as the affairs of the IAU. Then they would end up in England, where Struve would give his George Darwin lecture at a meeting of the Royal Astronomical Society.[78] Just before they left for

Figure 47   Bengt
Strömgren (1953).
Courtesy of Copen-
hagen University
Observatory.

Zurich, Hutchins called Struve to Chicago to meet Strömgren, who was in the United States and just then passing through the city. Evidently the chancellor had been in correspondence with the Copenhagen director to find out if he would return to Yerkes again, this time as Struve's successor. He was the only acceptable candidate in the latter's eyes, though he later insisted he had not recommended Strömgren for the post but had merely named him in response to Hutchins's questions, "in a spirit of trying to find an acceptable solution." It was a fine distinction. Evidently the Danish astrophysicist was willing at least to consider accepting the directorship if Struve decided to leave Yerkes.[79]

Although Strömgren's possible availability cleared the way for Struve to accept the Berkeley offer, he still had not decided whether to take it as he and Mary traveled through Europe. In Belgium he discussed the pros and cons with his close friend Swings as they walked in the woods picking mushrooms. During his trip Struve received letters from Page and from Chandrasekhar, making it all too clear that the Yerkes council members were still at odds with him on many matters he considered the director's prerogative. He would have been less than human if he did not contrast their critical actions unfavorably to the adulation he was receiving, with his honorary degree in Liège, his gold medal in London, and the friendly, cooperative messages from Shane.[80]

After his return to Yerkes in mid-October, Struve immersed himself in his work while trying to reach his final decision on the Berkeley offer. By the beginning of December he had nearly done so; he wrote Bowen that it was "more than probable" that he would soon he asking for observing time at Mount Wilson. Finally, on December 15 Struve took the plunge. He sent Bartky a short, noncommittal letter of resignation and Hutchins a longer one. In it he said that his administrative position had been "greatly weakened," that he could not carry through "reforms which [were] needed" or his own research, and that "the future of Yerkes [was] bleak." Under the circumstances, Struve wrote, the only solution was for him to resign and go to Berkeley, where although the "financial conditions" were "only slightly better . . . than my present position," he would face "a great challenge" to build up the department and "an opportunity" to do more research than he could at Yerkes and McDonald.[81] This time he almost really meant it. He still did not notify any of his faculty members, his colleagues elsewhere, or anyone in California of this resignation, and he prepared an eight-page typed memorandum reminding himself, in bitter terms, of all the details behind the bland generalities of his letter to Hutchins. This document can only have been intended as a "talking paper" he could take to Chicago if summoned by the chancellor. It consisted mostly of a list of complaints against what he called "the completely obstructionist attitude of Kuiper and, to a considerable extent, Chandrasekhar, Hiltner, and even Morgan." Basically his accusations boil down to the charge that they had different ideas than he did about what to do with the telescopes, research funds, and scientific visitors at Yerkes and McDonald and would no longer accept his unilateral decisions as binding. But parts of this memorandum are terms or conditions under which Struve would stay at Yerkes; they amounted to freedom from administrative and teaching duties, a director satisfactory to himself (Strömgren), and financial support, office space, and equipment for himself and his assistants on campus equivalent to those he would be getting at Berkeley. Surely by now he knew it could not be, but he could not prevent himself from setting them down.[82]

When Struve went to Chicago this time, however, on January 4, 1950, he found Strömgren with Hutchins in the chancellor's office. The Danish astrophysicist was on his way to Pasadena to lecture as a visiting professor at Caltech for one quarter and had stopped off in Chicago again. Hutchins made it clear that he was accepting Struve's resignation, and Strömgren, by his presence, was demonstrating that he was receptive to an offer, although he had not committed himself. The chancellor said he wanted Struve to retain some connection with the University of Chicago, and the onetime graduate student, faculty member, and director proposed that he be made a permanent research associate or visiting professor, coming back to McDonald to observe with the eighty-two-inch instead of with the sixty-inch and one-hundred-inch at Mount Wilson, and also occasionally to Yerkes for conferences. He would need $2,000 to $2,500 per year in traveling expenses from Chicago and a house permanently reserved for his use on Mount Locke. These demands were quite unrealistic, and no doubt not at all what Hutchins had in mind. Chandrasekhar, who had met separately with the chancellor that same day, would be the acting chairman until a new director was in place. He also quickly agreed that Struve had really resigned, and in a long discussion with him back at Yerkes that evening he ensured that Struve would no longer take part in actions concerning the future of the Astronomy Department.[83]

Only then did Struve write the Berkeley dean that "no insuperable obstacles . . . remained . . . to [his] accepting the offer" and that he would like to come out in February to work out the final "details" (or "package," in present-day terms). They turned out to involve bringing Phillips with him from Yerkes as a junior faculty member, Lillian Ness, his longtime secretary, and Alice Johnson, his scientific assistant. The Berkeley administration met all his demands, and he accepted the position of professor and chairman of its astronomy department, effective June 30, 1950. From there he went on by train to Pasadena, where Bowen, delighted to learn that Struve would soon be "a near neighbor," was "extremely cordial" and assured him that his observing requests would receive "sympathetic consideration."[84] The University of California issued a press release immediately after its Board of Regents meeting on February 25, announcing Struve's appointment as Einarsson's successor, and the news was out at last.[85]

In Pasadena Struve also saw Strömgren, who told him that he would probably accept the Yerkes directorship but still had a few details of his own to work out with Bartky and Hutchins. These included fixing up a house for him and his family at Williams Bay and also a house at McDonald Observatory. The same problems that had plagued Struve's accepting the Yerkes directorship in 1932 were back, complicating Strömgren's decision in 1950. Struve's heart was still at Yerkes, which he thought could be "saved" only by Strömgren, and he implored Kuiper to assure the Danish astrophysicist that he and the rest of the faculty would support him.[86]

Chandrasekhar was in effective charge of the department by February, depending heavily on Kuiper for advice on appointments. They both urged Strömgren to accept the Yerkes post, and Chandrasekhar invited him to stay in his house when he visited the observatory for a week at the end of March.[87] By then Chandrasekhar had been officially named acting chairman of the department to serve until a new director was chosen, had accepted, and had arrived. Strömgren was receptive but still did not commit himself. Struve, however, was convinced that he would take the job and that Yerkes Observatory would survive.[88]

Toward the end of April the astronomers at Yerkes had a second farewell dinner for Struve; this time none of the administrators from Chicago attended. He left for Lick Observatory and his Morrison Research Associateship on May 1, saying good-bye to Hutchins, Colwell, Harrison, and Vice President Emery T. Filbey in Chicago on the way to Midway Airport. True to astronomy to the last, he thanked them all, but especially Harrison for "the quiet and purposeful manner in which you conduct the difficult administrative problems that come to your office," and urged him to use it to "induce" the University of Texas to provide more money for McDonald Observatory.[89]

Struve worked at Lick throughout May, on his first love, measuring and analyzing spectrograms of $\beta$ Canis Majoris, a star with variable radial velocity, which Meyer had taken but never used before his death.[90] In June Struve went back to Williams Bay to clean out his office and turn over his responsibilities to Chandrasekhar, still the acting chairman. Then Otto and Mary Struve drove west at the end of June, arriving at their new home in Berkeley on the last day of the month. He spent most of July at Lick, completing his research work there, then went down the mountain to Berkeley. The Struve era had ended at Yerkes Observatory.[91]

CHAPTER TWELVE

# Epilogue: To the Centennial, 1950–1997

In the summer of 1950 Otto Struve, settled in his new base in California, was back at his first love, observing spectroscopic binaries and variable radial-velocity stars with the telescopes at Lick and Mount Wilson Observatories, where he was a welcome visitor. The situation was quite different at Yerkes Observatory. Parting, especially after a long, close association, can be traumatic. Subrahmanyan Chandrasekhar, the acting chairman, and Gerard P. Kuiper, his closest associate, were left in charge of Yerkes and McDonald Observatories. Struve believed they had driven him out; they believed they were saving the observatory and were suspicious of the former director's actions and motives. Yet Yerkes Observatory did not seem the same with his house and office empty.[1] When Struve resigned, Chancellor Robert M. Hutchins, who had worked so closely with him for so long, had said he hoped the former director could retain a connection with the University of Chicago, and Dean Walter Bartky had seconded him, though only weakly. Struve had converted these wishful thoughts into a dream of a longtime research associate's appointment for himself, under which he would come back to McDonald Observatory from time to time to observe, his expenses paid and a house on Mount Locke permanently assigned to him and his wife, as they had had while he was director. He now had an observing assistant, D. Harold McNamara, a recent Berkeley Ph.D., whom he wished to send to Texas to observe for him if he could not go himself.[2]

Chandrasekhar, when he heard indirectly of Struve's proposal, protested strongly to Bartky that it would be impossible to reserve a house just for Struve and not let anyone else use it. The housing shortage on the mountain was acute. The dean then notified Struve that he would be assigned a full allotment of observing time with the eighty-two-inch reflector during his first year away, and that Chicago would pay his traveling expenses to McDonald Observatory, but that he would have to occupy one of the two houses that other senior astronomers would also use when he was not there. Bartky made no promises beyond 1950–51. Struve could not go for his first observing period at McDonald under this new arrangement in August, so he sent McNamara in his place. He had excellent weather and obtained a large number of good spectrograms for Struve's programs. Chandrasekhar objected to the substitution, maintaining that the former director should have gone himself. Struve replied with barely suppressed hostility that he would give up the rest of his McDonald observing time and resign his research associateship. Chandrasekhar accepted his resignation at once, and neither Struve nor McNamara went back to Texas to observe again.[3]

About this same time Chandrasekhar heard from Hutchins that he had received a letter from the former director, commenting unfavorably on the impact a "loyalty oath" controversy was having on the University of California faculty. Believing that Struve might be angling for a chance to resign his Berkeley appointment and return to Chicago, Chandrasekhar advised strongly against taking him back. It would be "nothing short of a calamity," for in his last two years at Yerkes, and particularly in his last few months, Struve had "succeeded in losing the regard of all his colleagues and it was positively a relief to see him go." At about this same time Kuiper was replying to Jan H. Oort, who had written that Struve's departure from Yerkes was "tragic" and that he could see no way it could possibly help the observatory. Kuiper maintained instead that it would free Yerkes from a two-faced dictator, who had publicly played the role of a second George Ellery Hale but to his close colleagues revealed himself as a tyrant and a despot. In Kuiper's opinion his departure had cleared the way toward a bright future for Yerkes.[4]

Those letters were the low points of Chandrasekhar's and Kuiper's relations with Struve, written in the hot flush of anger soon after he had deserted them and they had helped speed his departure. But time and distance healed the wounds. Struve and Chandrasekhar had been close for years and admired one another's scientific work; they soon were back on friendly, reasonably intimate terms.[5] Kuiper and Struve were not so close, but they shared many opinions about science and other scientists. They too became reconciled, although Kuiper's basic opinion of his former director had not changed; he merely expressed it in much more restrained form.[6]

Figure 48   Yerkes staff, students, and visitors (1950). In front row, from left, Lyman Spitzer Jr., Paul W. Merrill, Subrahmanyan Chandrasekhar, Otto Struve, Dirk Brouwer, Gerard P. Kuiper, Otto Heckmann, and Nicholas U. Mayall. In second row Guido Münch is between Chandrasekhar and Struve, T. D. Lee between Struve and Brouwer, and Karl-Otto Kiepenheuer between Kuiper and Heckmann. W. Albert Hiltner is at far right end of second row, and Su-shu Huang is behind and to right of Lee. In second row from rear, at left, are Stewart Sharpless, Irene L. Hansen (now Osterbrock), and Donald E. Osterbrock. Thornton L. Page is tall man near center of rear row, William W. Morgan is second to right of him, and Aden B. Meinel is at right of that row. Courtesy of Yerkes Observatory.

At Berkeley Struve worked hard to build up a strong astrophysics department. Louis G. Henyey, who had preceded him, and John G. Phillips, who accompanied him, formed the Yerkes nucleus of his new team in California. Lillian Ness became the new departmental secretary and remained his faithful deputy. Alice Johnson, Struve's research assistant, decided to return to Williams Bay, but Su-shu Huang, who had done his Ph.D. work at Yerkes with Chandrasekhar, soon came to Berkeley on a Guggenheim Fellowship and stayed as a research associate. He worked very closely with Struve, supplying theoretical treatments for many of the problems on stellar rotation and close binary stars in which they both specialized. Jorge Sahade, the Argentine astronomer who had collaborated with Struve at McDonald and Yerkes, also spent long periods working with him in California.[7]

As an experienced academic, Struve quickly grasped the differences between the Universities of California and Chicago: more "red tape" and inertia at Berkeley, but more resources if they could be tapped, complicated by the existence of Lick Observatory, a powerful, prestigious, well-supported research astronomy department at an off-campus site. He worked successfully to reduce the teaching loads in the Berkeley department and to provide more money for research leaves, instruments, and equipment.[8]

In 1953 Struve received offers to become director of Perkins Observatory, now operated by Ohio State University, and of Harvard College Observatory. He said he would consider each from the viewpoint of whether he could do more for astronomy there than at Berkeley. At Ohio State there was little potential for building up a research department. Struve let word of the offer seep out, however, and C. Donald Shane notified University of California president Robert G. Sproul of it, warning him of the "great misfortune" it would be for Lick and for Berkeley if the star chairman were to leave. After a hurried series of conferences and memos among the top administrators, Struve got quick action, the promise of much-needed space in a new building (which was built as Campbell Hall) for the rapidly growing astronomy department. Then he declined the Ohio State offer.[9]

Harvard was not so easy. When Harlow Shapley retired in 1952, both Donald H. Me zel and Bart J. Bok considered themselves his logical successor as director. But Harvard University President James B. Conant had other ideas, for a visiting committee chaired by J. Robert Oppenheimer had reported that none of the Harvard astronomers was "first rank" on the national scene.[10] Conant named Menzel temporary acting director while seeking a new face from outside. Oort apparently declined the post before Dean McGeorge Bundy offered it to Struve. The Berkeley chairman considered it very carefully and notified his dean that he was doing so. He visited Harvard and talked with all the senior faculty members at the observatory. This time Henyey and Harold F. Weaver, the first faculty member Struve had hired in the Berkeley department from outside Yerkes (although he had been a postdoctoral fellow there for one year), went to Chancellor Clark Kerr to warn him of the new offer. They, and Struve, made it clear that to keep him at Berkeley would take more funds for astrophysics, to bring in top-level visiting scientists from abroad. Albrecht Unsöld, whom Struve described as "probably the most distinguished theoretical astronomer in the world," had come in the fall of 1950, and Struve wanted more like him. Another hurried series of high-level conferences ensued. Sproul told Kerr that he was "most anxious to keep Struve here, and would do anything . . . within reason to hold [him at Berkeley]." [11]

Then, once again, Struve declined the outside offer and remained in California. He stated that there were two or three "obstacles" Harvard could not meet, but he probably had made up his mind from the beginning to stay at Berkeley and had set his demands to Bundy so high that he knew they would be unacceptable. He was enjoying building up his new research department, and there was too much history to overcome at Harvard, where the observatory faculty, staff, and graduate students were split into opposing camps supporting Menzel and Bok. Apparently Sproul and Kerr came through with a $100,000 war chest for astronomy to keep Struve at Berkeley, and he specified that it should be used to support astrophysics, not "the bottomless pit of orbital work in the solar system." [12]

His department continued to grow and flourish, and everyone who had known Struve in his later years at Yerkes noted how much more relaxed and approachable he now was at the University of California. In 1956, while observing with the sixty-inch telescope at Mount Wilson, he suffered a bad fall, breaking several ribs and cracking two vertebrae. He was hospitalized for more than five weeks and had to wear a body cast for another month, but he seemed to recover and returned to work. [13]

In 1959, however, Struve did leave the University of California, to become director of the nascent National Radio Astronomy Observatory (NRAO) at Green Bank, West Virginia. There were no senior American radio astronomers, and Struve felt it was his duty to accept the post, to become a mentor and role model for the young Ph.D.'s

Figure 49 Otto Struve with slide rule (1955). Courtesy of National Research Council of Canada.

who were the hope of the next generation. Although they revered him at Green Bank, he soon came to realize that with no experience or ideas based on his own work in the field, he could not be a true scientific leader. Disappointed in himself, he decided to resign. Struve was suffering from hepatitis, probably first contracted during his years of privation in the Russian and White armies and in Turkey.[14] After three years at Green Bank, he retired from the NRAO directorship in January 1962, planning to spend part of each year at the Institute for Advanced Study in Princeton and part as a visiting professor at Caltech. He wanted to return to the kind of research he knew best, with telescopes, spectrographs, and gifted theoretical collaborators. But he had little more than a year to live, spending most of it in Pasadena and a few months in Princeton before he was hospitalized, returned to Berkeley, and died on April 6, 1963, at age sixty-five.[15]

In spite of his illness and decreasing strength, in the last five years of his life Struve managed, with the help of his assistants from Berkeley, to complete and publish an elementary textbook on astronomy and a more advanced one, both strongly stressing astrophysics. He also gave an invited series of lectures on astrophysics at the Massachusetts Institute of Technology, emphasizing the contributions radio astronomy would soon be making, and converted them into a third book. He was a compulsive worker and writer to the last.[16]

When Struve left Berkeley in 1959, Henyey succeeded him as chairman of the department. He continued to build its astrophysical strength. Under him and Weaver, his successor as chairman in 1964, it became one of the top astronomy and astrophysics departments in the United States, fulfilling Struve's dream. His longtime secretary, Lillian Ness, who had accompanied him to the NRAO, stayed in Green Bank as librarian after his retirement until she retired herself in 1965 and returned to Seattle, near her birthplace.[17]

Bengt Strömgren did accept the Yerkes directorship and returned to Williams Bay in January 1951.[18] But even before he arrived, Hutchins had announced his own resignation and was to leave the University of Chicago that summer. He had tired of his job, become increasingly at odds with many faculty members, and was under fire from some members of the Board of Trustees for antagonizing wealthy potential contributors to the university. The chancellor's brilliant speeches, always enlivened with humorous attacks on the status quo, made him a hero to the students but a dangerous radical to conservative businessmen, industrialists, and corporate lawyers. Hutchins had been looking for a way out since the end of World War II, and in 1951 he found it as an executive of the new Ford Foundation. Struve visited him at his office in Pasadena a few times, always trying to get support for international activities in astronomy, but with less success than in the old days at Chicago.[19]

There were many similarities between the former boy president and the former boy director, both burned out in their jobs in middle age. They were both tall, striking figures, Hutchins handsome, Struve austere and somewhat forbidding. No one ever ignored either of them. Both followed in their fathers' footsteps, but far from their birthplaces, and much more successfully than their parents in the eyes of the world. Both transformed their high administrative jobs, so that their fathers would hardly have recognized them. Both married wives who initially seemed "above" them, Maude Hutchins a member of New York society, and Mary Struve a native-born American. Neither woman could share her husband's successes, and both came to resent and sabotage them. Hutchins divorced his wife and married his secretary; Struve could never do that, but in his later years Mary Struve

Figure 50 Vesta
Hutchins and Chan-
cellor Robert M.
Hutchins at a farewell
dinner just before he
left the University of
Chicago (1951).
Courtesy of the De-
partment of Special
Collections, Univer-
sity of Chicago
Library.

became more and more a recluse, and he was much closer psycholog-
ically to Lillian Ness than to her.[20]

Strömgren was a great scientist but was far more considerate than
Struve, especially in seeking advice from his senior colleagues and
trying to heed it. Yerkes became a more relaxed research center. As
director, Strömgren was able to put into practice some of his own
scientific ideas, especially the interference-filter spectral classification
system, based on photoelectric photometry, that bears his name. Using
quantitative measurements, it was a brilliant amalgam of William W.
Morgan's empirical, almost artistic concepts with the physical theory
of stellar atmospheres Strömgren knew so well. Milder tensions soon
arose between the new director, Chandrasekhar, and Kuiper, however, some related to teaching, others to the division of resources and

to Strömgren's directorial style. In 1957 he resigned to become the first professor of astrophysics at the Institute for Advanced Study, where he could concentrate entirely on research.[21]

In 1952 Morgan had suffered a nervous breakdown and was hospitalized. He resigned the managing editorship of the *Astrophysical Journal*, and Chandrasekhar, who had become associate managing editor in 1945, succeeded him. The theoretical astrophysicist went on to a long and extremely fruitful editorship. Already making frequent trips to the campus, now as managing editor he drove to Chicago weekly, no matter what the weather, to spend one or two long days at the University of Chicago Press office and in discussions, research conferences, and teaching in the Physics Department. He had begun collaborating with Enrico Fermi when the latter supervised the Ph.D. thesis of future Nobel laureate Tsung-Dao Lee on the internal structure of white-dwarf stars. Soon Chandrasekhar and Fermi published two joint papers on magnetohydrodynamics and the turbulent magnetic fields in the interstellar gas clouds in our Galaxy.[22] The astrophysicist increasingly found the atmosphere of the Physics Department more stimulating for the kind of research he wanted to do than that of Yerkes Observatory. If he needed to stay overnight in Chicago, Chandrasekhar would go to the International House on the campus, where Struve had arranged a reservation for him on his first visit in 1936. In 1959 Chandrasekhar rented a small apartment at the edge of the campus so that he and his wife could have a regular base whenever they wished to stay in the city, and in 1964 they moved from Williams Bay to an apartment overlooking Lake Michigan. A few years later they moved to another apartment, close to International House.

As managing editor of the *Astrophysical Journal* for nineteen years, Chandrasekhar was an outstanding success, very much in the mold of Struve as director of Yerkes. He knew which way the *Journal* should go, demanded the impossible of authors, referees, and editorial staff, and got it. He could be harsh in his judgments, but they were always motivated by his deep commitment to making and keeping the *Astrophysical Journal* the best journal of the field in the world. He succeeded completely, riding the crest of the explosive post-*Sputnik* growth of space science and astronomy, channeling it into expanding the size and readership of the *Journal*, beginning the supplements and separate letters sections, and welcoming every new observational, experimental, or theoretical subfield of astrophysics as it came into being. With his tremendous prestige, Chandrasekhar's decisions could not be seriously questioned by authors, referees, editors at the Press, or even University of Chicago officials. The new chancellor in 1951, Lawrence A. Kimpton, was a successful money raiser and corporate manager, but he had few intellectual pretensions. Yerkes Observatory and the Astronomy Department never again had the close relationship with him or his successors that Struve had enjoyed with Hutchins. But Chandrasekhar, with his combination of great scientific achievements, recognition, presence, and background, was able to cut through the layers of academic bureaucracy and deal directly with successive chancellors, presidents, and provosts. In 1971, when after nineteen years as managing editor he felt that it was time for him to lay down the burden and for the University of Chicago to give up the ownership of the *Journal,* he personally negotiated the agreement under which Provost Edward H. Levi turned it over to the American Astronomical Society. Chandrasekhar continued to write and publish research papers and books, increasingly mathematical as the years went on, long after his retirement. He died in 1995 in the University of Chicago hospital near his home, after a heart attack. Chandrasekhar's contributions to astrophysics were immense.[23]

After Strömgren's resignation, Kuiper became director of Yerkes Observatory in 1957, this time with the real power. All of Struve's predictions came true; Kuiper was a great scientist who found it very difficult to distinguish between his own interests and those of his department and of astronomy. He sincerely believed that the research he was doing was the most important there was. Why else would he be doing it? It followed that all the resources of the university should be devoted to it. Why waste them on less important projects? He was always surprised if his colleagues did not see eye to eye with him, and his reign was brief and stormy. Nevertheless he had become an outstanding planetary astrophysicist who reawakened the whole field of solar-system research and made it once more a growing, vital scientific area. Kuiper was an encyclopedist in the best European tradition. The book he had edited titled *The Atmospheres of the Earth and Planets,* based on the 1947 Yerkes symposium, was revised and updated in 1952. This led to a four-volume *The Solar System* that Kuiper edited, published between 1953 and 1963, and then to the nine-volume *Stars and Stellar Systems: Compendium of Astronomy and Astrophysics,* which appeared between 1959 and 1975, the last volume two years after his death. These works were teaching and reference tools for a generation of astronomers. All of this Kuiper had clearly envisioned at least as early as 1950.[24]

The University of Chicago Press published all these volumes, but Kuiper had departed for the University of Arizona in early 1960, as founding director of its Lunar and Planetary Laboratory (LPL).[25] Morgan succeeded him as Yerkes director, a position he had never really wanted but felt he had to accept to preserve the observatory he loved. Morgan believed to the end of his life that there was a special mystique or ambiance connected with Yerkes Observatory that helped him to do research, and that should have helped others. After his initial discovery of the nearby spiral arms in our Galaxy, he improved and extended his map of their location, based on his spectral classification, done with the forty-inch refractor and with the photographic spectrograph built in the Yerkes shop for the MKK spectral atlas. But much of his later work on the spectra of globular clusters, and on the forms and spectra of galaxies, in which he made many important new discoveries on stellar populations, was based on spectrograms and direct photographs he obtained at Lick, Mount Wilson, and Palomar Observatories. He and Harold L. Johnson together defined and developed the UBV (ultraviolet-blue-visual) photometric system, based on his own deep knowledge of stellar spectra and on Johnson's photoelectric measurements made in Arizona and California. Always Morgan returned to Yerkes to work up the data, analyze it, and describe his results in papers. He gave up the directorship in 1963, remained chairman of the department until 1966, and retired in 1971, but he continued working in research until ill health stopped him. In 1994 he died at his home, a few hundred feet from the observatory where he had spent his entire scientific career.[26]

W. Albert Hiltner succeeded Morgan as director in 1963. He had become an expert in photoelectric photometry, the field Struve had assigned him to in 1946. Hiltner's outstanding discovery, made jointly with John S. Hall of Amherst College and the Naval Observatory in 1949, was the polarization of the light of distant stars by interstellar dust in our Galaxy. It was completely unexpected and revealed that the dust particles are nonspherical and are aligned by a magnetic field. Hiltner made nearly all his observations at McDonald Observatory; Hall, who had begun working with him there, made further measurements at the Naval Observatory in Washington.[27] In 1966 Hiltner handed the Yerkes directorship over to C. Robert O'Dell, who had been on the University of Chicago faculty for only two years. Thus the string of Struve's Yerkes staff members who followed him as directors was broken at last. Hiltner left Yerkes in 1970 to become chairman and director of the University of Michigan Observatory, where he had a long and successful career until his retirement in 1985. He died in Michigan in 1991.[28]

One of the longest-lived Yerkes astronomers, certainly the one with the longest observing career, was George Van Biesbroeck. He had retired in 1945 but continued observing asteroids and comets photographically with the Yerkes twenty-four-inch reflector and double stars with the forty-inch and the McDonald eighty-two-inch for many more years. That observing he did as an unpaid, volunteer emeritus professor, but in the 1950s he began to draw a salary again part time, measuring and analyzing photographic plates in a program organized by Kuiper to amass statistics on the numbers of asteroids. In 1963 the old astronomer moved to Tucson to go back to work under Kuiper, by then the director of the LPL. Van Biesbroeck had observed with the eighty-two-inch McDonald telescope on his eighty-second birthday; in Arizona he observed with the Kitt Peak eighty-four-inch and the Steward Observatory ninety-inch on those respective birthdays; and he liked to say, with a twinkle in his eye, that he would do the same with the 158-inch Mayall telescope. That was not to be however; he died in Tucson at age ninety-four.[29]

Just as Van Biesbroeck was observing double stars with the eighty-two-inch reflector, the University of Chicago was losing control of it. Its original agreement with the University of Texas to operate McDonald Observatory, signed in 1932, was written to expire thirty years later, in 1962. After World War II nearly all state universities, even those that had once been entirely teaching institutions, converted themselves into research centers. Texas, with its strong sense of state pride and its then abundant oil money, was no exception. In 1952 it appointed its first new astronomy faculty member since Harry Y. Benedict's time, Frank N. Edmonds, a Yerkes Ph.D. He was a theorist who had done his thesis under Chandrasekhar, but in 1960 Gerard de Vaucouleurs, an observer who specialized in planetary and galaxy research, became the second member of the new University of Texas astronomy department. McDonald Observatory was an obvious asset to use in building it up. Under the new ten-year agreement, negotiated in 1962, Texas took over responsibility for operating the observatory, with its own director and faculty. Chicago and Texas astronomers would use the telescopes equally, with each university guaranteed at least 25 percent of the observing time. Initially Texas would be hard pressed to use that much, but as its astronomy faculty grew the Yerkes astronomers' share would obviously fall to that limit. In 1963 Harlan J. Smith, a Harvard Ph.D., newly appointed professor of astronomy at Austin, became the first University of Texas director of McDonald Observatory.[30]

The University of Chicago presence at McDonald gradually dropped off, although for many years its astronomers continued to use a part of the observing time there. Their university reimbursed Texas for its proportionate share of the cost of operating the telescope. When Chandrasekhar moved to the campus in Chicago, the center of gravity of the department began to shift there. A forty-inch telescope in a midwestern climate was no longer a first-rate research instrument, except for the specialized programs it was especially suited to—photographic parallax and proper-motion determinations and Morgan's spectral classification work. Both programs depended on continuity and on large amounts of observing time with a moderate-aperture telescope. Before long most of the faculty members and all the teaching, undergraduate and graduate alike, were concentrated on the campus. There were obvious advantages from the point of view of the university as a community of scholars: interactions with other departments, faculty members, and students. There were less obvious disadvantages related to absolute concentration on astronomical research, so much a feature of Struve's years at Yerkes, and a perennial shortage of space on the campus.

As Yerkes Observatory approaches its centennial, there are only four faculty members based there, plus several research associates and fellows and a number of highly skilled technicians. No students take classes at Williams Bay, but several are usually in residence, working on research with faculty members stationed there. More students come in summer to gain observing experience. The faculty members get their observational data as guest observers at McDonald and at Palomar, at national observatories in Arizona and Chile, in Antarctica, on high-altitude airplanes, and above the earth's atmosphere, from the Hubble Space Telescope. Since 1994 they have also been able to observe with the new Astrophysical Research Consortium (ARC) 3.5-meter telescope at Apache Point, New Mexico, a larger reflector on an even higher, drier, and clearer site than McDonald Observatory. Chicago is one of five universities that together make up ARC and operate the telescope.

With its large, well-designed building, thanks to George Ellery Hale's foresight, Yerkes has superb precision shops, where engineers and instrument makers design and construct spectrographs, photometers, imaging cameras, and infrared detector packages for all these observing locations. The University of Chicago is one of the leading institutions of the Center for Astrophysical Research in Antarctica, with headquarters at Yerkes Observatory. It still has one of the outstanding astronomical libraries of the world, with sets of observatory publications and scientific journals going back to the earliest years of astrophysics.

The legacy of the first fifty-three years of Yerkes Observatory is the astrophysics of stars and nebulae. That is what Hale planned it for, and that is what Edwin B. Frost, Struve, and their staff members used it for. As a result of their work, and the work of other astronomers at Mount Wilson Observatory (started as a Yerkes Observatory expedition), Palomar Observatory (which had been in operation just two years before Struve left Yerkes), Harvard College Observatory, Lick Observatory, and many smaller observatories, scientists understood stars and nebulae much better in 1950 than they had in 1897. Astronomers at Yerkes and McDonald did very little research on galaxies, but knowledge of the stars and nebulae that are their building blocks went into understanding them.

Another legacy was the idea of a cooperative observatory, used by astronomers from more than one university, which Struve proposed as a way of building up McDonald Observatory and operating it effectively. Only Indiana University, and to a smaller extent the University of Minnesota, adopted the plan and put money into it before World War II nearly stopped astronomical research and ended the cooperative plan, but Struve's idea guided John B. Irwin's vision of a national photoelectric observatory that led to the founding of Kitt Peak National Observatory.[31] After Kitt Peak's success, Cerro Tololo Interamerican Observatory and the European Southern Observatory were natural extensions of the idea, as were the smaller combinations of institutions formed to operate a single observatory, such as the Wisconsin-Indiana-Yale-National and the ARC groups, closer to Struve's original plan.

The tradition of Yerkes Observatory, going back to Edward E. Barnard, was hard work in a poor climate. Being ready to observe whenever the skies cleared even marginally was a necessity to get results in the upper Midwest. Doing so was drilled into all the graduate students who observed at Yerkes in Frost's day and Struve's, and they took it with them wherever they went.

The history of Yerkes Observatory illustrates very clearly how important single human beings can be in deciding the direction of scientific research, and in the success or failure of specific scientific institutions. Of course the objective situation is very important too: the state of the telescope-building art, understanding of physics and astronomy, the stage of development of detectors, and so forth. But within those constraints, which apply everywhere, specific human beings who understand them and their limitations, who are able to form ideas on how to use or surmount them, and who have the drive to carry them out can and do affect the history of astronomy.

The most important single individual in the history of Yerkes Observatory was Charles T. Yerkes. Scientists like to think of scientists as the movers and shakers of science, but astronomy without telescopes can be barren indeed, as many of Hale's contemporaries could testify. Without "the Titan" there would have been no Yerkes Observatory with its "largest and best . . . telescope in the world." At the end of the previous century, and in the first half of this one, America overtook Germany, Great Britain, Italy, and France, the cradles of astrophysics, and became the world leader in astronomical research. That leadership was based on two unique advantages: numerous fine mountaintop observing sites with clear, stable, good-seeing skies in California, Arizona, and, as we now recognize, Hawaii; and a handful of very wealthy men and women who for one reason or another were willing to provide large sums of money to build observatories. One of them was Yerkes; some of the others were James Lick, Catherine Wolfe Bruce, Andrew Carnegie, John D. Rockefeller and his offspring, and William J. McDonald. Certainly a large part of their reasons involved vanity, but that does not make their telescopes any less useful. Astronomers can never forget their donors.

Yerkes would never have given the money to build the forty-inch telescope had it not been for Hale and William Rainey Harper. Although they differed by twelve years in age and forty pounds or more in bulk, Harper and Hale were similar bundles of nervous energy and organizational ability. Each could attract others to work effectively with him and could separate wealthy capitalists from their money for a good cause. Harper had many previous successes under his belt when he met Yerkes; Hale's only experience in money raising was persuading his own father to finance his projects. But in his first six years under Harper's tutelage he learned to cajole, flatter, state the main objective to the potential donor clearly, forcefully, and succinctly, and above all, never to give up until he had completed the "scheme" and "secured" the necessary funds. By the time Hale's father died in 1898, the Yerkes director was ready to strike out on his own. He succeeded brilliantly in convincing Carnegie's agents to provide the funds that enabled him to use his scientific assets, the sixty-inch mirror he owned and the Snow horizontal telescope he controlled, to found his own independent observatory on Mount Wilson. His mentor Harper, dying in Chicago, resented Hale's breakaway move and tried to block it, but in the end he realized he had taught the younger man well, and he forgave him.

Their successors, Frost and Harry P. Judson, were also similar in having been chief lieutenants who inherited the main job and then kept it long past their prime. Both tried to maintain the status quo but demonstrated that not to advance and change with the times inevitably means to fall behind. The radial-velocity program conceived in 1897 was largely irrelevant by the 1920s; the right successors to Burnham in 1917 and Barnard in 1924 were probably not the astronomers whose work was most nearly like theirs; and Hale's Kenwood twelve-inch telescope of 1891 was not the cutting-edge technology Yerkes Observatory most needed in 1927. Yet it is wrong to be too hard on Frost. He was a fine human being and a humanitarian, and in his talks to schoolchildren, to their parents in church groups, and to general audiences on the campus in Chicago, he probably did more to popularize astronomy than Hale, Struve, and all the other astronomers who ever worked at Yerkes Observatory (except Barnard).

Three of Frost's actions had far-reaching positive consequences for Yerkes Observatory that went far toward redeeming all his inaction and weak appointments. One was his humanitarian action in saving Struve from the White army relief camps in Turkey and bringing him to Yerkes Observatory. The Yerkes director had no idea what an outstanding research worker Struve—then twenty-four years old and with no real graduate training or observational experience—would turn out to be. Frost simply knew the young man was the scion of a great astronomical family. That was enough for him; he wrote letters, called wealthy and influential Chicago friends, and pulled strings until he got the young officer out of Turkey and into the United States, then gave him a job that enabled him to get an education and a staff position. Frost's second important action was his insisting in 1926 that Lick Observatory must not be allowed to hire Struve away from Yerkes. By then the director realized just how good the young Russian was and how much the future of Yerkes Observatory depended on him. Dean Henry G. Gale followed Frost's recommendations, or at least enough of them to keep the eventual savior of the observatory on its staff. This incident alerted Gale to Struve's importance; from then on the dean himself took charge of pushing him upward and binding his commitment to the University of Chicago.

Frost's third important action—like the first, done with no thought of getting an advantage for himself or his observatory, but simply for the good of astronomy—was sending sound advice to Benedict in 1926 on how to get McDonald's bequest and how to spend it to build "the greatest observatory in the South." Benedict had written several directors to ask their counsel; Frost alone, perhaps because he was doing so little else, had the time and interest to reply at length. His letters opened the connection with Benedict that Struve and Hutchins exploited to propose the agreement to build and operate McDonald Observatory, which profited both universities so much.

Struve and Hutchins were both supreme activists, never at rest, always seeking ways to improve their observatory and their university and to solve the problems of astronomy and of the world. Hutchins, who believed in debate, discussion, philosophy, and a type of education most scientists considered impractical and rather useless, nevertheless wanted his university to have the best faculty and to be the best in the world in every subject. He recognized that meant concentrating on research, at least in science, and in Struve he found the person who wanted that and nothing more. As director, chairman, and distinguished professor, Struve never fought the president and chancellor as most of the other science chairmen did. He simply proposed appointments and programs that would improve astronomical research at the University of Chicago. He took whatever Hutchins gave him, and he did improve the department tremendously. In its glory years and just after them, Yerkes Observatory was probably the outstanding university astronomy research department in the world and was second only to Mount Wilson and Palomar Observatories as a research institution. Struve sought the best young prospects he could find, no matter what country they came from, no matter what color their skin. That mode of proceeding appealed to Hutchins, and he supported Struve all the way. It was not how most other observatory directors operated (except Shapley), but that no doubt made it all the more attractive to Hutchins.

The young faculty members Struve chose were excellent; Chandrasekhar, Morgan, Kuiper, Strömgren, and among those who left for greener fields, Pol Swings, Henyey, and Jesse L. Greenstein, were all awarded many important medals and lectureships for their research, as was Struve himself. Yet in the end Yerkes Observatory was too small a stage to hold them all; they clashed, and one by one they left. But is not that the eternal story of life, and is it not the way science must be done? Hale left before he aged, but Frost stayed far too long before he gave way to Struve. It was hard for Frost to give up the directorship in 1932, but long past the time he should have done it. It was even harder for Struve to give it up in 1950, but probably it was time for him to do so. Strömgren was able to flourish briefly at the University of Chicago, and Chandrasekhar for a much longer time, in ways they could not have done had Struve still been director. Berkeley might not have had the outstanding department it has today had Struve had not followed Henyey there and built it up as he did. Certainly Greenstein created another outstanding department at Caltech, and Swings at the Institut d'Astrophysique in Liège. Astronomy thrives on changes that seem harsh when they occur.

The tragedy for Frost and for Struve was that in their time a director was supposed to be an outstanding scientist who kept the post until he retired or died. The only way to get rid of a living director was through what seemed to the outside world to be a knockdown fight. With the appointments of Morgan and Hiltner Yerkes Observatory went over to the modern system of a director appointed for a limited term, with the possibility of reappointment for a second one. It provides for change without revolution. But neither Morgan or Hiltner was willing to commit all his time and effort to finding support, either from a wealthy donor or from a government agency, to build a new, large telescope in a good climate, even as they saw control of McDonald Observatory slipping away to the University of Texas.[32] Without the apparently assured lifetime job Struve had had as director, they were not willing to bet their careers on getting a telescope as he had done. Everlasting directors are not in fashion in our world today, but weak directors and diffused responsibility have brought new problems unknown in Hale's, Frost's, and Struve's time.

As we have seen, Frost was far less creative than his predecessor or his successor. He accomplished less in research than they did, and under his direction the observatory fell behind other astronomy departments in prestige and power. Most astronomers know something about Hale's career and Struve's, but few even recognize Frost's name. Yet is being a great scientist the goal to which all other human qualities must give way? Frost was a warm, friendly person, honored and respected by the citizens of Williams Bay and Lake Geneva. There are a Frost House, Frost Woods, Frost Park, Frost Drive, and Frost Circle in present-day Williams Bay. Its older inhabitants, the children and grandchildren of his contemporaries, honor his memory. There is no Hale Street or Struve Avenue; few in the town except the astronomers would know their names. Frost, at the end of his life, wrote a rather shallow autobiography, in which every page radiates his happiness in himself, his family, his friends, and nature; Struve, at the end of his life, sick and disappointed, wrote ceaselessly about astronomy. Frost died in Billings Hospital surrounded by his wife and children; Struve, childless, died alone in a hospital in Berkeley. Which was better?

Finally, like Sherlock Holmes following up the clue of the dog that did not bark in the night, we might speculate about two appointments that were not made. What would have happened if James E. Keeler, in 1898, had not declined the professorship Harper offered him, but instead had turned down the Lick directorship and gone to Yerkes to take over the stellar and nebular spectroscopic work as Hale had urged him to do? Frost would then not have been a member of the early Yerkes staff. Very probably Keeler would have died in 1900 anyhow, whether he was at Yerkes or Lick; the immediate cause of his death was a cerebral thrombosis, but it seems to have been a consequence of heart disease and lung cancer, presumably the result of years of cigar smoking. If so, Hale would almost certainly have tried to recruit Carl Runge, just as he did in 1898, and if he had declined, would have gone to Frost next. At that time there simply were no other American spectroscopists with any experience and credibility at all. But by 1900 Frost might have felt too well established at Dartmouth to leave his New England home. When Hale moved on to Mount Wilson, he almost certainly would have taken Keeler with him to use the sixty-inch if he was still alive. Who then would have become the second Yerkes director? There were many traditional astronomers who were undoubtedly eager to get their hands on the largest operating telescope in the world; probably Harper would not have promoted Barnard or Sherburne W. Burnham, neither of whom had any real academic credentials, to the post. Most likely it would have been someone like William J. Hussey, then at Lick Observatory, who became the director at the University of Michigan a few years later, or George C. Comstock, already director at the University of Wisconsin. Yerkes Observatory would have lost its astrophysical tradition with Hale. It is a sobering thought and makes Frost's conservative policy of continuing the past a positive virtue.

As a second hypothetical situation, what if Hutchins had heeded the advice of Frost, the senior Yerkes staff members, and Hale and appointed Joel Stebbins director of the observatory in 1932 rather than Struve? Stebbins would not have demanded a large reflector in a good climate, and there probably would have been no Chicago-Texas agreement to build McDonald Observatory. He would have shifted his base from Madison and the 15-1/2-inch at Washburn Observatory to Williams Bay and the forty-inch, but probably he would have done his most significant observing at Mount Wilson, just as he actually did. He had already become a research associate of the Carnegie Institution of Washington in 1931; he built and tested his frequently updated and improved photoelectric photometers in Wisconsin; then he used them to measure interstellar extinction, globular clusters, and galaxies in summer visits to the sixty-inch and the one hundred-inch.[33] From Yerkes, with no teaching duties, he could have gone anytime during the year. Almost certainly Stebbins would have offered Struve the assistant directorship, but the disappointed young spectroscopist might have accepted the Harvard position instead. Whether he stayed at Yerkes or departed for Cambridge, Struve might have proposed the same kind of arrangement with Texas as the one that occurred, but without a strongly committed director behind it, it seems doubtful that a really useful large McDonald Observatory reflector at a good, clear-sky site would have resulted. Probably Struve would have had to do his observing, from Harvard or from Yerkes, on an unsatisfactory, guest observer basis at Mount Wilson. He could not have accomplished nearly as much as he did. Neither would Kuiper, Hiltner, Swings, Greenstein, or any of the other observers who made so many discoveries with the eighty-two-inch reflector. Morgan would have prospered under Stebbins's directorship, but Kuiper would not have been appointed to the Yerkes staff and probably would have gone to the observatory in Java where a job was waiting for him. After a year or two at Harvard Chandrasekhar might have returned to India to stay, and Strömgren might never have left Copenhagen.

These little "what-if" stories are highly speculative. But they do illustrate how important single individuals were in the history of American astronomy in the big-telescope era. A very small number of people were involved; their personal strengths and weaknesses played a crucial part in what did in fact happen. Yerkes Observatory was fortunate indeed to have three outstanding scientists, each committed to astrophysics, at its head for the first fifty-three years of its existence. They brought it down to us as a living, vital, productive research institution.

In the world at large, America had taken the lead in observational astronomy when the Lick Observatory thirty-six-inch refractor went into operation in 1888. The climate in the western United States was so much more favorable than that in Germany or England, and a few

American millionaires were so much more generous (or ostentatious) than their counterparts abroad, that the Old World was never in the race again until well after the period of this book, when European consortia and nations began to build and operate really big reflectors in Chile, Hawaii, and the Canary Islands. The Yerkes forty-inch was marginally larger than the Lick telescope but had fewer clear nights with good seeing; it was soon surpassed by the Mount Wilson sixty-inch, then the Mount Wilson one-hundred-inch, before Struve got the McDonald eighty-two-inch built and into operation. Generally speaking, though not always, the most important discoveries are made with the largest available telescopes at any particular time (or in the most recently exploited new wavelength region in our time). More photons from a star, nebula, galaxy, or unknown object mean more information, and hence the possibility of better understanding.

But understanding only comes to intelligent, active, skilled, well-trained astronomers and astrophysicists. The best telescopes require the best scientists to use them, no matter what language they speak, no matter what their religion, heritage, or skin color. Hale realized this to some extent and put it into operation at Mount Wilson, at least with northern Europeans; under his tutelage Walter S. Adams followed the same line, though less aggressively. Frost did the same at Yerkes, in the persons of Struve and Van Biesbroeck. As director himself, Struve followed it all the way, and the great success of Kuiper, Strömgren, and Chandrasekhar as research scientists showed that it worked. America's ethos had changed too, and after World War II most other universities, observatories, and research centers followed Struve's lead and sought the best astronomers and astrophysicists all over the world. Women astronomers' battle for recognition had just begun in 1950, though; Struve, like Frost, had welcomed them as graduate students and as postdoctoral research associates but never hired a woman as a faculty member. He was ahead of his time, but not that far; in all of America only Cecilia Payne-Gaposchkin had a really top-flight faculty research appointment, at Harvard, and hers came late in her career.

Before World War II America was much weaker than Europe in theoretical astrophysics. Except for Henry Norris Russell, there were almost no astrophysical theorists in astronomy departments or observatories. At Mount Wilson, Hale again took the lead in signing up outstanding Europeans, such as Jacobus Kapteyn (more of an analyzer of observational data on galactic structure and dynamics) and Sir James Jeans, as permanent research associates and paying them to come to Pasadena regularly as theoretical consultants. Frost was not attracted to this idea (and had no means of raising the money in any case), but Struve, from his own experience, clearly recognized the great need for theorists, both as researchers and as teachers. He encouraged Henyey, and brought Strömgren and Chandrasekhar to the Yerkes permanent staff from abroad. Strömgren's appointment was widely applauded; he was an applied theorist who used a telescope himself and shaped his theoretical work toward interpreting observational data and drawing conclusions from them. However, several directors and senior American astronomers including Shapley, advised Struve against adding Chandrasekhar to the Yerkes staff. Even though at that time most of his work was also fairly closely related to interpreting measured data on the masses, luminosities, and radii of stars, Shapley thought there was no place for such a "pure" theorist at an observatory. He wanted to keep Chandrasekhar at Harvard on a Society of Fellows appointment limited to two years, not give him a Harvard College Observatory permanent appointment. But Struve was right and Shapley was wrong, as proved by Chandrasekhar's great research successes, confirmed by so many prizes, awards, and outside offers. He made theory not only acceptable but desirable and even essential at American observatories, and they flourished because of it.

With the McDonald Observatory eighty-two-inch reflector, the second largest telescope in the world until the Palomar two-hundred-inch went into operation in 1948, Struve, Kuiper, Hiltner, Swings, Gerhard Herzberg, and their colleagues did outstanding research on stars, interstellar matter, planets, and comets. Morgan and Philip C. Keenan showed that even an old forty-inch in a poor climate, intelligently used on a well thought out program, could continue to produce important results that also drew recognition throughout the world. They and their former postdoctoral fellows and graduate students were well poised for the next round of frontier science—the drive to understand stellar evolution, galaxies, and the universe—as the 1950s dawned.

# ABBREVIATIONS USED IN NOTES AND BIBLIOGRAPHY

## Names

| | | | |
|---|---|---|---|
| WSA | Walter S. Adams | FE | Ferdinand Ellerman |
| AA | Arthur Adel | CTE | Christian T. Elvey |
| RGA | Robert G. Aitken | JWF | James W. Fecker |
| HWB | Horace W. Babcock | ETF | Emery T. Filbey |
| EEB | Edward E. Barnard | PF | Philip Fox |
| SBB | Storrs B. Barrett | EBF | Edwin B. Frost |
| WB | Walter Bartky | HGG | Henry G. Gale |
| HYB | Harry Y. Benedict | CPG | Cecilia Payne- |
| GVB | George Van Biesbroeck | | Gaposchkin |
| NTB | Nicholas T. Bobrovnikoff | HMG | Harry M. Goodwin |
| | | JLG | Jesse L. Greenstein |
| BJB | Bart J. Bok | GEH | George Ellery Hale |
| ISB | Ira S. Bowen | WRH | William Rainey Harper |
| JAB | John A. Brashear | RWH | R. W. "Pat" Harrison |
| SWB | Sherburne W. Burnham | LGH | Louis G. Henyey |
| EDB | Ernest D. Burton | GH | Gerhard Herzberg |
| JWC | John W. Calhoun | WAH | W. Albert Hiltner |
| SC | Subrahmanyan Chandrasekhar | RMH | Robert M. Hutchins |
| | | JAH | J. Allen Hynek |
| ECC | Ernest C. Colwell | AHJ | Alfred H. Joy |
| AHC | Arthur H. Compton | HPJ | Harry P. Judson |
| CCC | Clifford C. Crump | JEK | James E. Keeler |
| TD | Theodore Dunham | PCK | Philip C. Keenan |
| ASE | Arthur S. Eddington | LAK | Lawrence A. Kimpton |
| SE | Sturla Einarsson | GPK | Gerard P. Kuiper |

| | | | |
|---|---|---|---|
| MM | Max Mason | FS | Frank Schlesinger |
| PWM | Paul W. Merrill | TJJS | Thomas Jefferson |
| GWM | George W. Moffitt | | Jackson See |
| JHM | Joseph H. Moore | CDS | C. Donald Shane |
| WWM | William W. Morgan | HS | Harlow Shapley |
| FRM | Forest Ray Moulton | RGS | Robert G. Sproul |
| RSM | Robert S. Mulliken | HJLS | H. J. Lutcher Stark |
| LN | Lillian Ness | JS | Joel Stebbins |
| JHO | Jan H. Oort | HTS | Harlan T. Stetson |
| DEO | Donald E. Osterbrock | JMS | James M. Stifler |
| TLP | Thornton L. Page | BS | Bengt Strömgren |
| JSP | John S. Plaskett | ES | Elis Strömgren |
| DMP | Daniel M. Popper | MLS | Mary Lanning Struve |
| HPR | Homer P. Rainey | OS | Otto Struve |
| GWR | George Willis Ritchey | PS | Pol Swings |
| FR | Franklin E. Roach | RJT | Robert J. Trumpler |
| HR | Hans Rosenberg | AU | Albrecht Unsöld |
| FER | Frank E. Ross | FW | Frederic Woodward |
| SR | Svein Rosseland | WHW | William H. Wright |
| HNR | Henry Norris Russell | | |

## Archives

AIP   American Institute of Physics Microfilm, Sources for History of Modern Astrophysics
  *Jan Hendrik Oort Selected Correspondence 1927–1956*

ASC   Special Collections, University of Arizona Library, Tucson
  *Bart J. Bok Papers*
  *Gerard P. Kuiper Papers*

CAH   Center for American History, University of Texas at Austin
  *President's Office Records 1907–1968*

CIT   California Institute of Technology Archives, Pasadena
  *Ira S. Bowen Papers*

HCO   Harvard College Observatory Archives, Harvard University Archives, Pusey Library, Cambridge, Massachusetts
  *Harlow Shapley Papers*

HHL   Henry Huntington Library, San Marino, California
  *George Ellery Hale Collection*
  *Mount Wilson Observatory Collection*
  *Walter S. Adams Papers*
  *Ira S. Bowen Papers*

*Alfred H. Joy Papers*
*Paul W. Merrill Papers*

HPM    George Ellery Hale Papers, Microfilm Edition, Carnegie Institution of Washington and California Institute of Technology, Pasadena

KK    Kevin Krisciunas Private Collection, Hilo, Hawaii

LC    Library of Congress, Washington
*Simon Newcomb Papers*
*Thomas Jefferson Jackson See Papers*

RL    University of Chicago Archives, Special Collections, Joseph Regenstein Library
*Thomas C. Chamberlin Papers*
*William Rainey Harper Papers*
*Robert M. Hutchins Papers*
*Presidents' Papers*

SLO    Mary Lea Shane Archives of the Lick Observatory, University Library, University of California, Santa Cruz
*Directors' Papers*
*Nicholas U. Mayall Papers*

UCB    Bancroft Library, University of California, Berkeley
*Otto Struve Papers*
*University of California Archives*

UWA    University of Wisconsin Archives, Memorial Library, Madison
*Department of Astronomy Papers*

VU    Special Collections, Vanderbilt University Library, Nashville
*Edward Emerson Barnard Barnard Papers*

YOA    Yerkes Observatory Archives, Williams Bay, Wisconsin

## Journals

*AJ*       *Astronomical Journal*
*AN*       *Astronomische Nachrichten*
*ApJ*      *Astrophysical Journal*
*ARA&A*    *Annual Review of Astronomy and Astrophysics*
*BAAS*     *Bulletin of the American Astronomical Society*
*BMNAS*    *Biographical Memoirs of the National Academy of Sciences*
*CMO*      *Contributions of the McDonald Observatory*
*HCO*      *Harvard College Observatory*
*JHA*      *Journal for the History of Astronomy*
*JOSA*     *Journal of the Optical Society of America*

JRASC    *Journal of the Royal Astronomical Society of Canada*
MNAS    *Memoirs of the National Academy of Sciences*
MNRAS  *Monthly Notices of the Royal Astronomical Society*
PA       *Popular Astronomy*
PAAS    *Publications of the American Astronomical Society*
PASP    *Publications of the Astronomical Society of the Pacific*
PYO     *Publications of the Yerkes Observatory*
QJRAS   *Quarterly Journal of the Royal Astronomical Society*
S&T     *Sky and Telescope*
YO      *Yerkes Observatory*

# NOTES

## Chapter One   Birth, 1868–1897

1. This account of Harper and Hale's persuading Yerkes to give the money to build "his" observatory, and much of the rest of the chapter, is based on Wright (1966) and Osterbrock (1984b, 1993a), each of which gives copious references to original sources and to the secondary literature. Wright states the date Harper and Hale called on Yerkes as Tuesday, October 2, 1892, but October 2 was a Sunday, when they would not have considered calling and he would not have been in his office. Almost certainly their call occurred on Friday, October 7, the date of the telegram quoted, T. W. Goodspeed to F. T. Gates, Oct. 7, 1892, RL.

2. Much of the material on Harper and the beginnings of the University of Chicago is based on Goodspeed (1916) and Storr (1966), which give many references to the original sources.

3. Walter S. Adams, "Ferdinand Ellerman," *PASP* 52 (1940): 165.

4. Osterbrock (1993a).

5. This account of the World Congress of Astronomy and Astro-Physics is is based on Osterbrock (1980, 1985a), both of which give many references to the original sources. See also Moore, Maschke, and White (1896).

6. The stories of the selection of the Williams Bay site for Yerkes Observatory, and of the later dedication, are based on Osterbrock (1985a), which contains many references to the original sources. The quotation is from the *Lake Geneva Herald*, Dec. 8, 1893.

7. Regarding the *Astronomical Journal* see Comstock (1922), and also "Preamble[s]" to the *AJ*, vols. 1 (1851), 6 (1861), and 7 (1888).

8. The material on the founding of the *Astrophysical Journal* is from Osterbrock (1984b, 1995a), works that contain many references to the original sources.

9. In addition to the sources of note 1, see *Williams Bay Observer*, May 20, 27, 1897.

10. *Williams Bay Observer*, June 3, July 9, 1897.

11. See note 6; *Williams Bay Observer*, Oct. 22, 1897.

12. *Williams Bay Observer*, Oct. 29, 1897.

## Chapter Two   Infancy, 1897–1904

1. Osterbrock (1993a) gives full references to these attempts and others, earlier and later, by Hale to detect the solar corona outside eclipse.

2. Wright (1966) and especially Osterbrock (1993a) give many details of the move from Kenwood to Williams Bay. See also *Lake Geneva Herald,* Sept. 13, 20, 1895; and *Williams Bay Observer,* May 28, Nov. 5, 26, Dec. 3, 10, 17, 31, 1896, Jan. 14, 21, 1897.

3. George E. Hale, "On the Presence of Carbon in the Chromosphere," *ApJ* 6 (1897): 412.

4. George E. Hale and Ferdinand Ellerman, "On the Spectra of Stars of Secchi's Fourth Type, I," *ApJ* 10 (1899): 87; George E. Hale, Ferdinand Ellerman, and J. A. Parkhurst, "The Spectra of Stars of Secchi's Fourth Type," *PYO* 2 (1904): 251.

5. Record Book, School District Clerk, Joint District 7, Towns of Walworth, Delevan, Geneva, and Linn, Walworth County, Wisconsin, and later of Associated Rural School District no. 1, Williams Bay School Board Office.

6. Osterbrock (1993a) gives full references.

7. Burnham (1900).

8. Osterbrock (1993b).

9. SWB to GEH, July 9, 1894, Mar. 8, 1895, YOA.

10. *Williams Bay Observer,* Mar. 19, May 21, Dec. 3, 24, 1896; EEB to WRH, Dec. 28, 1895, Apr. 25, 1896, RL.

11. Hale (1897b).

12. Osterbrock (1984a,b) and especially Sheehan (1995) give Barnard's career in much more detail, with full references to the original sources. The former of these describes Holden's battles with Burnham as well.

13. Osterbrock (1984b) gives full references.

14. Frost (1933); Struve (1938).

15. Edwin B. Frost, "The Bruce Spectrograph of the Yerkes Observatory," *ApJ* 15 (1902): 1; Edwin B. Frost and Walter S. Adams, "Radial Velocities of Twenty Stars Having Spectra of the Orion Type," *PYO* 2 (1904): 143.

16. D. Brouwer, "Frank Schlesinger, 1871–1943," *BMNAS* 24 (1945): 105.

17. Hale (1897b).

18. GEH to JAB, June 1, July 17, 22, 27, [189]2, HHL.

19. GEH to JAB, Sept. 13, [189]2, HHL.

20. Osterbrock (1993a) gives full references.

21. C. H. Rockwell to GEH, Oct. 26, 1897, JEK to GEH, Nov. 1, 1897, E. C. Pickering to GEH, Nov. 8, 1897, YOA.

22. George E. Hale, "The Harvard Conference," *ApJ* 8 (1898): 193; H. R. Donagle, "Photographic Flashes from Harvard Observatory," *PA* 6 (1898): 481.

23. GEH to S. Newcomb, Jan. 5, 16, 1899, LC.

24. [W. J. Hussey], "Third Conference or First Meeting," *BAAS* 1 (1910): 73.

25. Joel Stebbins, "The American Astronomical Society, 1897–1947," *PA* 55 (1947): 490.

26. Osterbrock (1993a) gives full references.

27. WSA to EBF, June 11, July 2, July 23, Aug. 26, 1900; WSA to GEH, Dec. 31, 1900, Mar. 13, 1901, YOA.

28. Wright (1966).

29. GEH to WRH, Sept. 19, 1898, YOA.

30. GEH to N. R. Ream, Jan. 19, 1899, YOA.

31. GEH to WRH, May 9, 1899, YOA.

32. JEK to GEH, Apr. 19, 1899; GEH to JEK, May 10, 1899, YOA.

33. GEH to WRH, Apr. 1, 1902, RL.

34. GEH to J. S. Billings, Jan. 24, 1902, GEH, "A Great Reflecting Telescope," [Jan. 20, 1902], YOA.

35. GEH to E. C. Pickering, Aug. 15, 1902, YOA.

36. GEH to W. W. Campbell, Oct. 4, 1902, GEH to WRH, Oct. 15, 1902, GEH to C. D. Walcott, Oct. 20, 30, 1902, YOA.

37. Wright (1966).

38. GEH to [I. F.] Blackstone, Feb. 14, 1903; Blackstone to GEH, Feb. 16, 1903; YOA.

39. GEH to J. S. Ames, Oct. 13, 1903; GEH to C. G. Abbott, Oct. 21, 1903; GEH to A. Agassiz, Nov. 28, 1903; YOA.

40. GEH to HMG, Jan. 22, 1903, HHL.

41. GEH to WRH, Feb. 11, Mar. 6, 25, 1903, RL.

42. GEH to WSA, Dec. 12, 1903, YOA; GEH to HMG, Dec. 24, 1903, HHL.

43. [GEH] to H. E. Snow, Jan. 30, 1904; G. S. Isham to [GE]H, Jan. 4, 25, 1904; HPM.

44. GEH to HMG, Feb. 28, May 26, 1904, HHL.

45. *Transactions of the International Union for Cooperation in Solar Research* 1 (1906): 1; Walter S. Adams, "The History of the International Astronomical Union," *PASP* 61 (1949): 5.

46. GEH to HMG, Oct. 14, Dec. 21, 1904, HHL.

47. GEH to WRH, Jan. 7, 17, 25, Feb. 11, Mar. 11, 27, 1905; WRH to GEH, Jan. 16, Feb. 17, Mar. 17, 1905; RL; GEH to HMG, Feb. 22, 1905, HHL.

## Chapter Three    Near Death, 1904–1932

1. EBF to GEH, Jan. 8, 20, Feb. 3, [20], Mar. 14, 1904; GEH to EBF, Jan. 12, 1904; HPM.

2. GEH to WRH, Feb. 13, 1904; WRH to GEH, Feb. 25, 1904; RL; GEH to WRH, Mar. 7, 1904, HPM.

3. EBF to GEH, Apr. 29, June 2, 16, 1904; GEH to EBF, May 26, 1904; HPM; EBF to WRH, June 4, 1904; WRH to [E]BF, June 5, 1904; RL.

4. Frost (1933); Storr (1966).

5. GEH to EBF, Nov. 21, 1904, Jan. 17, Feb. 17, 18, 1905; EBF to GEH, Dec. 27, 1904, Jan. 5, 24, Feb. 11, 24, 1905; HPM; EBF to WRH, Jan. 31, 1905, RL.

6. R. S. Woodward to EBF, Feb. 20, 1905; EBF to GEH, Feb. 28, Mar. 2, 16, Apr. 3, 1905; GEH to EBF, Mar. 21, 1905; HPM.

7. GEH to EBF, Feb. 26, Mar. 11, Apr. 7, 1905; EBF to GEH, Mar. 6, 15, 19, 1905; T. W. Goodspeed to [EBF], Mar. 20, 1905; HPM.

8. GEH to [G. S.] Isham, Oct. 27, Dec. 11, 26, 1905; Isham to [GE]H, Dec. 5, 18, 1905; EBF to GEH, Oct. 26, Nov. 15, Dec. 26, 1905; GEH to EBF, Nov. 5, 21, Dec. 10, 21, 1905; W. Heckman to EBF, Apr. 23, 1906, HPM.

9. EBF to GEH, May 24, 1905, HPM.

10. GEH to EBF, Jan. 5, 14, Apr. 26, May 16, June 18, 1905; EBF to GEH, Jan. 14, May 8, 1905; HPM; GWR to EEB, Dec. 21, [19]04, Oct. 3, 1905; VU.

11. Susan F. Schultz, "Thomas C. Chamberlin: An Intellectual Biography of a Geologist and Educator" (Ph.D. thesis, University of Wisconsin, Madison, 1976); S. G. Brush, "A Geologist among Astronomers: The Rise and Fall of the Chamberlin-Moulton Cosmology," *JHA* 9 (1978): 1; T. C. Chamberlin to EBF, Dec. 26, 1905, RL.

12. WRH to GEH, Mar. 17, 1905, RL; EBF to GEH, Apr. 1, 20, 1905, Jan. 26, 1906.

13. *PYO* 2 (1904): 1.

14. Burnham (1900).

15. Osterbrock, Gustafson, and Unruh (1988).

16. Joy (1958).

17. Burnham (1906).

18. Frost (1921); SWB to H. Crew, June 23, [19]14, SLO.

19. Sheehan (1995).

20. EBF to HGG, Mar. 15, 1923; HGG to EBF, Oct. 25, 1923; JS to EBF, Nov. 12, 1923; YOA.

21. F. H. Seares to EBF, June 1, 1923, YOA.

22. WSA to H. S. Pritchett, June 18, 1923, MWO.

23. JS to EBF, Nov. 12, 1923; EBF to A. O. Leuschner, June 15, 1923; Leuschner to EBF, June 19, 1923; YOA.

24. EBF to HNR, Nov. 22, 1923; HNR to EBF, Nov. 28, 1923; YOA.

25. Nicholson (1961), Morgan (1967), and Osterbrock (1994) are partial references; the rest is based on an as yet unpublished study of Ross's career by Osterbrock.

26. FER to [EB]F, Oct. 24, 1923, Jan. 9, 10, 24, 1924; EBF to FER, Nov. 28, 1924; YOA.

27. EBF to J. H. Tufts, Jan. 14, 1924; EBF to FER, Mar. 31, 1924; FER to EBF, Apr. 2, 1924; EBF to HGG, Apr. 4, 1924; YOA.

28. W. S. Eichelberger to [EB]F, Apr. 5, 1924; J. H. Tufts to EBF, Apr. 9, 1924; HGG to EBF, Apr. 10, May 6, 12, June 26, July 10, 1924; EBF to FER, Apr. 11, 22, 1924; FER to [EB]F, Apr. 24, May 19, June 29, 1924; EBF to HGG, May 3, 1924; EBF to J. S. Dickerson, June 19, 1924; J. S. Dickerson to EBF, June 23, 1924; YOA.

29. FER to [EB]F, Aug. 18, Sept. 21, 27, 29, 1924, YOA.

30. FER to [WH]W, July 3, 1924, SLO.

31. FER to [WH]W, Oct. 21, 1925, SLO.

32. Osterbrock (1993a); Frank E. Ross and Mary R. Calvert, *Atlas of the Northern Milky Way* (Chicago: University of Chicago Press, 1934).

33. FRM to EBF, Jan. 12, 1916; EBF to FRM, Jan. 14, 1916; GEH to EBF, Jan. 22, 1916; EBF to GEH, Oct. 15, 1917; YOA; Frost (1933).

34. W. Weaver, "Max Mason," *BMNAS* 37 (1964): 205.

35. [HG]G to [WS]A, Oct. 21, [19]08, HHL; [HGG] to [SB]B, Apr. 14, 1913; HGG to [WSA], [June 19, 1931]; YOA.

36. Crew (1943); T. (1944).

37. Goerge E. Hale, Walter S. Adams, and Henry G. Gale, "Preliminary Paper on the Cause of the Characteristic Phenomenon of Sun-Spot Spectra," *ApJ* 24 (1906): 185.

38. [HGG] to J. E. Raycroft [Apr. 15, 1909], to J. L. Wirt, Apr. 20, May 3,

1909, to R. D. Salisbury, Apr. 20, 1909 to F. D. Nichols, May 14, 1909; J. L. Wirt to HGG, May 26, 1909; RL; [HG]G to [WS]A, [June 15, 1910], HHL.

39. [HG]G to [WS]A, [Jan. 28], [Feb. 12], Apr. 3, 1912, HGG to WSA, July 10, 1912, HHL; [WS]A to [HGG], Oct. 6, 1913, Jan. 10, 1915; HGG to WSA, Jan. 20, 1915, [Apr. 19, 1928], July 8, 1930, May 22, 1931, June 25, 1937; RL.

40. EBF to J. F. Moulds, Jan. 16, 1928, Sept. 29, 1928, YOA; Wolfmeyer and Gage (1976).

41. EBF to HPJ, Sept. 28, 1916; [EB]F and [EE]B to Yerkes Observatory, June 8, [1918] (telegram); EBF to D. E. Felt, June 22, 1918; D. E. Felt to EBF, June 20, 1918; YOA. See also Edwin B. Frost, "Total Solar Eclipse June 8, 1918," *PA* 26 (1918): 458.

42. EBF to W. Wrigley Jr., Aug. 30, Sept. 29, 1922; Wrigley to EBF, Sept. 5, 30, 1922; EBF to HPJ, Sept. 20, 1922; YOA.

43. JS to EBF, Jan. 15, July 31, 1923; EBF to [J]S, Jan. 19, 1923; YOA.

44. EBF to Associated Press, Aug. 13, 1923, YOA.

45. J. Crelinsten, "Physicists Receive Relativity: Revolution and Reaction," *Physics Teacher* 118 (1980): 187–93.

46. RGA to EBF, July 9, 23, 1923; EBF to RGA, July 18, 1923; YOA.

47. EBF to HGG, June 7, 1927; HGG to FW, June 10, 1927; EBF to MM, July 2, 1927; YOA.

48. EBF to MM, Mar. 5, 1928; EBF to JWF, Jan. 21, Apr. 19, 1927, July 14, Nov. 9, 1928, Jan. 29, 1929, Sept. 9, 1930, Mar. 31, 1931; OS to JWF, Dec. 21, 1932, Apr. 10, 1933; YOA. These are only a small sample from the vast files on the ill-fated twelve-inch twin-telescope project.

49. EBF to MM, Apr. 5, 1928; MM to EBF, Apr. 10, 1928; YOA.

50. HGG to RMH, Dec. 30, 1931, RL; see also chapter 5.

51. JHM to [J]S, Feb. 24, 1928; JS to R. H. Curtiss, Jan. 13, 1927; UWA.

52. Morgan (1938).

53. Donald E. Osterbrock, "Science, Religion and Money: Perkins Observatory in the Great Boom and the Great Depression, 1919–1936," *BAAS* 24 (1992): 1167. This is the abstract of a longer paper, presented orally but still unpublished, that gives complete references, including letters, a "history," and interviews of several survivors who knew Crump.

54. EBF to CCC, Sept. 27, 1915, Feb. 2, 1916, May 18, 1918, Dec. 2, 1920, July 23, Oct. 17, 1923, Jan. 14, 1924; CCC to EBF, May 2, 1916, Jan. 3, 1924; YOA.

55. EBF to S. [C. Bartlett], Sept. 17, 1928; EBF to CCC, Sept. 17, 1928; YOA.

56. EBF to [F]W, Apr. 25, May 24, 1930; FW to EBF, June 23, 1930, YOA; SBB to FW, June 27, 1930; EBF to FW, June 30, 1930, RL.

57. AHC to RMH, Oct. 18, 1929, RL.

58. FER to [WH]W, Dec. 27, 1930, SLO; FER to [WS]A, Feb. 5, 1931; WSA to FER, Feb. 16, 1931; HHL.

59. H. D. Curtis to [GE]H, Mar. 24, 1900; GEH to WRH, Mar. 27, 1900; GEH to H. D. Curtis, June 11, 1900; JS to GEH, Dec. 17, 1900; G. C. Comstock to GEH, Dec. 22, 1900; YOA.

60. WRH to GEH, Dec. 19, 1902; GEH to WRH, Dec. 20, 1902; RL.

61. Hughes (1925).

62. Osterbrock, Brashear, and Gwinn (1990).

63. See chapter 4.

64. Osterbrock (1986).

65. EBF to B. A. Wooten, Aug. 3, Oct. 22, 1926, July 2, 1927, Dec. 1, 1928; B. A. Wooten to EBF, Aug. 12, 1926; WWM to EBF, Aug. 19, Sept. 3, 1926, Nov. 1, 1929; EBF to WWM, Aug. 24, 1926; EBF to HGG, Aug. 4, Aug. 28, 1926; YO; WWM transcript, University of Chicago.

66. EBF to HGG, Apr. 17, 1931; HGG to EBF, Apr. 21, 1931; YOA.

67. E. F. Carpenter to RGA, July 16, 1929, Apr. 11, 1930, SLO; PCK to WSA, Dec. 27, 1929, HHL; EBF to FW, Mar. 31, 1930; FW to EBF, Apr. 4, 1930, 1930; YOA.

68. Author's interviews of PCK, Sept. 20, Sept. 23, 1993, Tucson, Arizona. (I was in "the battleship" myself many times in the years 1949–52, visiting fellow graduate students and a postdoctoral fellow who lived there then, and it had not changed much!)

69. Record Book, School District Clerk, Joint District 7, Towns of Walworth, Delavan, Geneva, and Linn, Walworth County, Wisconsin, and later of Consolidated Rural School District no. 1, Williams Bay School Board Office; Lake Geneva News, 1919–48; Frost (1933).

70. [CCC], [brief, almost daily reports on EBF's condition, based on telephone calls from Mary H. Frost, at Billings Hospital, for posting on bulletin board], Oct. 28–Nov. 27, 1931; OS, [memo to staff], [Nov. 22, 1931]; EBF "to the Staff of the Yerkes Observatory," Nov. 23, 1931; OS to EBF, Nov. 18, Nov. 30, Dec. 4, 1931; YOA.

71. FER to RGA, Nov. 23, 1931, SLO; EBF to HGG, Dec. 21, 1931, YOA.

## Chapter Four   The Savior, 1897–1931

1. The family tree is traced in detail in A. H. Batten, "The Struves of Pulkova—a Family of Astronomers," *JRASC* 71 (1977): 345.

2. A. H. Batten, *Resolute and Undertaking Characters: The Lives of Wilhelm and Otto Struve* (Dordrecht: Reidel, 1988).

3. L. Goldberg, "Otto Struve," *QJRAS* 5 (1964): 284; K. Krisciunas, "Otto Struve, 1897–1963," *BMNAS* 61 (1992): 351. These are the two best biographies of our subject, the former especially for the science, the second especially for the details of his personal life. Krisciunas gives many references to early letters tracing his escape from Russia, his travails in Turkey, and his rescue by Frost. Most of the account of Struve's life up to the time he reached Yerkes Observatory is based on these two sources, particularly the latter. Some additional material on his World War I service is from OS to D. Donnell, Nov. 22, 1940, YOA.

4. NTB, "My Life, Part I, 1896–1924" (372-page typed manuscript), [ca. 1960], SLO.

5. The quotations are from EBF to HPJ, Jan. 19, Apr. 14, 1921; EBF to A. Kaznakoff, Oct. 10, 1921; YOA.

6. GVB to RGA, Oct. 8, 1921, SLO.

7. EBF, Aug. 1919, "A Program of Instruction and Research at Yerkes Observatory" (typed program), YOA.

8. EBF to I. L. Carter, June 19, 1922, EBF to Graduate Office, June 20, July 5, 1922; EBF to [HG]G, June 15, 1923; YOA.

9. Otto Struve, "On the Spectroscopic Binary 13 γ Ursae Minoris," *PA* 31 (1923): 90.

10. [Joel Stebbins], "Twenty-eighth Meeting," *PAAS* 5 (1927): 335.

11. GVB to [RG]A, Aug. 4, 1923, SLO.

12. EBF to HGG, June 8, July 13, 1923; OS, "A Summary of My Work at the Yerkes Observatory October 1921–December 1923," [Dec. 8, 1923]; YOA.

13. EBF to Graduate Office, Jan. 10, 1923, YOA.

14. Otto Struve, "A Study of Spectroscopic Binaries of Short Period" (Ph.D. thesis, University of Chicago, 1923), YO Library; Otto Struve, "On the Nature of Spectroscopic Binaries of Short Period," *ApJ* 60 (1924): 167.

15. EBF to Graduate Office, Nov. 8, 1923; EBF, "Report on Final Examination for Degree of Ph.D." [of OS], Dec. 8, 1923, YOA.

16. EBF to J. H. Tufts, Apr. 17, Apr. 30, 1923; J. H. Tufts to EBF, Apr. 24, May 17, 1923; RL; EBF to HGG, Nov. 22, 1923, YOA; Edwin B. Frost, "Yerkes Observatory" [annual reports for 1922–23 and 1923–24], *PAAS* 5 (1927): 135, 241.

17. Edwin B. Frost, "Yerkes Observatory" [annual report for 1924–25], *PAAS* 5 (1927): 340.

18. Frost (1933).

19. OS, "Abridged Record of Family Traits," [1938], National Academy of Sciences Archives; EBF to M. M. Lanning, Aug. 20, 1923; Lanning to [EB]F, Aug. 22, 1923; YOA; EBF to RGA, Nov. 24, 1922, May 3, 1923, SLO.

20. Naomi Greenstein, "Reminiscences Prompted by Questions from Kevin Krisciunas" (manuscript), Jan. 21, 1988, KK.

21. Marriage license no. 7080, [p.] 248, May 21, 1925, Muskegon County, Michigan.

22. OS, "Biographical Information, University of California President's Form no. 1501," Apr. 3, 1950, UCB.

23. GVB to RGA, May 2, 1926, SLO.

24. Edwin B. Frost, Storrs B. Barrett, and Otto Struve, "Radial Velocities of 368 Helium Stars," *ApJ* 64 (1926): 1.

25. Otto Struve, "On the Calcium Clouds," *PA* 33 (1925); 639; continued in *PA* 34 (1926): 1.

26. Joel Stebbins, "Thirty-fourth Meeting," *PAAS* 5 (1927): 279.

27. Otto Struve, "Changes in the Bright Lines in 15 κ Draconis," *PAAS* 5 (1925): 294.

28. Otto Struve, "On the Motions of the Calcium Clouds in Space," *PAAS* 5 (1925): 294.

29. Payne-Gaposchkin (1984).

30. Otto Struve, "Stellar Atmospheres: A Contribution to the Study of High Temperature Ionization in the Reversing Layers of Stars," *ApJ* 64 (1926): 204.

31. [RGA] to EBF, Mar. 19, 1926; EBF to RGA, Mar. 24, 1926; SLO.

32. EBF to HGG, Mar. 25, Apr. 5, 1926, YOA.

33. Osterbrock, Gustafson, and Unruh (1988, 169).

34. OS to WSA, Mar. 5, Apr. 15, 1926; WSA to OS, Mar. 10, 1926; HHL.

35. PWM to OS, Jan. 30, Feb. 12, 1926; OS to PWM, Feb. 3, 5, 15, 1926; HHL.

36. AHJ to EBF, July 13, 1926, HHL; OS to EBF, Sept. 13, 1926, YOA.

37. OS to PWM, Nov. 10, 27, 1926; PWM to OS, Nov. 20, 1926; HHL.

38. Otto Struve, "Interstellar Calcium," *ApJ* 65 (1927): 163.

39. WSA to OS, Dec. 18, 1926, HHL; OS to RGA, Jan. 28, 1927; [RGA] to OS, Feb. 1, 3, 1927, SLO.

40. JSP to OS, Dec. 22, 1926, HHL.

41. JSP to OS, Dec. 28, 1926, HHL.

42. OS to JSP, Jan. 5, 1927; WSA to OS, Jan. 17, 1927; OS to WSA, Jan. 23, 1927; HHL.

43. See note 38 above.

44. [Joel Stebbins], "Thirty-seventh Meeting," *PAAS* 6 (1931): 1; Otto Struve, "On the Effect of Distance on the Intensity of Detached Calcium Lines," *PAAS* 6 (1931): 16.

45. HS to OS, Nov. 22, 1926; OS to HS, June 22, 1927; HCO.

46. Bok (1978); Gingerich (1990).

47. HS to OS, June 24, 1927, HCO.

48. [R. S. Dugan], "Thirty-eighth Meeting," *PAAS* 6 (1931): 19; Otto Struve, "The Spectrum of Nova Aquilae 1927," *PAAS* 6 (1931): 34.

49. OS to HS, Sept. 13, Oct. 21, 1927; HS to OS, Oct. 25, 1927; HCO; OS to WSA, Oct. 24, 1927, HHL.

50. Otto Struve, "Further Work on Interstellar Calcium," *ApJ* 67 (1928): 353; Otto Struve, "Intensities of Calcium Lines in Early Type Stars," *HCO Bulletin,* no. 857 (1928): 11.

51. OS to HS, Feb. 17, Mar. 1, 1928; HS to OS, Mar. 1, 1928; HCO.

52. See note 20 above.

53. Douglas (1957); Kilminster (1966).

54. Arthur S. Eddington, "Diffuse Matter in Interstellar Space," *Proc. Roy. Soc. Lond.,* ser. A, 111 (1926): 424.

55. OS, "Plan for Study," [Nov. 11, 1927], HHL.

56. [WSA], "Form Sent to Guggenheim Foundation—Confidential Report on Candidate for Fellowship," Dec. 21, 1927, HHL; HS to H. A. Moe, Feb. 1, 1928, HCO.

57. OS to [H]S, Aug. 6, 1928, HCO.

58. Otto Struve, "The Determination of Stellar Distances from Intensities of the Detached Calcium Line K," *MNRAS* 89 (1929): 567.

59. "Report of the Meeting of the Royal Astronomical Society," *MNRAS* 89 (1929): 221; seating plans from the privately published records of the RAS Club by permission.

60. JHO to [O]S, Nov. 21, 1928; OS to JHO, Dec. 15, 1928; AIP.

61. EBF to OS, Dec. 19, 1928, Apr. 18, 1929; OS to [EB]F; Mar. 16, Apr. 18, 1929; EBF to G. Struve, Jan. 27, 1930; YOA.

62. OS to [H]S, Aug. 6, 1928, HCO; B. P. Gerasimovich and Otto Struve, "Physical Properties of a Gaseous Substratum in the Galaxy," *ApJ* 69 (1929): 1.

63. Struve (1957); McCutcheon (1991).

64. G. Shajn and Otto Struve, "On the Rotation of the Stars," *MNRAS* 89 (1928): 222.

65. Struve (1958); McCutcheon (1991).

66. G. Tikhoff to O[S], Sept. 11, 1928; OS to [EB]F, Oct. 13, 1928; EBF to W. W. Husband, Commissioner General of Immigration, Nov. 9, 1928; YOA.

67. OS to EBF, Mar. 16, 1929, YOA; Otto Struve, "The Stark Effect in Stellar Spectra," *ApJ* 69 (1929): 173.

68. Otto Struve, "Pressure Effects in Stellar Spectra," *ApJ* 70 (1929): 85; Otto Struve, "The Stark Effect as a Means of Determining Absolute Magnitudes," *ApJ* 70 (1929): 237.

69. Edwin B. Frost, Storrs B. Barrett, and Otto Struve, "Radial Velocities of 500 Stars of Spectral Class A," *PYO*, pt. 1 (1929): 1.

70. OS to WSA, Dec. 9, 1929; WSA to OS, Dec. 24, 1929; HHL.

71. WSA to EBF, Mar. 17, 1925, HHL.

72. Donald E. Osterbrock, "Walter Baade, Observational Astrophysicist (1): The Preparation," *JHA* 26 (1995): 1.

73. OS to WSA, Apr. 1, May 14, 1930; WSA to OS, Apr. 5, 1930; HHL.

74. OS to WSA, Aug. 3, 1930; PWM to OS, July 24, Aug. 12, 1930; OS to PWM, Aug. 3, 15, 1930; HHL.

75. EBF to HGG, Apr. 17, 1929, YOA; Christian T. Elvey, "The Contours of Hydrogen Lines in Stellar Spectra," *ApJ* 71 (1930): 191; Christian T. Elvey, "The Rotation of Stars and the Contours of Mg$^+$ 4481," *ApJ* 71 (1930): 221.

76. WWM to EBF, Nov. 1, 15, 1929, Feb. 26, 1930, YOA.

77. William W. Morgan, "Studies in Peculiar Stellar Spectra. I. The Manganese Stars," *ApJ* 73 (1931): 104; "II. The Spectrum of BD −18°3789," *ApJ* 74 (1931): 24; William W. Morgan, "On the Occurrence of Europium in A-Type Stars," *ApJ* 75 (1932): 46.

78. OS, "Budget Recommendations for Year 1932–33, Dept. of Astronomy and Astrophysics," [Nov. 30, 1931], YOA.

79. Ledoux (1987).

80. PS to [WS]A, Apr. 8, 1931; P. C. Gilpin to WSA, Apr. 15, 1931; WSA to Gilpin, Apr. 21, May 2, 1931; [WSA] to PS, Apr. 30, 1931; HHL.

81. P. C. Gilpin to WSA, Apr. 27, 1931, HHL; PS to [EBF], July 2, [19]31, YOA.

82. Otto Struve and Pol Swings, "On the Interpretation of the Emission Lines in Stars of Early Spectral Class," *ApJ* 75 (1932): 161; Pol Swings and Otto Struve, "The Bands of CH and CN in Stellar Spectra," *Physical Review* 39 (1932): 142.

83. P. C. Gilpin to WSA, May 5, 1931; PS to [WS]A, Jan. 1, 1932, Oct. 14, 1935, Jan. 6, 1936; WSA to PS, Oct. 22, 1935; HHL; PS to EBF, Jan. 2, 1932, YOA.

84. PS to OS, Aug. 29, 1932, Jan. 8, May 24, 1933; OS to PS, Sept. 23, 29, 30, 1932, June 18, 1933; YOA.

## Chapter Five    The Boy President, 1929–1932

1. EBF to OS, May 4, 1929, YOA.

2. The material on Robert M. Hutchins and on the University of Chicago, its administration, and its trustees in this chapter and throughout the book, where not otherwise referenced, is largely from Dzubak (1991), McNeill (1991), and to a lesser extent Ashmore (1989).

3. HGG to EBF, Jan. 28, Oct. 26, 1931; EBF to HGG, Jan. 30, 1931; YOA.

4. RMH to "Chairmen of Departments," Nov. 17, 1930, YOA.

5. RMH to EBF, May 4, 1931; RMH to W. H. Clark, May 4, 1931; EBF to RMH, May 11, 1931 [three letters same date]; YOA.

6. EBF to [F]W, July 28, 1931, RL.

7. See chapter 3, notes 56, 57; RMH to AHC, Oct. 24, 1929, RL.

8. J. A. Firbank to HHS, June 29, 1931; HHS to J. A. Firbank, July 3, 1931; HHS to RMH, July 3, 1931; [F]W to RMH, July 16, 1931; FW to EBF; Aug. 21, 1931, RL.

9. A. E. Whitford, "Joel Stebbins, 1878–1966," BMNAS 49 (1978): 293.

10. J. H. Tufts, "Memorandum to the President," Jan. 19, 1926, RL.

11. JS to EBF, Nov. 3, 6, 21, Dec. 22, 28, 1928, EBF to JS, Nov. 5, Dec. 21, 1928, YOA; [JS] to EBF, Nov. 11, 1928, UWA.

12. EBF to JS, Jan. 2, 16, 1929; JS to EBF, Jan. 17, Feb. 6, 1929; GWM to JS, Jan. 29, Feb. 2, 1929; YOA; [JS] to EBF, Feb. 5, 6, 19, 1929, UWA.

13. JS to EBF, Apr. 16, 23, May 15, 16, June 10, 15, 1929, YOA.

14. EBF to F. B. Jewett, Nov. 22, 1929; JS to EBF, Dec. 11, 18, 1929, Jan. 25, Mar. 27, Oct. 17, 1930; EBF to [J]S, Jan. 23, 1930; EBF to FW, Jan. 11, 1930; JS to CTE, Nov. 5, 1930; YOA.

15. EBF to [GE]H, Mar. 12, 1929, HPM; EBF to [WS]A, Mar. 12, 1929, HHL. Hale was in England when he received this letter, and he endorsed it "Ans. Apr. 5 from London, strongly approving choice of Stebbins, both for Y.O. & Journal." No reply has turned up from Adams (Frost no doubt kept both their letters in his personal files), but in his later letter to Dean Gale, referenced below, he strongly supported Stebbins.

16. EBF to RMH, May 11, 1931; HHS to FW, July 17, 1931; RL.

17. HGG to WSA, May 22, 1931, HHL.

18. WSA to HGG, June 1, 1931, HHL.

19. This attitude is explored in general, and in the case of Adams in particular, in Donald E. Osterbrock, "The Appointment of a Physicist as Director of the Astronomical Center of the World," JHA 23 (1992): 155.

20. HGG to WSA, June 10, 1931, HHL.

21. HGG to WSA, June 13, 1931, HHL; FW to HHS, July 20, 1931, RL.

22. HGG to RMH, Nov. 3, 1931, RL; GVB to RGA, Nov. 17, 1931, SLO.

23. OS to HS, Nov. 25, 1931; OS to W. de Sitter, Nov. 27, 1931, YOA.

24. HGG to EBF, Dec. 11, 1931; HGG to EBF, Feb. 29, 1932, RL.

25. FR to DEO, May 30, 1991.

26. FER to [WH]W, Dec. 27, 1930; FER to [RG]A, Nov. 23, Dec. 4, 28, 1931; [RGA] to FER, Dec. 22, 1931, SLO; FER to WSA, Apr. 13, 1931, HHL.

27. HGG to EBF, Feb. 3, 18, 1932, EBF to HGG, Feb. 8, 1932, YOA.

28. HS to OS, Feb. 8, 1932; OS to HS, Feb. 18, 1932, HCO.

29. HS to OS, Feb. 25, 1932, OS to H. H. Plaskett, Mar. 11, 1932, HCO.

30. OS to HS, Feb. 29, 1932, HCO.

31. HS to OS, Mar. 3, 1932 (telegram and letter, both of same date), HCO.

32. HS to HGG, Mar. 3, 1932, HCO; HS to EBF, Mar. 3, 1932, YOA.

33. OS to HS, Mar. 8, 1932, HCO.

34. HGG to HS, Mar. 7, 1932, HCO.

35. HS to OS, Mar. 13, 1932, HCO; HNR to HGG, Mar. 9, 11, 1932; HNR to OS, Mar. 11, 1932; RL.

36. L. E. Dickson to [HY]B, [ Feb. 27, 1932], CAH; HGG to HNR, Mar. 14, 1932, RL.

37. OS to HS, Mar. 18, 1932, HCO; A. H. Farnsworth to EBF, Mar. 26, 1932, EBF to A. H. Farnsworth, Mar. 31, 1932, YOA.

38. HHS to RMH, Dec. 22, 1931, RL.

39. HGG to WSA, Mar. 22, 1932, WSA to HGG, Mar. 24, 1932, HHL.

40. Evans and Mulholland (1986) is the main source for the story of the beginnings to McDonald Observatory. Their book contains a wealth of information from within the University of Texas, but its authors did not understand the situation at Yerkes Observatory and the University of Chicago at the time.

41. HYB to TJJS, Aug. 28, 1905, LC. Benedict's becoming a member of the AAS is from PAAS 5 (1927): 280.

42. HYB to EBF, Feb. 17, 1926; EBF to HYB, Feb. 25, 1926; YOA.

43. HYB to RGA, Mar. 26, 1926; RGA to HYB, Apr. 2, 1926; SLO; HYB to EBF, Mar. 27, 1926; EBF to HYB; Apr. 1, 1926, YOA; fifteen other letters from other directors and senior astronomers to HYB, Mar. 30–June 17, 1926, CAH.

44. HYB et al. to W. L. W. Splawn, May 21, 1927; J. W. Kuehne to HYB, July 29, [Sept. 5], 1927; HYB to TJJS, June 17, 1929; CAH.

45. HYB to CCC, Feb. 1, 1932; CCC to HYB, Feb. 10, 1932, YOA. Struve (1947a, 1962). The second of these last two references, written by Struve just a few months before his death, which he knew was approaching, was his frankest published statement about the beginnings of McDonald Observatory. He wrote it because "most of the people involved . . . are no longer living." He himself was fully rational when he wrote it, but he did not include any dates in this popular article.

46. See note 33 above.

47. HYB to J. S. Phenix, Feb. 4, 1932; F. M. Little to RMH, Mar. 22, 1932; CAH; see also HYB to RJT, Jan. 29, 1932, SLO.

48. OS to HS, Mar. 29, Apr. 3, 1932, HCO.

49. HS to OS, Mar. 31, Apr. 6, 1932, HCO.

50. OS to HS, Apr. 10, 1932, HCO; EBF to OS, Dec. 19, 1928, May 4, 1929, YOA.

51. Struve (1940, 1947a); Evans and Mulholland (1986, 26, 33); RMH to HYB, Apr. 4 (telegram), Apr. 18, 1932; HYB to OS, Apr. 7, 1932 (telegram); OS to HYB, Apr. 12, 1932 (telegram); HYB, "Joint Operation of the McDonald and Yerkes Observatories" (memo), Apr. 22, 1932; CAH.

52. "Appointment Certificate," signed by F. W. Hunnewell, secretary, Apr. 14, 1932, HCO.

53. OS to HS, Apr. 19, 1932, HCO.

54. HGG to EBF, Apr. 19, 1932, YOA; HGG to WSA, Apr. 19, 1932; WSA to HYB, May 7, 1932; HHL.

55. WSA to HYB, May 7, 1932, HHL.

56. RMH to GEH, Feb. 11, 1932; GEH to RMH, Feb. 16, 1932; GEH to [HG]G, Apr. 26, [1932]; RL; HYB to GEH, May 2, 5, Sept. 29, 1932; GEH to OS, July 7, 1932; GEH to HYB, Sept. 7, Oct. 5, 1932; HPM.

57. Evans and Mulholland (1986, 26–29). The full text of the agreement, except for the date it was finally signed in late September 1932, is attached to GEH to HYB, Oct. 5, 1932, HPM.

58. R. L. Moore to [HY]B, May 9, 1932, CAH; OS to HNR, May 10, 1932; HNR to OS, May 17, 1932; YOA.

59. OS to HS, May 10, 1932; HS to OS, May 14, 1932; YOA; OS to HS, July 11, 1932, HCO.

60. OS to FS, May 10, 1932; FS to OS, July 22, 1932; YOA.

61. OS to WSA, May 11, 1932; WSA to OS, May 13, 23, 1932; HHL.

62. Christian T. Elvey, "The Climate at the McDonald Observatory," *CMO*, no. 1 (1940): 18.

63. OS to HGG, Mar. 18, Apr. 30, 1932, YOA.

64. FR to DEO, May 30, 1991.

## Chapter Six   The Boy Director, 1932–1936

1. A. F. Caftan to EBF, May 31, 1932; A. F. Caftan to OS, Dec. 30, 1932, YOA.

2. EBF to W. H. Garrett, Apr. 5, 1932, YOA.

3. Frost (1933, 259–61).

4. OS to HGG, May 25, 1932; OS to RMH, May 25, 1932; J. P. Moulds to H. E. Servis, June 2, 1932; RL.

5. HGG to OS, July 9, 18, 1932; OS to HGG, July 21, 1932; YOA.

6. OS to HGG, Aug. 15, 1932; HGG to OS, Aug. 19, 1932; YOA; Frost (1933, 262–63).

7. Frost (1933, 262–63).

8. Philip Fox, Joseph H. Moore, Walter S. Adams, and Edwin B. Frost, "Eclipse Reports," *PA* 40 (1932): 459.

9. PF to EBF, Oct. 14, 1931; EBF to PF, Oct. 15, 1931; YOA.

10. PF to OS, Nov. 4, 14; Dec. 17, 1932; OS to PF, Nov. 11, Dec. 19, 1932; PF to P. J. Byrne, Nov. 30, 1932; YOA.

11. *Chicago Tribune*, May 26, 28, 1933; Otto Struve, "Arcturus and the Century of Progress," *University of Chicago Magazine* 25 (1933): 312; Frost (1933, 256).

12. OS to R[G]A, May 13, 1935, YOA; *Williams Bay Leaves*, May 23, 1935.

13. S. B. Nicholson, "Frank Elmore Ross, 1874–1960," *PASP* 73 (1961): 182; William W. Morgan, "Frank Elmore Ross," *BMNAS* 39 (1967): 391; HGG to OS, Apr. 21, 1933, YOA.

14. OS to HGG, Mar. 21, 1933; HGG to OS, Mar. 23, 1933; YOA; OS and FER to "Dear Colleague," Mar. 25, 1933, HCO.

15. OS to HGG, June 6, 1933, YOA.

16. S. Arend, "In memoriam Georges-Achille Van Biesbroeck," *Ciel et Terre* 90 (1974): 321; P. Muller and P. Baize, "G. A. Van Biesbroeck (1880–1974)," *L'Astronomie* 88 (1974): 306.

17. OS to HGG, Jan. 7, July 3, Aug. 9, Nov. 10, 1933; HGG to OS, Jan. 9, July 7, Aug. 7, 1933; YOA.

18. OS to GEH, July 13, 1932, GEH to OS, Sept. 7, 1932, HPM.

19. V. Gushee to C[CC], Feb. 28, 1932, CCC to V. Gushee, Mar. 2, 1932, YOA.

20. Clifford C. Crump, "β Cephei," *ApJ* 79 (1934): 246; Cliford C. Crump, "The Radial Velocity of δ Ceti," *ApJ* 79 (1934): 351. The dates when the author had completed the papers were published at the ends of them, the standard practice at that time.

21. M. D. Welch to OS, Feb. 15, 23, 1940; OS to M. D. Welch, Feb. 20, 1940; YOA.

22. PCK, interview with author, Oct. 8, 1986.

23. CCC to OS, Aug. 11, 1932, Mar. 2, 1933, YOA. The second of these, stating that Crump would not be back to observe on Saturday night, was endorsed by Struve "file."

24. CCC to JS, Apr. 4, 1933; JS to CCC, Apr. 7, 1933, UWA.

25. Ripon College press release, May 16, 1957, Earlham College Archives, Richmond, Indiana; T. D. Hamm, Earlham College, to DEO [Aug.] 14, 1991.

26. OS to HGG, May 8, Nov. 22, 1933, YOA.

27. J. S. Dickerson to OS, Nov. 23, 1932; OS to J. S. Dickerson, Dec. 8, 15, 1932; YOA.

28. OS to HGG, June 14, 1933; OS to W. B. Harrell, Mar. 26, Apr. 7, July 7, 1936; W. B. Harrell to OS, Apr. 2, July 3, 1936; OS to L. R. Flook, July 3, 18, 23, 1936; WWM to L. R. Flook, July 16, 1936; YOA.

29. Otto Struve and Christian T. Elvey, "Preliminary Results of Spectrographic Observations of 7 ε Aurigae," *PYO* 7, pt. 2 (1930): 1; see also chapter 9.

30. Otto Struve, "17 Leporis: A New Type of Spectrum Variable," *ApJ* 76 (1932): 85; Otto Struve, "On the Absorption Lines of Hydrogen in Be Stars," *ApJ* 76 (1932): 309; Otto Struve, "Notes on Be Stars," *ApJ* 76 (1932): 210.

31. HTS to RGA, Sept. 15, 1928, F. B. Stetson to RGA, Apr. 4 (telegram), 29, 1929; RGA to W. L. Smyser, May 14, 1929; SLO; HTS to EBF, Feb. 4, Oct. 23, 1929; R. H. Curtiss to EBF, June 15, 1929; YOA.

32. Nicholas T. Bobrovnikoff, "Astronomy at Ohio State University and Perkins Observatory" (manuscript), Jan. 2, [19]69, Department of Astronomy Records, Ohio State University; HTS to [O]S, Apr. 15, 1932, YOA.

33. OS to HTS, May 6, 22, 1932; HGG to [O]S, May 20, 1932; HGG to HTS, May 20, 1932; HTS to HGG, June 9, 1932; OS to HS, July 11, 1932; YOA.

34. OS to HGG, June 10, 1932; OS to HTS, June 18, July 15, 30, 1932; OS to W. R. Rayton, July 19, 1932; YOA.

35. Donald E. Osterbrock, "Franklin E. Roach, 1905–1993," *BAAS* 26 (1995): 1608.

36. OS to HTS, Sept. 22, 1932; HTS to OS, July 18, Aug. 6, Sept. 23, 30, 1932; OS to RMH, Oct. 1, 1932; FR to OS, Oct. 9, 22, 1932; [untitled agreement, Oct. 10, 1932]; RMH to E. D. Soper, Oct. 11, 1932 Soper to RMH, Oct. 14, 1932; YOA.

37. OS to HNR, Oct. 8, 1932; HTS to OS, Oct. 17, 25, 29, 1932; OS to HTS, Oct. 26, 1932; OS to FR, Oct. 25, 26, 27, 1932; FR to OS, Oct. 27, Nov. 3, 1932; OS to HGG, Nov. 3, 1932; YOA.

38. OS to FR, Nov. 12, 16, Dec. 6, 7, 20, 21, 1932; FR to OS, Nov. 12, 15, Dec. 10, 13, 19, 1932; YOA.

39. FR to [CC]C, Nov. 29, 1932; FR to OS, Dec. 3, 1932; HTS to OS, Dec. 6, 1932; YOA.

40. OS to FR, Dec. 14, 1932; OS to [HT]S, Jan. 3, 1933; YOA; Otto Struve,

"The Width of Bright Hα in γ Cassiopeiae," *ApJ* 77 (1933): 66; Otto Struve, "An Emission Line of Hydrogen in the Spectrum of Rigel," *ApJ* 77 (1933): 67.

41. FR to GWM, Nov. 28, 1932; OS to FR, Jan. 15, 1933; YOA Actually the correct factor is not the square of the aperture, as Roach and Struve thought, but the first power, more like 1.7 than 3, but it was clear that the sixty-nine-inch should have been faster.

42. OS to HTS, May 3, 20, July 3, Aug. 23, 1933; HTS to OS, May 4, July 8, 14, Aug. 31, 1933; OS to FR, May 3, 1933; YOA.

43. NTB, note 32 above.

44. FR to OS, Oct. 17, 31, Nov. 25, 1933; OS to HGG, July 3, Oct. 31, 1933; HTS to [O]S, Dec. 18, 1933; YOA.

45. E. D. Soper to OS, Sept. 9, 12, Oct. 31, 1933; OS to E. D. Soper, Sept. 11, 1933; OS to HNR, Oct. 2, 1933; OS to HTS, Oct. 24, 1933; YOA.

46. AHC to OS, Oct. 17, 1933, YOA.

47. OS to HTS, Nov. 28, Dec. 19, 1933; HTS to OS, Nov. 29, Dec. 21, 1933; FR to OS, Nov. 25, Dec. 17, 1933; E. D. Soper to OS, Dec. 19, 27, 1933; NTB to E. D. Soper, Dec. 20, 1933; OS to E. D. Soper, Dec. 22, 1933; NTB to O[S], Nov. 24, 1933; YOA. At times of crisis, Bobrovnikoff wrote to Struve in Russian, to preserve secrecy, although Struve frowned on this practice and never followed it himself. This and other letters Bobrovnikoff wrote in Russian were kindly translated for me by R. A. McCutcheon.

48. NTB to O[S], Jan. 17, Feb. 6, June 6, 1934; FR to [O]S, Jan. 25, Feb. 7, 28, Mar. 23, 30, May 19, 21, 1934; JAH to OS, May 14, 1934; YOA.

49. OS to HGG, Dec. 19, 1933, Mar. 16, May 9, 1934; OS to FR, Jan. 8, 1934; FR to OS, May 3 (telegram), 1934; NTB to OS, May 3, July 20, Dec. 18, 1934; OS to NTB, Jan. 25, Feb. 28, May 5, 7, 1934; YOA.

50. OS to HGG, Aug. 7, Nov. 10, 1933; May 9, Aug. 6, 17, 1934; HGG to OS, Jan. 13, 1934; YOA; Franklin E. Roach, "A Study of Stellar Spectra in the Region 6562–7593," *ApJ* 80 (1934): 233.

51. See note 40 above.

52. OS to HGG, May 9, 1934, YOA.

53. JS to F. Aydelotte, May 8, 1933, UWA; H. D. Curtis to RGA, Mar. 3, 1934; H. D. Curtis to [WH]W, Jan. 29, 1936; SLO; NTB to OS, Nov. 30, 1934, Mar. 4, July 6, 1935; A. W. Smith to OS, May 15, July 25, 1935; YOA.

54. OS to HGG, Nov. 23, 1933; HGG to OS, Nov. 25, 1933; YOA.

55. A[S]E to [RMH], Feb. 1, 1934; OS to A[S]E, Feb. 19, 1934; RMH to A[S]E, Feb. 20, 1934; A[S]E to OS, Mar. 4, 1934; OS to HGG, Apr. 4, 1934; YOA; DEO interviews with WWM, Nov. 22, 1989, Oct. 28–30, 1991.

56. OS to G. Enders, Mar. 21, 1934; Enders to [O]S, Mar. 25, 1934; OS, "To Whom It May Concern," Mar. 28, 1934; YOA.

57. Lillian Ness's date of birth is from records in the possession of her niece, Laurel A. Andrew, Olympia, Washington. The first letters in the YOA that Ness typed for Struve, and that thus bear her initial "n," are dated May 9, 1934.

58. OS to HGG, June 23, July 13, 1934, Sept. 23, 1935; [J]AH to OS, Jan. 14, Feb. 8, 1935; OS to LGH, Feb. 16, Mar. 8, 14, 18, 1935; LGH to OS, Feb. 26, Mar. 15, 1935; YOA.

59. OS to HGG, July 6, 1934; HGG to OS, July 7, 1934; M. R. Calvert to OS, Jan. 23, 1936; YOA.

60. OS to HGG, Nov. 9, 1933, HGG to OS, Nov. 15, 1933; YOA.

61. OS to HGG, Feb. 16, May 3, Oct. 9; HGG to ETF, Oct. 10, 1935; OS, "Notes concerning Discussion with President Hutchins and Dean Gale October 8, 1935," [Oct. 9, 1935]; HGG to OS, Oct. 10, 1935; YOA.

62. [J.] Van Biesbroeck, "La naissance d'un observatoire," *Ciel et Terre* 39 (1934): 97–98. The first letter is clearly by her; the other two letters published with it, by G. Van Biesbroeck, are straightforward descriptions that make no such claims as his wife did.

63. OS to HGG, Jan. 17, 1935, YOA; Otto Struve and William W. Morgan, "The Spectrum of Nova Herculis," *PAAS* 8 (1936): 124.

64. Otto Struve and Theodore Dunham, "The Spectrum of the B0 Star τ Sco," *ApJ* 77 (1933): 321.

65. Fermi (1968).

66. OS to W. Gleissberg, Nov. 24, 1933, YOA.

67. E. R. Murrow to JMS, Nov. 16, 1933; OS to HGG, Jan. 8, 1934; JMS to W. Weaver, Feb. 16, 1934; [H]R to OS, Apr. 3, May 4, 1934 (telegram); YOA.

68. OS to FS, June 29, 1933; OS to ES, Oct. 14, Dec. 9, 1933; YOA.

69. OS to HGG, May 10, 1934, Oct. 21, 1935, YOA.

70. HGG to OS, Apr. 8, 1935; OS to HGG, Apr. 13, May 3, 1935; JMS to HGG, Apr. 29, 1935; OS to JMS, June 26, 28, Sept. 17, Oct. 19, 1935; JMS to OS, June 27, 1935; S. A. Goldsmith to JMS, Oct. 18, 1935; JMS to S. A. Goldsmith, Oct. 21, 1935; YOA.

71. J. Whyte to C. Razovsky, Jan. [15], 1936; JMS to OS, Jan. 28, May 19, 1936; OS to JMS, Feb. 7, 1936; OS to HGG, Apr. 29, June 2, 1936; Whyte to JMS, May 14, 1936; JMS to Whyte, May 19, 1936; OS to Comptroller, June 11, 1936; YOA.

72. GPK to RJT, Dec. 13, 1936, ASC.

73. OS to FS, May 27, 1935; FS to OS, June 4, 1935; FS to H. E. Shantz, June 10, 1935; HR to [O]S, Jan. 7, 1937; YOA. Rosenberg wrote this last letter to Struve, a recommendation for a graduate student at Yerkes Observatory, in German.

74. Hans Rosenberg, "Darkening at the Limb and Color Index of an Eclipsing Variable (U Cephei)," *ApJ* 83 (1936): 67; OS to ES, June 3, 1937; YOA.

75. HS to OS, Jan. 4, 1939, YOA; W. Gleissberg, "Hans Rosenberg," *Publica tons of the Istanbul University Observatory*, no. 13 (1940): 1.

## Chapter Seven    Resurrection on the Campus and at Yerkes, 1893–1937

1. See, for instance, Struve (1947a).

2. See chapter 2.

3. This treatment of See is based largely on Osterbrock (1984b), which gives full references to the original source material. Two of the most important letters are TJJS to E. B. Hurlbert, Aug. 20, [18]92, RL (the first letter of application), WRH to TJJS, [June 27, 1892], (the agreement formalizing See's later appointment), RL. Nearly all the significant letters are in RL.

4. G. D. Purinton to WRH, July 26, 1892, RL.

5. TJJS to E. B. Hurlbert, Aug. 30, [18]92, RL.

6. GEH to WRH, Jan. 28, 1894, RL.

7. T. J. J. See, "Theory of the Determination by Means of a Single Spectroscopic Observation, of the Absolute Dimensions, Masses and Parallaxes of Stellar Systems Whose Orbits Are Known from Micrometrical Measurements, with a Rigorous Method for Testing the Universality of the Law of Gravitation," *AN* 139 (1895): 17.

8. T. J. J. See, "The Study of Physical Astronomy," *PA* 2 (1895): 249, 289, 337.

9. Williams G. Hoyt, *Lowell and Mars* (Tucson: University of Arizona Press, 1976); William L. Putnam, *The Explorers of Mars Hill: A Centennial History of Lowell Observatory* (Flagstaff, Ariz.: Lowell Observatory, 1994).

10. WRH to TJJS, Nov. 6, [18]97, Feb. 8, 25, 1898, LC; TJJS to WRH, Mar. 4, 1898, RL.

11. J. Lankford, "A Note on T. J. J. See's Observations of Craters on Mercury," *JHA* 11 (1980): 129; C. J. Peterson, "A Very Brief Biography and Popular Account of the Unparalleled T. J. J. See," *Griffith Observer* 54, no. 7 (1990): 2; C. J. Peterson, "Visionary and Dangerous Genius: The Life and Career of the Unparalleled T. J. J. See" (unpublished book-length manuscript, 1995).

12. TJJS to HYB, Dec. 17, 1927, Nov. 10, 1930, Sept. 17, 1932, CAH.

13. Struve (1947a).

14. R. T. Chamberlain, "Thomas Crowder Chamberlin, 1843–1928," *BMNAS*, 15 (1932): 307; S. F. Schultz, "Thomas C. Chamberlin: An Intellectual Biography of a Geologist and Educator" (Ph.D. thesis, University of Wisconsin, Madison, 1976); Stephen G. Brush, "A Geologist among Astronomers: The Rise and Fall of the Chamberlin-Moulton Cosmogony," *JHA* 9 (1978): 1.

15. Osterbrock, Brashear, and Gwinn (1990).

16. C. E. Gasteyer, "Forest Ray Moulton," *BMNAS* 41 (1970): 341. This is the main source for the section on FRM.

17. Forest R. Moulton, "The Limits of Temporary Stability of Satellite Motion, with an Application to the Question of the Existence of an Unseen Body in the Binary System F. 70 Ophiuchi," *AJ* 20 (1899): 33; T. J. J. See, "Remarks on Mr. Moulton's Paper in AJ 461," *AJ* 20 (1899): 56; [S. C. Chandler, editorial note], *AJ* 20 (1899): 56. See also Lankford (1980), cited in note 11 above.

18. See Hoyt (1976), cited in note 9 above.

19. Forest R. Moulton, "Capture Theory and Capture Practice," *PA* 20 (1912): 67; H. Jefferys, "The Planetesimal Hypothesis," *Science* 69 (1929): 245; Forest R. Moulton, "The Planetesimal Hypothesis," *Science* 69 (1929): 246.

20. WB to OS, Jan. 1, 1934, Sept. 4, Dec. 16, 1935; OS to WB, Jan. 4, Apr. 11, 1934, Aug. 13, Nov. 15, 1935; YOA.

21. OS to WB, Aug. 17, 1934, YOA.

22. OS to WSA, Oct. 19, 1935, YOA.

23. FW to OS, Aug. 5, 1935; OS to FW, Aug. 18, 1935; YOA.

24. OS to HGG, Oct. 2, 1935, YOA.

25. D. H. DeVorkin, "The Harvard Summer School in Astronomy," *Physics Today* 37, no. 7 (1984): 48.

26. FER to [O]S, Aug. 18, 30, [19]35; OS to FER, Aug. 21, 1935; YOA.

27. G. W. Preston, "Olin C. Wilson (1909–1994)," *PASP* 107 (1995): 97.

28. OS to O. C. Wilson, Aug. 12, 28, Sept. 9, 27, 1935; O. C. Wilson to OS, Sept. 3, 23, 1935; OS to WSA, Aug. 12, 14, Oct. 2, 1935; WSA to OS, Aug. 24, 1935; YOA.

29. Osterbrock, Gustafson, and Unruh (1988); Cruikshank (1993).

30. [GPK] to HNR, Feb. 19, 20, 1935, ASC; RGA to H. A. van Coenen Torchiana, Mar. 29, 1935; WHW to RGS, July 3, Aug. 21, 1935; SLO; OS to GPK, Aug. 12, 193;, YOA.

31. [GPK] to OS, Aug. 20, 1935, ASC; OS to BJB, Sept. 17, 30, 1935; BJB to OS, Sept. 25, 1935; YOA.

32. [OS], "Notes concerning Discussion with President Hutchins and Dean Gale—October 8, 1935," [Oct. 9, 1935]; WSA to OS, Oct. 7, 1935; YOA.

33. OS to GPK, Oct. 9, 1935; OS to HGG, Oct. 9, Nov. 5, 1935; HGG to OS, Oct. 10, 1935; YOA; GPK to OS, Nov. 1, 7, 1935, Apr. 18, 26, 1936; OS to GPK, Nov. 6, 1935; ASC.

34. [OS, background paper on SR, undated, summer 1935]; OS to SR, Aug. 16, 1935; SR to OS, Sept. 11, 1935; YOA.

35. G. A. Bliss to OS, Oct. 4, 12, 1935; OS to Bliss, Oct. 7, 9, 1935; YOA.

36. OS to SR, Oct. 19, 1935 (two letters); SR to OS, Nov. 19, Dec. 5, 1935, YOA.

37. OS to HNR, Aug. 28, 1935; HNR to OS, Sept. 5, Nov. 17, 1935; HS to OS, Oct. 2, 1935; OS to HGG, Oct. 15, Dec. 16, 1935; YOA.

38. OS to SR, Dec. 19, 1935, YOA.

39. M. Rudkjøbing, "Bengt Georg Daniel Strömgren (1908–1987)," *QJRAS* 29 (1988): 282.

40. OS to GPK, Dec. 19, 1935, Jan. 6, 1936; OS to HGG, Dec. 31, 1935; OS to BS, Jan. 4 (cablegram), 6, 11, 1936; YOA.

41. BS to OS, Jan. 18 (cablegram), Feb. 5, Mar. 6, 1936; OS to BS, Feb. 18, Apr. 9, 1936; BS to RMH, Mar. 6, 1936, YOA.

42. HNR to OS, Jan. 3, 1936; OS to GPK, Jan. 6, 1936; GPK to OS, Jan. 7, Feb. 6, 1936; YOA.

43. Wali (1991). Much of the description of Chandrasekhar's youth, education, family, and experiences in England is from this excellent biography. However, the details of his negotiations with Struve and Hutchins, his friendship with Kuiper, and his early experiences in America are from the letters cited below.

44. OS to HGG, Sept. 23, 1934, Feb. 3, 1936, YOA.

45. OS to RMH, Jan. 13, 1936; RMH to OS, Jan. 15, 1936; RL.

46. GPK to OS, Jan. 7, 28, 1936, YOA.

47. OS to RMH, Jan. 20, 1936, RL; OS to SC, Feb. 6, 1936; SC to OS, Feb. 8, 1936; YOA.

48. OS to HS, Jan. 6, 1936, HCO, and letters from HNR, HS, and GPK to OS listed above.

49. OS to HGG, Feb. 3, 1936, YOA.

50. OS to GPK, Jan. 23, 1936; GPK to OS, Jan. 31, Feb. 8, 11, 23, 1936, YOA.

51. OS to GPK, Feb. 14, 25, 1936; OS to SC, Feb. 25, 1936; OS to HS, Feb. 25, 1936; YOA; OS to RMH, Feb. 26, 1936, RL.

52. HS to OS, Feb. 28, 1936; SC to OS, Feb. 28, 1936; GPK to OS, Mar. 1, 1936; YOA.

53. OS to R. Whipple, Mar. 3, 1936; R. Whipple to OS, Mar. 4, 1936; OS to SC, Mar. 3, 1936; YOA.

54. OS to HGG, Mar. 4, 1936, RL.

55. SC to OS, Mar. 7, 14 (telegram), 1936; OS to D. Barbier, Mar. 12, 1936;

YOA; SC to [GP]K, Mar. 17, 1936; OS to RMH, Mar. 23, 1936; RL; OS to GPK, Mar. 4, 12, 1936; GPK to OS, Mar. 20, 1936, ASC.

56. OS, "Discussion on March 17 [1936] with Professor Bliss and Dr. Bartky concerning appointment of Chandrasekhar" (memo), Mar. 17, [1936]; YOA.

57. HGG to [ET]F, Mar. 23, 1936, RL; OS to RMH, Jan. 22, 1938; HGG to OS, Feb. 11, 1938; YOA.

58. RMH to SC, Mar. 23, 1936; SC to RMH, Mar. 24, 1936 (both radiograms); RL.

59. OS to SC, Mar. 23, 1936, YOA.

60. SC to OS, Mar. 31, Apr. 8, 1936; YOA; SC to [GP]K, Apr. 8, 1936; [GPK] to SC, Mar. 11, 15, May 11, 1936; ASC.

61. OS to ES, Mar. 4, 1936; ES to OS, Mar. 18, 1936; BS to OS, Apr. 6, 1936; YOA.

62. OS to J. [S.] Hall, Mar. 10, 1936, YOA.

63. OS to AHC, Apr. 2, 1936; AHC to OS, Apr. 3, 1936; YOA.

## Chapter Eight   Birth of McDonald Observatory, 1933–1939

1. OS to HS, Aug. 16, 1932, YOA.

2. OS to RGA, Dec. 15, 1932, YOA. See also Evans and Mulholland (1986), one of the main sources for this entire chapter.

3. OS to HGG, Sept. 28, Oct. 27, Dec. 19, 1932, YOA.

4. HS to OS, Mar. 9, May 15, 1932; OS to HS, May 19, 1933; HCO.

5. Osterbrock (1993a).

6. GWR to OS, Sept. 9, 1932; HYB to OS, Feb. 10, 17, 1933; YOA.

7. Minutes of University of Texas Board of Regents meeting, Feb. 16, 1933; OS to HYB, Feb. 15, 1933; OS to HJLS, [Feb. 21], [Mar. 6] (telegram), Mar. 7, May 5, 1933; HJLS to OS, Mar. 5, 6 (telegram), 1933; YOA.

8. HJLS to L. C. Haynes, Feb. 26, 1933, CAH; OS to HJLS, Mar. 14, 1933, YOA.

9. OS to WSA, June 26, July 5, 1933; WSA to OS, June 28, 1933; OS to JSP, June 26, 1933; JSP to OS, June 29, 1933 (two letters of same date); YOA.

10. OS to HGG, July 29, 1933; OS to HNR, Aug. 1, 1933; RMH to OS, Aug. 12, 1933; B. H. Jester to RMH, Feb. 12, 1934; YOA.

11. HJLS to OS, Mar. 5, 1934; HJLS to P. E. Bliss, Mar. 5, 1934; OS to HJLS, Mar. 7, 1934; YOA.

12. OS to HGG, Oct. 31, 1933, Jan. 3, May 18, Oct. 8, 1934, YOA.

13. OS to WSA, Feb. 21, 1933; OS to HGG, June 4, 14, Aug. 17, 1934; HGG to OS, June 13, 1934; YOA; FR to DEO, May 30, 1991.

14. OS to BJB, Sept. 30, 1935; OS to HGG, Nov. 5, Dec. 31, 1935; OS, "Memo concerning discussion with President Hutchins and Dean Gale, Oct. 8, 1935" [Oct. 8, 1935]; YOA.

15. OS to GPK, Jan. 23, 1936, YOA.

16. Otto Struve, Christian T. Elvey, and Franklin E. Roach, "Reflection Nebulae," *ApJ* 84 (1936): 219; Otto Struve, "On the Interpretation of the Surface Brightness of Diffuse Nebulae," *ApJ* 85 (1937): 194.

17. OS to HNR, Feb. 23, 1933; OS to WSA, Jan. 8, 1934, Jan. 14, 1935; WSA to OS, Jan. 15, 1934; OS to HGG, Jan. 17, Feb. 25, 1935, Feb. 2, 1938; HGG to OS, Jan. 29, 1938; YOA.

18. See, e.g., Osterbrock (1994).

19. OS to GPK, Jan. 8, 1936; OS to HGG, Oct. 25, 1937; YOA.

20. FER to OS, Aug. 18, [19]35; M. L. Steffen to OS, May 17, 1937; OS to HGG, May 19, Sept. 17, 1937; YOA.

21. OS to HGG, Mar. 14, May 3, Aug. 16, 1935; HGG to OS, Aug. 2, 1935, YOA.

22. OS to RMH, Mar. 15, Sept. 13, 1937; M. L. Steffen to OS, Mar. 20, 1937; OS to M. [L.] Steffen, Mar. 22, 1937; HGG to OS, June 11, 1937; YOA; OS to RMH, June 26, 1937; HGG to GWM, July 1, 1937; GWM to HGG, July 8, 1937; RL.

23. OS to GWM, Feb. 28, 1939; GWM to OS, Mar. 6, 1939; R. S. Perkin to OS, Mar. 28, 1940; OS to Perkin, Mar. 30, 1940; YOA.

24. OS to HS, Mar. 10, 1936, YOA; FR to DEO, May 30, 1991.

25. Donald E. Osterbrock, "Franklin E. Roach, 1905–1993," BAAS 26 (1994): 1608.

26. BJB to OS, Jan. 12, Mar. 1, 1936; OS to BJB, Feb. 14, 1936; HS to OS, Feb. 4, 1936; OS to HS, Feb. 15, 1936; YOA.

27. HS to OS, Feb. 17, 28, Mar. 12, 1936; OS to C. K. Seyfert, Mar. 21, 1936; YOA.

28. BJB to OS, Feb. 5, 1937; OS to BJB, Feb. 11, 1937; OS to HGG, Oct. 14, 1937; YOA.

29. Jesse L. Greenstein, "An Astronomical Life," ARA&A 22 (1984): 1.

30. Otto Struve, "A New Slit-Spectrograph for Diffuse Galactic Nebulae," ApJ 86 (1937): 613–19.

31. Jesse L. Greenstein and Louis G. Henyey, "The Spectra of the North America Nebula and of the γ Cygni Nebula," ApJ 86 (1937): 620; OS to HGG, Dec. 7, 1937; HGG to OS, Dec. 9, 1937.

32. Otto Struve, George Van Biesbroeck, and Christian T. Elvey, "The 150-Foot Nebular Spectrograph of the McDonald Observatory," ApJ 87 (1938): 559; OS to BJB, Jan. 17, Feb. 26, 1938, YOA.

33. OS to HGG, Mar. 7, 21, Aug. 8, 1938; YOA; Otto Struve and Christian T. Elvey, "Observations Made with the Nebular Spectrograph of the McDonald Observatory," ApJ 89 (1939): 119; ibid. II, ApJ 89 (1939): 517.

34. OS to WSA, Feb. 26, 1938; WSA to OS, Mar. 2, 1938; YOA.

35. WHW to RGS, Apr. 16, 1936; WHW to OS, Feb. 4, 15, 1939; OS to WHW, Feb. 8, 1939; ISB to WHW, Feb. 9, 1939; SLO; OS to GPK, Jan. 19, 1939, YOA.

36. OS to BJB, Nov. 23, 1936, YOA.

37. OS to BJB, Apr. 3, 1938; OS to HGG, Apr. 4, 1938; YOA.

38. OS to HGG, May 24 (two letters), Aug. 6, Oct. 13, 16, 1938; E. L. McCarthy to [O]S, June 4, 1938; OS to Dean of Graduate Students, Rochester, July 16, 1938; OS to E. L. McCarthy, Aug. 9, Sept. 30, 1938, YOA.

39. OS to E. L. McCarthy, Oct. 17, 1938, YOA. The tests are described more formally, and the comparison of the four mirrors is made quantitatively in John S. Plaskett, "The 82-Inch Mirror of the McDonald Observatory," CMO 1 (1939): 24.

40. OS to HGG, Dec. 5, 1938; HGG to OS, Dec. 6, 1938; YOA.

41. Otto Struve, "Stars with Extended Atmospheres," Proceedings of the American Philosophical Society 8 (1939): 211.

42. OS to HGG, Jan. 18, 1939; GVB to [O]S, Feb. 22, 1939; [OS] to WWM, Mar. 4, 1939; YOA.

43. GVB to [O]S, Jan. 17, 1939, YOA; George Van Biesbroeck, "The Coudé Spectrograph," *CMO* 1 (1939): 103.

44. OS to WWM, May 21, 26, 1939; OS to [G]VB, May 23, 1939; GVB to [O]S, May 29, 1939; YOA.

45. Otto Struve and Franklin E. Roach, "The Ultraviolet Spectra of 17 Leporis and P Cygni," *ApJ* 90 (1939): 727.

46. Otto Struve, "The Ultraviolet Spectra of A and B Stars," *ApJ* 90 (1939): 699.

47. H. W. Morelock to RMH, Feb. 24, Apr. 4, 1939; RMH to Morelock, Mar. 2, Apr. 18, 1939; RL.

48. OS to BJB, Nov. 12, Dec. 20, 1938; [OS], "Faculty Conference—September 29, 1938," [Sept. 30, 1938]; Announcement of Nineteenth Annual Meeting, AAAS Southwest Division, May 2–5, 1939; RL.

49. JWC to [HYB], 1924; JWC to W. L. W. Splawn, Apr. 22, 1925; CAH; OS to RMH, May 11, 1937; JWC to RMH, June 13, 1938; OS to HPR, Mar. 7, 1939; HPR to OS, Apr. 19, 1939; YOA.

50. OS to RMH, Nov. 28, 1938, Feb. 10, Mar. 13, Apr. 17, 1939; RMH to OS, Dec. 12, 1938, Feb. 16, Mar. 17, May 1 (telegram), 4 (telegram), 1939; W. V. Morganstern to [RM]H, Apr. 17, May 1, 1939; JWC to RMH, May 26, 1939; RL.

51. "Program of Symposium on 'Galactic and Extragalactic Structure' at the McDonald Observatory, May 4–8, 1939," [Feb. 3, 1939], SLO; "Program for the Dedication of the McDonald Observatory, *PA* 47 (1939): 141.

52. [RJT] to H[Y]B, Jan. 13, 1932, SLO; BJB to OS, Nov. 17, 1938, Jan. 4, 1939; OS to BJB, Dec. 20, 1938; YOA.

53. OS to HGG, Mar. 8, May 27, June 7, 22, 1939; HGG to OS, June 2, 1939; HWB to OS, May 23, 1939; OS to HWB, May 17, 1939; YOA.

54. OS to [GP]K, May 29, 1939, YOA.

55. OS to JHM, Oct. 6, Nov. 2, 1939; JHM to OS, Oct. 23, 1939; YOA.

56. PS to OS, Jan. 10, 193[4], Feb. 15, May 10, June 12, July 1, 1935; OS to PS, Jan. 23, 1934, Jan. 15, 1935; YOA.

57. PS to OS, Apr. 24, Aug. 16, Nov. 19, 25, Dec. 15, 20, 25, 28, 1935, YOA.

58. OS to HGG, Jan. 10, Feb. 8, Mar. 8, 1939; PS to OS, Jan. 8, Feb. 3, Mar. 19, Apr. 11, May 7, July 7, 1939; OS to PS, Jan. 21, June 8, 1939; YOA.

59. [OS] to [BJ]B, Sept. 13, 1939; OS to HGG, Sept. 27, 1939; YOA.

## Chapter Nine   An Extraordinarily Fine Group, 1936–1942

1. [GPK] to SC, May 11, 1936, ASC; Cruikshank (1993).

2. Elizabeth Ross to [O]S, Feb. 10, [19]36; OS to Elizabeth Ross, Feb. 12, July 28, 1936; YOA.

3. [GPK] to RJT, Dec. 13, 1936; RJT to [GP]K, July 16, 1937, ASC.

4. BS to OS, Aug. 15, Sept. 20 (telegram), Oct. 15, Nov. 12, 1936; OS to BS, Sept. 18, 28, Oct. 13, 1936; YOA.

5. OS to BS, Nov. 20, 1936; BS to OS, Nov. 26, 1936; WWM to BS, Dec. 14, 1936, YOA.

6. SC to OS, Apr. 8, 1936, YOA; SC to GPK, Apr. 8, 1936, ASC.

7. SC to OS, June 12, July 9, 27, 1936; OS to FW, June 22, 1936; OS to SC,

June 29, July 3 (two letters), 1936; H. H. Moore to OS, July 1, 1936; OS to American Consul General, London, July 14, 1936; D. G. Dwyne, Consul General, London, to OS, July 24, 1936; YOA.

8. Wali (1991).

9. OS to SC, Nov. 11, Dec. 2, 1936; SC to OS, Nov. 17, Dec. 4, 1936; YOA.

10. [GPK] to RJT, note 3 above.

11. SC to OS, Dec. 16, [1936], SC to WWM, Dec. 16, 1936 (both telegrams); OS, "To Whomsoever This Letter May Be Presented," Dec. 28, 1936; YOA; Wali (1991).

12. OS to RMH, Dec. 24, 30, 1936; BS to OS, Dec. 28, 1936, Jan. 1 (two letters), 27, 1937; OS to BS, Dec. 30, 1936; OS to ES, Jan. 15, 1937; OS to N. Bohr, Jan. 15, 1937; ES to OS, Mar. 11, 1937; YOA; HGG to F. Barkley, Jan. 7, 1937; RL.

13. BS to OS, Feb. 21, Mar. 26, 28, 1937; OS to BJB, Mar. 5, 1937, YOA.

14. OS to BJB, Jan. 6, 1936; BJB to OS, July 7, 1936; OS to ETF, Feb. 25, 1937; M. L. Steffens to OS, Mar. 20, 1937; YOA.

15. Otto Struve, Gerard P. Kuiper, and Bengt Strömgren, "The Interpretation of Epsilon Aurigae," *ApJ* 86 (1937): 570.

16. HNR to WSA, Jan. 18, 1937, HHL.

17. OS to M. Irwin, May 15, 1937; OS to R. D. Hemens, May 15, 1937, YOA.

18. R. D. Hemens to OS, May 21, 1937; OS to R. D. Hemens, [June 11], 1937; R. P. Rohrer to OS, June 9, 1937; OS to R. P. Rohrer, Sept. 2, 1937; OS to M. Irwin, Sept. 20, 1937; OS to RMH, Dec. 1, 1937; OS, "Conference on December 1, 1937" (memo to file), Dec. 2, 1937, YOA.

19. OS to ETF, Jan. 3, 1938; OS to HGG, Jan. 5, 193[8]; OS to R. D. Hemens, Feb. 17, 1938; R. D. Hemens to OS, Apr. 5, 1938; G. J. Laing to OS, Apr. 14, Apr. 20, 1938; OS to G. J. Laing, Apr. 15, 1938; OS, "Conversation with Mr. Hemens" (memo to file), Apr. 20, 1938.

20. OS to R. D. Hemens, Feb. 19, 1938; OS to M. Alexander, May 10, 1938; YOA.

21. Subrahmanyan Chandrasekhar, *An Introduction to the Study of Stellar Structure* (Chicago: University of Chicago Press, 1939).

22. OS to C. F. Huth, Jan. 15, 1938; OS to RMH, Jan. 22, 1938; RMH to OS, Jan. 26, 1938; WWM to OS, Feb. 24, 1938; YOA.

23. HGG to OS, Feb. 11, 17, 1938; OS to HGG, Feb. 12, 14, 1938; OS, [untitled memo to file on his discussion with HGG on Feb. 16], Feb. 17, 1938; YOA. After Gale had retired, Struve denied he had ever agreed to keep Chandrasekhar off the campus; the dean had ordered him to do so (OS to RMH, July 12, 1941; YOA).

24. Record of the "Innominates," University of Chicago, 1917 to 1947, RL.

25. OS to C. R. Moore, Dec. 1, 1939; C. R. Moore to OS, Dec. 5, 1939, Jan. 5, 1940, YOA.

26. Bengt Strömgren, "On the Hydrogen and Helium Content of the Interior of the Stars," *ApJ* 87 (1938): 520.

27. BS to [O]S, Nov,. 30, 1938, YOA; Bengt Strömgren, "The Physical State of Interstellar Hydrogen," *ApJ* 89 (1939): 526.

28. OS to BJB, Jan. 21, Aug. 6, 1937, YOA.

29. William W. Morgan, "On the Spectral Classification of the Stars of Types A to K," *ApJ* 85 (1937): 380.

30. OS to WSA, Oct. 12, 1937; WSA to OS, Oct. 19, 193; HHL.

31. William W. Morgan, "On the Determination of Color Indices of Stars from a Classification of Their Spectra," *ApJ* 87 (1938): 460.

32. WWM to [O]S, Sept. 27, 1939, YOA.

33. BJB to [O]S, Aug. 6, 1937, Feb. 20, 1938.

34. OS to WWM, Jan. 20, Sept. 25, 1939; WWM to [O]S, Jan. 14, May 23, 1939; OS to [L]N, Jan. 12, May 22, 28, 1939; HGG to OS, May 18, 1939; YOA.

35. OS to RMH, Apr. 20, 1937; OS to HGG, June 22, 1937; YOA.

36. OS to HGG, Jan. 7, 27, 1937; HGG to OS, Jan. 25, 1937; M. L. Steffen to OS, Feb. 16, 1937; YOA.

37. OS to RMH, Apr. 20, Oct. 7, 1937; RMH to OS, May 7, [1937], (telegram); OS to HGG, June 21, Aug. 6, Oct. 14, 1937; OS, "Re Conference with Dean Gale and President Hutchins 6/24/37" (memo), [June 25, 1937]; YOA.

38. HGG to OS, June 22, 1937, Oct. 10, 1938; OS to HGG, Jan. 4, Oct. 7, 1938; OS to M. Steffen, Jan. 14, 1938; YOA; HGG to WSA, July 27, 1937; HHL.

39. OS to HGG, Jan. 15, 1938; BS to [ML]S, Feb. 26, 1938; BS to OS, Feb. 26, May 12, 1938; BS to [CT]E, May 3, 1938; YOA.

40. OS to [ES]E, Mar. 22, 1938; OS to HGG, Mar. 23, 1938, YOA; Subrahmanyan Chandrasekhar, G. Gamow, and M. A. Tuve, "The Problem of Stellar Energy," *Nature* 141 (1938): 982; Karl Hufbauer, *Exploring the Sun: Solar Science since Galileo* (Baltimore: Johns Hopkins University Press, 1991).

41. OS to BS and [Sigrid] Strömgren, Mar. 26, 1938 (telegram), YOA.

42. BS to OS, Aug. 28, Sept. 28, 1938; OS to BS, Sept. 9, 1938; YOA.

43. PS to OS, Sept. 7, Nov. 11, Dec. 25, 1938, YOA.

44. OS to WWM, July 5, [19]38; WWM to OS, July 7, 1938; YOA. This is just one sample, an exchange in which Struve sent Morgan, from McDonald, a letter containing fifteen numbered instructions to carry out, without any general guidance or thanks for previous work, and Morgan replied briefly on what he had done on most of them without ever going beyond the literal words of his orders.

45. OS to BS, Oct. 11, 1938; OS to HGG, Oct. 14, 1938; [OS], "Memorandum for discussion with President Hutchins, October 1938," [Oct. 15, 1938]; OS to PS, Oct. 29, 1938; YOA.

46. HGG to OS, Oct. 19, Nov. 9, 1938; OS to RMH, Oct. 21, 1938; OS to HGG, Oct. 22, 1938; YOA.

47. BS to OS, Nov. 23 (cablegram), 29, 1938, YOA.

48. OS to RMH, Nov. 28, 1938; OS to PS, Dec. 12, 1938; YOA.

49. BS to OS, Jan. 21, Feb. 15, Apr. 20, May 26, Aug. 23, Dec. 23, 1939; OS to BS, May 10, Oct. 16, 1939; YOA; H. S. W. Massey and D. R. Bates, "The Continuous Absorption of Light by Negative Hydrogen Ions," *ApJ* 91 (1940): 202.

50. OS to ETF, Feb. 4, 1939; ETF to SC, Feb. 4, 1939; YOA; A. J. Shaler, ed., *Les novae et les naines blanches* (Paris: Hermann, 1941).

51. Subrahmanyan Chandrasekhar, *Eddington: The Most Distinguished Astrophysicist of His Time* (Cambridge: Cambridge University Press, 1983); Subrahmanyan Chandrasekhar, *Truth and Beauty: Aesthetics and Motivations in Science* (Chicago: University of Chicago Press, 1987); Wali (1991).

52. K. Wurm to OS, Jan. 22, [19]38, YOA.

53. W. Baade to OS, Nov. 28, 1938; [WW]M to OS, Jan. 17, 1939; OS to WWM, Jan. 19, 1939; AU to OS, Jan. 3, 1939; OS to AU, Jan. 6, 21, Feb. 4, 1939; OS to U.S. Consul, Kiel, Jan. 20, 1939; YOA; OS to AU, Dec. 17, 1938; OS to WSA, Jan. 20, 1939; WSA to OS, Feb. 19, 1939; HHL.

54. OS to AU, Jan. 21, Mar. 8, 28, 1939; AU to OS, Feb. 23, Mar. 22, 1939; YOA.

55. GPK to L. Spitzer Jr., Feb. 7, 1948, YOA.

56. [OS] to G[VB], May 23, 1939; LN to AU, June 1, 1939; AU to Yerkes Observatory, June 12, 1939 (telegram); YOA.

57. OS to AU, Aug. 16, 1939; AU to [O]S, Aug. 22, 1939; OS to B. H. Uhl, District Director, Immigration Service, Nov. 4, 1939; YOA; Albrecht Unsöld, "Quantitative Spectralanalyse des BO Sternes τ Scorpii," pt. 1, Zeitschrift für Astrophysik 24 (1942): 1.

58. OS, "Conference with Dean Gale—December 1, 1937" (memo), Dec. 1, 1937; OS to HGG, Dec. 4, 1937; HGG to OS, Dec. 10, 1937; RSM to OS, Dec. 6, 1937; OS to RSM, Dec. 7, 1937; YOA.

59. OS to AA, Dec. 16, 1937; HNR to OS, Dec. 18, 1937; OS to R. Wildt, Dec. 22, 1937; OS to G. H. Shortley, Dec. 18, 1937; OS to RSM, Jan. 4, 18, 1938; RSM to OS, Jan. 8, 1938; NTB to OS, Jan. 25, 1938; OS to HGG, Apr. 4, 1938; YOA.

60. There are letters dated 1938 in YOA to or from all those named, and many more, indicating attendance at the symposium.

61. "Recent Progress on the Interpretation of Molecular Spectra and the Study of Molecular Spectra in Celestial Objects," ApJ 89 (1939): 283.

62. OS to D. H. Menzel, July 2, 1941; OS to RSM, Nov. 21, 1941; YOA.

63. OS to HGG, Mar. 18, 1939; OS to RMH, Sept. 4, 1939; OS to ETF, Apr. 1, 1940; YOA.

64. OS to ETF, Feb. 7, 1939; OS to HGG, Feb. 8, Oct. 9, Nov. 14, 1939; OS, "Conference with Mr. Filbey, June 5, 1940," June 6, 1940; WWM to [O]S, Sept. 20, 1939, Jan. 29, Aug. 16, 1940; OS to R. R. McMath, Jan. 5, 1940; OS to PCK, Feb. 2, 1940; PCK to [O]S, Feb. 4, June 12, 1940; OS to D. M. Wiggins, June 11, 1940; YOA.

65. OS to G. Breit, Oct. 3, 8, Nov. 7, 1938; G. Breit to OS, Oct. 7, 12, 1938; YOA.

66. G. Breit to OS, Nov. 13, 17, Dec. 21, 1938; OS to G. Breit, Nov. 14, 18, Dec. 23, 1938; YOA; G. Breit and E. Teller, "Metastability of Hydrogen and Helium Levels," ApJ 91 (1940): 215.

67. H. M. Pillans, "On the Spectrum of β Lyrae," ApJ 80 (1934): 51; Otto Struve, "The Puzzle of β Lyrae," Observatory 57 (1934): 265.

68. Otto Struve, "The Spectrum of β Lyrae," ApJ 93 (1941): 133; J. R. Gill, "Some Line Intensities in β Lyrae," ApJ 91 (1941): 104; Jesse L. Greenstein and Thornton L. Page, "The Spectrum of β Lyrae in the Visual Region," ApJ 91 (1941): 118; Gerard P. Kuiper, "On the Interpretation of β Lyrae and Other Close Binaries," ApJ 91 (1941): 133.

69. Otto Struve, "The Problem of β Lyrae," PASP 70 (1958): 1.

70. V. Rojansky to OS, Nov. 10, 22, 1939; OS to V. Rojansky, Nov. 16, 1939; YOA; V. Rojansky, "The Hypothesis of the Existence of Contraterrene Matter," ApJ 91 (1940): 257.

71. OS to G. Reber, Nov. 2, 1939; G. Reber to OS, Nov. 8, 1939; YOA; G. Reber, "Cosmic Static," *ApJ* 91 (1940): 621.

72. Louis G. Henyey and Philip C. Keenan, "Interstellar Radiation by Free Electrons and Hydrogen Atoms," *ApJ* 91 (1940): 625.

73. OS to HGG, Apr. 5, Sept. 9, 1938; HGG to OS, Apr. 13, 1938; HGG to FER, Sept. 25, 1938; YOA; RMH to HGG, Apr. 16, 1938; RMH to FER, Sept. 30, 1939; RL.

74. G. C. Moore to OS, Sept. 30, [19]39, YOA; FER to WHW, Oct. 5, [19]39, YOA; FER to WHW, Oct. 5, [19]39, SLO.

75. OS to C. E. Merriam, Apr. 21, 1938; OS to A. J. Carlson, June 10, 1938; OS to G. K. K. Link, June 25, 1938; OS to A. J. Dempster, July 27, 1938; RMH to OS, June 21, 1938; OS to RMH, June 27, 1938; YOA.

76. RMH to OS, Sept. 20, 26, 1939; OS to RMH, Sept. 24, Oct. 9, 1939; YOA.

77. RMH to ETF, Jan. 9, 1940; H. I. Schlesinger to ETF, Feb. 3, 1940; RL.

78. HGG to OS, Jan. 21, Apr. 26, 1939; OS to HGG, Jan. 24, Apr. 25, May 10, 1939; OS, [memo to file re meeting with RMH], June 13, [19]39; YOA.

79. OS to ETF, Jan. 12, 1940; OS, "Conference with Mr. Filbey" (memo to file}, Feb. 26, 1940; YOA.

80. G. A. Bliss to OS, May 29, 1940; OS to G. A. Bliss, June 1, 1940; YOA.

81. OS to ETF, Mar. 22, May 14, 1940; ETF to OS, Apr. 5, 1940; LN to [O]S, Apr. 26, 1940; SC to OS, May 1, 1940; YOA.

82. OS, "Conference with [RMH] on September 16, 1940" (memo to file), Sept. 18, 1940, YOA.

83. BJB to OS, Nov. 10, 1936, Mar. 16, [19]37, June 19, 1938, Jan. 10, 1939; OS to BJB, June 30, 1938, Jan. 18, 1939; YOA. These are only a few of the many letters along these lines that passed between them.

84. D. Levy, *The Man Who Sold the Milky Way: A Biography of Bart Bok* (Tucson: University of Arizona Press, 1993); J. A. Graham, C. M. Wade, and R. M. Price, "Bart J. Bok," *BMNAS* 64 (1994): 73.

85. OS, [notes for a meeting with Yerkes Observatory faculty, Nov. 5, 1940]; YOA.

86. OS to [BJ]B, Oct. 14, 1940; B[J]B to [O]S, Nov. 2, 1940; OS, [untitled memo for RMH and AHC], Nov. 4, [19]40; OS, "Memorandum" (to file), Nov. 4, [19]40; YOA.

87. OS to HS, Nov. 5, 1940; OS to AHC, Nov. 5, 1940; OS to WB, Nov. 6, 1940; OS to BJB, Nov. 6, 1940; YOA.

88. OS to BJB, Nov. 12, 20, Dec. 13, 1940; AHC to OS, Nov. 14, 1940; OS, "Plan for Mr. Bok's Appointment at the University of Chicago" (memo), Nov. 25, [19]40; OS to AHC, Dec. 2, 1940; ETF to AHC, Dec. 10, 1940; OS, "Memorandum regarding offer to Bart J. Bok," [Dec. 13], 1940; BJB to OS, Dec. 21, 1940; YOA.

89. OS to RMH, Dec. 27, 1940; OS to BJB, Jan. 21, 1941; BJB to OS, Feb. 10, 1941; YOA.

90. N. L. Bowen to OS, Mar. 27, 1940; OS to N. L. Bowen, Mar. 29, 1940; OS to RMH, June 6, 1940; OS to AHC, Nov. 27, 1940; AHC to OS, Nov. 25, 1940; OS to JS, Feb. 19, Mar. 8, 1941; YOA.

91. A. B. Wyse to OS, Feb. 17, 20, 1941; D. H. Menzel to OS, June 28, 1941;

OS to R. Minkowski, Aug. 19, 1941; R. Minkowski to OS, Aug. 22, 1941 (telegram); YOA.

92. "Program of the Sixty-sixth Meeting of the American Astronomical Society, Yerkes Observatory, Williams Bay, Wisconsin, September 7–9, 1941"; OS to "Dear Fellow Alumnus," Feb. 14, 1941; J. L. Guion to OS, Feb. 19, 1941; NTB to OS, Feb. 21, 1941, JAH to OS, Feb. 24, 1941; YOA; D. B. McLaughlin, "The Sixty-sixth Meeting of the American Astronomical Society," *PA* 49 (1941): 401.

93. Otto Struve, "The Department of Astronomy of the University of Chicago," *Science* 94 (1941): 337.

94. OS to AHC, Aug. 13, 1940; AHC to OS, Aug. 1940; HNR to OS, Aug. 9, 1941; OS to HNR, Aug. 13, 1941; YOA.

**Chapter Ten    World War II, 1939–1945**

1. OS to BJB, Sept. 13, 1939; OS to HS, Sept. 27, 1939, YOA.

2. OS to RMH, Sept. 4, 1939; WWM to [O]S, Sept. 8, 1939; OS to [WW]M, Sept. 17, 1939; YOA.

3. OS, "Discussion at Trustees' dinner, January 11, 1939" (memo), [Jan. 12, 1939]; OS, [untitled memo], June 13, [19]39; RMH to OS, July 21, Aug. 8, 1939; OS to RMH, July 31, 1939; OS to HPR, Sept. 12, 1939; YOA; W. Weaver to OS, Oct. 10, 1939; ETF to RMH, Oct. 13, 1939; RL.

4. OS, "Report concerning the Yerkes Observatory," July 10, [19]39; OS, "Staff meeting August 31, '39," (memo), [Sept. 6, 1939]; [OS], "Summary of Plans for the Reorganization of the Yerkes Observatory," Oct. 10, [19]39; YOA.

5. ETF to OS, Nov. 27, 1939; OS to ETF, Nov. 28, 1939; LN to [O]S, Apr. 23, 26, 1940; OS to [WW]M, Apr. 25, 28, 1940; OS to [L]N, Apr. 25, 1940; YOA.

6. OS, "Collaboration in Astronomical Research" (eight-page typed memo), Nov. 17, [19]39; OS, "Summary of Collaboration Plans" (briefing paper for RMH to use in discussions), Nov. 27, [19]39; OS, "Plan for Astronomical Collaboration in Connection with the McDonald Observatory" (seven-page proposal), Mar. 16, 1940; YOA.

7. OS, "Conference with Dr. Weaver," Nov. [24], 1939; OS, "Memorandum—Conference with Mr. Hutchins," Jan. 11, 1940; V. Bush to OS, Apr. 15, 1940; V. Bush to [C.] Dollard, Feb. 19, 1940; YOA; WSA to OS, Apr. 24, 1940, HHL.

8. OS, "Visit to Harvard on November 21" (memo), Nov. 24, [19]39, YOA.

9. OS to JWC, Mar. 20, 1940; OS to HPR, Mar. 20, 1940; [OS] to RMH, Apr. 28, 1940; YOA.

10. OS, "Memorandum concerning possible collaboration with Yale University Observatory," Nov. 27, [19]39; OS, "Memorandum regarding the possibility of collaboration with the Observatory of the University of Michigan," Nov. 27, [19]39; RMH to OS, Mar. 28, 1940; S. A. Mitchell to OS, Apr. 9, 1940; H. D. Curtis to OS, July 15, 1940; YOA; Otto Struve, "Cooperation in Astronomy," *Scientific Monthly* 50 (1940): 142.

11. OS, "Conference with Dr. W. J. Luyten," Mar. 18, [19]40; OS to RMH, Mar. 20, 1940; F. K. Edmondson to OS, June 12, 1940; H. B. Wells to OS, July 17, 1940; OS to H. B. Wells, Dec. 26, 1940; OS to H. T. Briscoe, Feb. 14, 1941; H. T. Briscoe to OS, Mar. 29, 1941; OS to RMH, Apr. 3, 1941; YOA.

12. OS to [S]C, Jan. 17, 1941; SC to OS, Jan. 20, 1941; OS to AHC, Feb. 5, 1941, YOA; OS to RMH, Feb. 19, Mar. 12, 1941, RL.

13. SC to [O]S, May 20, 1941; OS to C. V. Raman, May 24, 1941; YOA.

14. GPK to [O]S, Feb. 3, 1941; OS to RMH, Mar. 12, 1941; YOA.

15. OS, "Discussion with Kuiper on May 31" (memo to self), [June 1, 1941]; GPK to OS, June 3, 16, 23, Nov. 16, 1941, Mar. 14 1942; OS to GPK, June 19, July 3, 1941; AHC to ETF, June 11, 1941; OS to AHC, June 19, Nov. 18, 1941; YOA; GPK to WHW, June 23, 1941, SLO.

16. D. P. Phemister, Q. Wright, and P. W. Moore to OS, Nov. 7, 1941; OS to D. P. Phemister, Nov. 24, 1941; YOA; OS to HS, July 11, Aug. 4, 1941, Feb. 4, 1942; HS to OS, July 8, Aug. 11, 1941; HCO.

17. OS to CTE, Dec. 10, 1941; OS, "Report concerning McDonald Observatory," Feb. 3, 1942; OS to RMH, Feb. 4, 1942; YOA; M. L. Humason to HS, Dec. 10, 1941; OS to HS, Feb. 10, 1942; HCO.

18. HS to OS, Nov. 8, 1941; OS to HS, Nov. 17, 1941; OS to CTE, Nov. 18, 1941; OS to RMH, Jan. 2, 1942; YOA. Otto Struve, "The Cosmogonical Significance of Stellar Rotation," PA 53 (1945): 201.

19. GPK to [O]S, May 27, 1943; OS to GPK, Aug. 26, Sept. 23, 1943; YOA; Donald E. Osterbrock, "Guido Münch: An Appreciation," in Structure and Dynamics of the Interstellar Medium, ed. G. Tenorio-Tagle, M. Moles, and J. Melnick (Berlin: Springer Verlag, 1989), xix.

20. RSM to OS, Nov. 19, 1941; OS to RSM, Nov. 21, 1941; Program, Conference on Spectroscopy, University of Chicago, June 22–25, 1942; YOA. Robert S. Mulliken, "Conference on Spectroscopy: Introduction to the Conference," Reviews of Modern Physics 14 (1942): 57.

21. Otto Struve, "Astronomy Faces the War," PA 50 (1942): 465.

22. RMH to OS, Apr. 4, 1941; OS to HS, June 13, Aug. 4, 1941; OS to CTE, Aug. 19, 1941; OS to RMH, June 18, 1942; YOA; OS to HS, Oct. 13, 1942, HCO.

23. OS to AHC, Jan. 27, 1942; OS to WB, May 29, June 3, 5, July 19, 1942; [OS], Navigation course announcement, June 2, 1942; OS to W. M. Storck, June 9, 1942; OS to GPK, July 19, Nov. 7, 1942; OS, GVB, J. Titus, certificate for course on "Practical Air Navigation" given at YO, Aug. 18, 1942; M. H. Stein to OS, Nov. 6, 1942; OS to J. Titus, Nov. 28, 1942; YOA.

24. AHC to Executive Officers, Depts. of Physical Sciences, July 23, 1940; OS to AHC, July 24, Dec. 30, 1940; OS, [announcement for posting at Yerkes Observatory], June 26, 1942; YOA; DeVorkin (1980). The last reference is the basis for this treatment of the Yerkes Optical Bureau during World War II. It contains many references to the original source material.

25. HS to OS, Aug. 6, 1942, Jan. 22, 1943; OS to HS, Nov. 24, 1942; HCO; OS to ETF, Aug. 15, 1942, YOA; OS to WSA, May 5, 1943, HHL.

26. SC to [O]S, May 22, July 24, Oct. 9, Nov. 12, 1941; OS to W. K. Jordan, July 2, 1941; OS to [S]C, Oct. 29, Dec. 5, 1941; OS to PS, Nov. 6, 1941; YOA.

27. OS, "Interview with Mr. Hutchins on December 22, 1942" (memo), [Dec. 23, 1942]; Wali (1991).

28. Osterbrock (1994).

29. OS to WSA, Mar. 18, 25, Apr. 13, June 8, 1942, May 21, 1943; WSA to OS, Mar. 21, Apr. 6, 1942, May 13, 25, 1943; YOA. Otto Struve and Pol Swings, "The Spectrum of α² Canum Venaticorum," ApJ 98 (1943): 361.

30. PS to OS, Aug. 3, 24, Sept. 9, 19, 21, 22, 1942; OS to PS, Aug. 6, 24, 29, 1942; YOA.

31. PS to OS, Feb. 14, Mar. 5, 10, Apr. 20, May 15, 20, 1942, YOA.

32. OS, [memo re PS], Nov. 8, [19]39; PS to OS, Apr. 4, May 13, 16, 26, 27, 31, 1940; YOA.

33. PS to OS, Jan. 5, Feb. 26, Oct. 7, Nov. 9, Dec. 9, 1941; OS to PS, Feb. 7, 1941; YOA.

34. OS to PS, Feb. 13, Apr. 24, 1942; PS to OS, Feb. 18, 1942; YOA.

35. PS to [IS]B, Nov. 27, Dec. 27, [19]42, CIT; PS to N. [U. Mayall], May 27, 1942, June 6, 1943; PS to JHM, Mar. 23, Sept. 8, [19]43; SLO; OS to HS, Apr. 26, May 16, 1943, HCO.

36. OS to ETF, July 29, 1942; OS to GPK, July 22, [GPK] to P[S], May 6, 1943, ASC; GPK to WSA, Sept. 19, Oct. 24, 1943, HHL.

37. GPK to [O]S, Jan. 6, 9, 1944, YOA; S. F. Kuiper to [H]S, Jan. 10, 1944; HS to GPK, Jan. 3, Mar. 22, 1944, HCO; GPK to [JH]M, Feb. 29, 1944, SLO. Gerard P. Kuiper, "Titan: A Satellite with an Atmosphere," ApJ 100 (1944): 378; Cruikshank (1993).

38. GPK to OS, Aug. 7, Sept. 9, [Oct. 10], Oct. 19, Dec. 10, 1944, Mar. 5, May 20, June 6, 30, July 26, 1945; F. E. Terman to OS, Dec. 4, 1945, YOA; GPK to WSA, Jan. 1, 17, Apr. 10, June 8, 1945; S. [F.] Kuiper to WSA, July 28, 1945; HHL; GPK to JHM, Jan. 1, 18, Apr. 12, 1945, SLO; GPK to HS, July 25, 1945, HCO.

39. TLP to OS, Aug. 11, 1945; OS to WB, Aug. 22, 1945, YOA.

40. OS, "Conference with President Hutchins on March 5" (memo), [Mar. 5, 1942]; OS to RMH, Dec. 24, 1942; YOA; OS to RMH, Dec. 26, 1942, Jan. 2, 1943; RMH to OS, Jan. 2, 15, 1943; RL; OS to WSA, July 20, 1942, HHL; OS to HS, Jan. 16, 1943, HCO.

41. OS to WSA, May 31, 1944, HHL; OS to AHC, Feb. 6, 1945; OS to WB, Apr. 21, 1945; YOA. Code (1992).

42. OS to RMH, Jan. 11, 1943, YOA; OS to HS, Jan. 31, 1944, HCO. Otto Struve, C. U. Cesco, and J. Sahade, "The Spectroscopic Orbit of BD Virginis," ApJ 100 (1944): 181; M. Schönberg and Subrahmanyan Chandrasekhar, "On the Evolution of the Main Sequence Stars," ApJ 96 (1942): 161.

43. OS to WSA, Nov. 14, 1944, Apr. 2, May 30, June 4, July 3, 1945; WSA to OS, Nov. 20, 1944, Apr. 6, May 26, 1945; RSM to WSA, July 11, 1945; HHL; OS to WB, June 4, 1945, YOA.

44. OS to RMH, Aug. 17, 1945, YOA.

45. Chicago Tribune, Apr. 21, 1945; OS to RMH, Apr. 24, 1945, YOA.

46. Minutes of meeting of Policy Committee of the Division of Physical Sciences, May 9, [1945], May 18, [19]45; OS to Policy Committee, May 18, 1945; RMH to H. I. Schlesinger, June 17, 1945; YOA. These are only a few of the many memos, records, and memoranda on this subject in the Yerkes archives.

47. OS to RMH, Aug. 27, 1945; [OS, handwritten pencil notes for conference with RMH, two sheets, both sides of paper, Aug. 29, 1945]; OS, [untitled memo, one typed page, on Aug. 27 conference with RMH], Sept. 21, 1945; YOA.

48. RMH, Summary of address to Academic Senate, July 2, 1945; ECC to All Members, Physical Sciences Division, Sept. 20, 1945; YOA. Chicago Sun, July 3, 1945.

49. OS to RMH, Apr. 16, [19]45; OS to HNR, July 27, 1945; OS to HS, Aug. 8, 1945; YOA. *Chicago Daily News,* Aug. 10, 1945.

## Chapter Eleven Golden Years, 1945–1950

1. PCK to [O]S, May 17, June 6, Oct. 12, 1943; OS to PCK, May 21, June 9, Sept. 20, 1943; OS to WB, Aug. 31, Sept. 13, 1945; WB to OS, Sept. 5, [1945] (telegram); YOA.

2. OS, [memo on conference with RMH, June 27, 1944], July 3, [19]44; OS to RMH, Jan. 11, May 2. 1945; YOA. I lived in the Van Biesbroecks' boarding-house for three years, 1949–52, and like the other graduate students and post-doctoral fellows who also stayed there, or simply took their meals with us, came to know many visiting astronomers from all over the world.

3. CTE to D[MP], Jan. 26, 1942; CTE to [O]S, Mar. 20, July 3, 1942; OS to CTE, Mar. 20, June 13, Aug. 18, 1942; OS to AHC, July 2, 1942; YOA.

4. CTE to [O]S, Aug. 20, 29, Sept. 2, 13, 22, 1942; OS to [CT]E, Aug. 24 (telegram and letter same date), Sept. 2, 15, 1942; OS to AHC, Sept. 24, 1942; YOA.

5. CTE to [O]S, Oct. 2, 1942, Jan. 19, May 25, 1943, Jan. 30, 1944; OS to CTE, Dec. 8, 1942, Feb. 14, Dec. 5, 1944; LN to [CT]E, Oct. 16, Dec. 31, 1942; OS to WB, Dec. 16, 1944; OS to AHC, Dec. 16, 1944; YOA.

6. OS to AHC, June 24, Oct. 8, Dec. 8, 1942; AHC to ETF, June 30, 1942; AHC to OS, Oct. 5, 1942; OS to CTE, May 20, 1943; OS to GPK, Dec. 7, 1943; OS to WB, July 27, 1945; YOA.

7. OS to E. Dershem, Feb. 22, 1946; OS to AA, Feb. 26, Apr. 7, 17, 1946; AA to OS, Apr. 15, May 4, 11, 27, 1946; [GPK], "Telephone conversation with Dr. Arthur Adel, June 26, 1946, 2 PM" (memo), June 26, 1946; GPK to AA, June 26, 1946; AA to GPK, June 29, 1946; YOA.

8. D. Hinds to [O]S, Sept. 28, 1946; WAH to [L]N, Sept. 27, [28], Oct. 30, 1946; GPK to [O]S, Oct. 1, 13, 1946; WAH to [O]S, Oct. 2, 1946; AA to OS, Oct. 11, 1946; OS to NTB, Feb. 10, 1947; YOA.

9. OS to AA, Oct. 31, 1946; AA to OS, Nov. 9, 1946; YOA. Otto Struve, "Reports of Observatories, 1946–1947: Yerkes and McDonald Observatories," *AJ* 53 (1948): 156; R. L. Walker, "Arthur Adel 1908–94," *BAAS* 26 (1994): 1600.

10. OS to GH, Aug. 30, Sept. 14, 27, Dec. 20, 1943; GH to OS, Sept. 4, Dec. 10, 29, 1943; OS to RSM, Sept. 12, 1943; RSM to OS, Sept. 29, 1943; OS to RMH, Oct. 9, 1943; RSM to AHC, Dec. 3, [19]43; AHC to OS, Dec. 15, 1943; YOA; A. E. Douglas and G. Herzberg, "CH+ in Interstellar Space and in the Laboratory," *ApJ* 94 (1941): 381.

11. OS to GH, Jan. 4, 29, 1944; OS to AHC, Jan. 6, 31, 1944; RSM to GH, Jan. 14, 21, 1944; AHC to OS, Jan. 14, 1944; GH to OS, Jan. 27, 1944; YOA.

12. GH to OS, Mar. 3, 23, June 23, 1944, Feb. 13, Apr. 13, [July 8] (telegram), Aug. 27 (telegram), 1945; OS to WB, Mar. 27, 1944; OS to R. J. Seeger, Mar. 30, 1944; OS to [S]C, July 5, 1945; WWM to OS, July 17, 1945; YOA.

13. OS to GH, Apr. 24, 1944, Apr. 24, 1944, Apr. 26, May 7, July 23, 1945, YOA; Otto Struve, "Reports of Observatories, 1945–1946: Yerkes and McDonald Observatories," *AJ* 52 (1947): 146; Gerard P. Kuiper, "Reports of Observa-

tories, 1947–1948: Yerkes and McDonald Observatories," *AJ* 54 (1948): 70; see also note 9 above. These three annual reports and the two subsequent ones listed in note 31 are the sources for much of the information concerning research, appointments, and resignations in the University of Chicago astronomy faculty throughout this chapter.

14. See chapter 10, note 47; SC to RMH, Aug. 24, 1945; RMH to SC (telegram), Aug. 27, 1945; RL; OS to HS, Jan. 23, 1946, YOA.

15. OS to RMH, Sept. 6, 1945, with attached three-page "Reorganization of the Astronomy Department" plan, YOA.

16. GPK to OS, July 5, Aug. 23, 1945; OS to GPK, Aug. 31, 1945; YOA.

17. OS to RMH, Oct. 5, 9, 1945; OS, "Conference with Mr. Hutchins on October 9, 1945" (memo), [Oct. 10, 1945]; YOA.

18. OS, "Conference with President Hutchins on July 3, 1945" (memo), July 5, [19]45; OS to WB, July 14, 1945; OS to [WA]H, Aug. 14, 1945; YOA.

19. OS to [WW]M, Jan. 17, Feb. 13, 1945; SC to [O]S, Jan. 29, 1945; OS to WAH, Oct. 10 (telegram), Oct. 11, Nov. 14, 1945; OS to RMH, Oct. 31, 1945; WWM to [O]S, Jan. 16, 1946; OS to [WW]M, Jan. 20, 1946; YOA. Morgan mentioned to me several times in 1973–80 that he had learned only indirectly, not from Struve, that Hiltner had become assistant director.

20. WB to OS, Oct. 11, 1945; LGH to OS, Nov. 24, 1945; OS to RMH, Dec. 17, 1945; RMH to [O]S, Dec. 19, 1945 (telegram); OS, "Items to be incorporated in an agreement between the University of Texas and the University of Chicago for the construction of a Coude spectrograph" (memo), Jan. 19, 1946; WAH to OS, May 11, 1946; OS to WAH, May 15, 1946; OS, "Conference with Mr. Hutchins on May 28, 1946," May 29, 1946; YOA.

21. OS to BS, June 5, 1945, Feb. 7, 1946; BS to OS, Dec. 8, 15, 1945; OS to RMH, Jan. 26, 28, 1946; YOA.

22. Otto Struve, "The Copenhagen Conference of the International Astronomical Union," *PA* (1946): 54, 327; [OS] to [P]S, Apr. 4, 1946; OS to BS, Apr. 16, 1946; YOA.

23. Otto Struve, "The Zeeman Congress," *PA* 55 (1947): 175; BS to OS, Apr. 14 (telegram), May 18, June 17, 29, Aug. 15, Sept. 10 (telegram), 1946; OS to BS, June 20, 1946; C. J. Gorter to OS, Aug. 19, 1946; YOA.

24. OS to RMH, Nov. 26, 1945, Aug. 5, 1946; OS to BS, Aug. 8, 1946; W. C. Johnson to LAK, Aug. 9, 1946; WB to OS, Aug. 19, 1946; YOA; OS, [biographical data, including his year-by-year salary at the University of Chicago, June 15, 1961], UCB.

25. Wali (1991).

26. RMH to OS, Aug. 16, YOA; OS to WB, Aug. 22, 29, 31, Sept. 4 (telegram), 1946; WB to OS, Aug. 31, 1946; YOA; Wali (1991).

27. OS to RMH, Sept. 9, 1946; LN to [WA]H, Sept. 23, 1946; SC to [O]S, Oct. 17, 1946; OS to [S]C, Oct. 24, 1946; HNR to OS, Jan. 22, 1947; YOA; SC to RMH, Oct. 3, 1946; RMH to HNR, Oct. 31, 1946; HNR to RMH, Nov. 4, 1946; RL; Wali (1991).

28. G. Shajn to OS, Aug. 15, Oct. 11, [Nov. 15], Dec. 15, 1946; OS to G. Shajn, Sept. 9, 1946; LN to G. Shajn, Oct. 21, 1946; GPK to OS, Nov. 29, 1946; YOA; G. Shajn and Otto Struve, "The Absorption Continuum in the Violet Region of the Spectra of Carbon Stars," *ApJ* 106 (1946): 86.

29. OS to WB, Jan. 4, Jan. 29, May 29, 1946; OS to GPK, Jan. 30, 1946; WB to OS, Feb. 5, 1946; YOA.

30. JAH to OS, May 15, 1946; OS to JAH, May 21 (telegram), 22, 1946; OS to WB, July 24, 1946; YOA.

31. Gerard P. Kuiper, "Reports of Observatories, 1948–1949: Yerkes and McDonald Observatories," *AJ* 54 (1949): 225; Subrahmanyan Chandrasekhar, "Reports of Observatories, 1949–1950, Yerkes and McDonald Observatories," *AJ* 55 (1951): 209. These two annual reports, together with the earlier ones cited in notes 9 and 13, contain brief descriptions of much of the research accomplished at Yerkes and McDonald in the years 1946–50. See also Gerard P. Kuiper, "Planetary and Satellite Atmospheres," *Reports on Progress in Physics* 23 (1950): 247.

32. OS to F. C. Hoyt, May 1, 1945; F. C. Hoyt to OS, May 30, 1945; YOA.

33. Garrison (1995).

34. OS to A. E. Whitford, Dec. 26, 1945; Feb. 27, 1946; A. E. Whitford to OS, Jan. 2, 194[6]; UWA; OS to JS, Jan. 29, 1946; OS to WB, Apr. 6, 1946; WAH to [O]S, May 28, 1946.

35. PS to [O]S, Feb. 21, June 10, Dec. 7, 1946; OS to PS, June 18, Dec. 19, 1946; P[S] to G[PK], Oct. 27, 1947; GPK to PS, Nov. 10, 1947; YOA.

36. OS to RMH, Sept. 9, 1946; OS to BS, Feb. 26, 1947; BS to OS, Feb. 26, 1947; YOA.

37. SC to [O]S, Nov. 5, 1946; OS to BS, Mar. 6, 1947; YOA; Bengt Strömgren, "On the Density Distribution and Chemical Composition of the Interstellar Gas," *ApJ* 108 (1948): 242.

38. H. C. Van de Hulst to GPK, Mar. 2, 1946; OS to GPK, Mar. 22, 1946; OS to WB, Mar. 23, 1946; H. C. Van de Hulst to OS, Oct. 25, 1946; GPK to OS, Apr. 21, 1948; YOA.

39. OS to GPK, Feb. 14, 1946; [O]S to WB, Mar. 28, 1946 (telegram), May 10, July 13, Dec. 19, 1946; WB to [O]S, Mar. 28, 1946 (telegram); YOA.

40. OS to GPK, Nov. 8, 1945; GPK to OS, Nov. 12, 1945; P. van de Kamp to OS, Jan. 2, 21, 1946; YOA.

41. OS to GPK, Feb. 25, 1946; OS to K. A. Strand, Aug. 12, Sept. 4, 1946; YOA.

42. OS to WB, Aug. 24, 1946; OS to M. R. Calvert, Feb. 7, 1947; YOA.

43. GPK to OS, June 21, 1947; GPK to PS, June 27, 1947; YOA.

44. [OS], "Present Difficulties: Discussed with Mr. Hutchins on Jan. 10, '47" (typed memo) [Jan. 11, 1947]; OS to [L]N, Jan. 31, 1947; YOA.

45. OS to HNR, Feb. 6, 1947; HNR to OS, Feb. 12, 1947; OS to PS, Mar. 3, 1947; PS to OS, Apr. 4, 1947; OS to BS, Apr. 2, 1947; BS to OS, Apr. 9, 1947; OS to WSA, Apr. 2, 1947; WSA to OS, Apr. 8, 1947; YOA.

46. OS to HS, Jan. 17, 1947; OS to WB, Mar. 24, 28, May 7, 1947; ECC to Members of the Department of Astronomy and Astrophysics, Apr. 28, 1947; LAK to SC, May 20, 1947; OS to [L]N, June 27, 1947; YOA.

47. Dzuback (1991); McNeill (1991).

48. Robert M. Hutchins, "The Department of Astronomy of the University of Chicago," *Science* 47 (1947): 195; Otto Struve, "Stellar Spectroscopy," *Science* 47 (1947): 204; and other articles in the same issue of *Science*.

49. OS to V. Bush, July 21, 1947; V. Bush to OS, Sept. 15, 1947; UCB.

50. These three papers, all published in 1947, are listed in the bibliography among Struve's publications.

51. "A Double Anniversary," *S&T* 7 (October 1947): 36; C. H. Gingerich, "The Seventy-seventh Meeting of the American Astronomical Society," *PA* 55 (1947): 399; "Abstracts: Of Papers Presented at the 77th Meeting, Evanston, Ill., Sept. 3–6, 1947," *AJ* 53 (1948): 105.

52. GPK to WB, Sept. 13, 1947, YOA; Kuiper (1949); Hynek (1951).

53. GPK to RMH, July 21, Aug. 5, 1947; RMH to GPK, July 31, Aug. 31, Oct. 17 (telegram), 1947; GPK to WB, Oct. 28, 1947; YOA; GPK to ECC, Oct. 21, 1947; OS to [RW]H, Nov. 1, 1947; RL.

54. GPK to LAK, May 20, 1947; OS to LAK, May 21, 1947; GPK to ECC, July 8, 1947; OS to GPK, Aug. 8, 11, 1947; OS to WB, Oct. 25, 1947; YOA.

55. [OS], "Memorandum Regarding the Needs of the Astronomy Department," [Nov. 1, 1947]; GPK to OS, Nov. 4, 1947; RMH to OS, Nov. 19, 1947; YOA.

56. RMH to OS, Feb. 10, 1948; OS to [GP]K, Nov. 3 (telegram), 22, Dec. 19 (telegram), 1948; GPK to OS, Nov. 23, 1948; YOA.

57. OS, "List of items discussed with Messrs. Colwell, Harrison and Bartky on January 22, 1948" (memo), [Jan. 23, 1948]; GPK to OS, Jan. 30, 1948; YOA.

58. [S]C to G[PK], Feb. 24, 1948; [GPK] to [S]C, Mar. 4, 1950; ASC; [S]C to G[PK], Feb. 5, Mar. 8, 1948; GPK to [S]C, Feb. 26, Mar. 9, 29, 1948; YOA.

59. [OS], "Conference with Vice-President Harrison and Dean Bartky at the Yerkes Observatory on April 24, 1948" (memo), [Apr. 25, 1948], YOA.

60. I was a student in this class, Astronomy 302, and attended the lecture, but I had no idea anything like this was going on.

61. MLS, "Events," [Jan. 18, 1949]; OS to BS, [Jan. 23, 1949]; UCB. The first of these is a very biased account by Mary Struve, but it includes a day-by-day chronology, written soon after the "events," that is probably correct in the main. The letter from Struve to Strömgren is much more even-handed and gives his own thinking at the time. It is a copy Mary Struve typed, not exactly verbatim, since she abbreviated many words and apparently left out a few, but otherwise it is probably close to what Struve wrote.

62. [OS] to [RM]H, Jan. 5, 1949, UCB.

63. RMH to OS, Jan. 8, 1949; OS to all Members of the Astronomy Department with Rank of Instructor and Higher, Jan. 10, 1949; YOA.

64. [O]S to "Dearest" [MLS], Jan. 5, 1949, UCB. Both this and the letter to RMH of note 62 are copies typed by Mary Struve, apparently verbatim except for the omission of a personal part of this letter. Struve had mailed her a handwritten copy of the letter he had sent to Hutchins; she transcribed the copy of it.

65. OS to HS, Jan. 18, 24, 1949, HCO; OS to [GP]K, Jan. 21, 1949; GPK to OS, Jan. 26, 1949; YOA.

66. D. E. Lilienthal to OS, Oct. 19, 1948; OS to BJB, Oct. 26, 1948; SLO; OS to RMH, Jan. 11, 1949, RL; OS to ISB, Jan. 11, 29, 1949; OS to WSA, Feb. 25, 1949; HHL. O[S] to [MLS], [Feb. 26, 1949], UCB. The last is a large fragment of a letter, with the top torn off, but it is signed "Otto" and is clearly to his wife. It can be dated from the fact that it was written just before his Vanuxem lectures at Princeton, March 1–3, 1949.

67. [S]C to GPK, Feb. 24, 1949; "Constitution for the Council of the Astronomy Department," [May 15, 1949]; "Council Minutes" of meetings May 17, 1949, through February 2, 1950; ECC to "All interested parties," Aug. 9, 1949; YOA.

68. WB to OS, May 11, 1949; OS to RMH, June 4, 1949; OS to RWH, July 25, 1949; YOA.

69. C. Donald Shane, "The Award of the Bruce Gold Medal to Dr. Otto Struve," PASP 60 (1948): 155; ISB to OS, Nov. 9, 1949; HHL; Struve (1950).

70. OS to LGH, Jan. 15, 1949, SLO.

71. [OS] to [P]S, Apr. 4 1946; OS to H. M. Pillans, Jan. 25, 1947; OS to M. R. Calvert, Feb. 7, 1947, YOA.

72. OS to [CD]S, Sept. 15, 1947, Mar. 1, 1948; R. M. Underhill to CDS, Nov. 24, 1948; CDS to SE, Dec. 30, 1948; SLO.

73. CDS, "Tell Einar" (notes for a telephone conversation with SE), Jan. 20, 1949; CDS to SE, Jan. 21, 1949; OS to LGH, Feb. 7, 1949; CDS to OS, Feb. 8, 1949; [S]E to [CD]S, Feb. 10, 1949; CDS to RGS, Feb. 16, 1949; SLO.

74. CDS to WSA, Mar. 21, 31, 1949; JS to CDS, Mar. 28, 1949; CDS to ISB, [Mar. 31, 1949]; CDS to H. Williams, Apr. 5, 6, 1949; SLO; OS to SE, Mar. 28, 1949; LGH to [O]S, Mar. 31, 1949; UCB.

75. ISB to OS, June 16, 26, 1947, Feb. 25, 1949; OS to ISB, June 23, Aug. 14, Sept. 17, 1949, July 8, 1948, Feb. 21, Mar. 28, 1949; WSA to OS, May 17, 1949; HHL.

76. OS to SE, Apr. 22, 1949, SLO.

77. CDS to OS, July 18, Aug. 16, Sept. 20, 1949; OS to CDS, Aug. 9, Sept. 2, 1949; SLO.

78. OS to SE, Aug. 19, 1949; SLO; "Mr. Struve's addresses" (itinerary Sept. 3–Oct. 16, 1949), [Sept. 1, 1949]: OS to [M.] Phillips, Sept. 9, 1949; YOA.

79. OS to BS, Jan. 23, 1949, UCB (see note 61); OS to [GP]K, Jan. 7, 1950, ASC.

80. TLP to OS, Sept. 20, 1949; SC to OS, Sept. 29, 1949; OS to [GP]K and [S]C, Oct. 1, 1949; YOA; OS to [CD]S, Sept. 25, 1949; CDS to OS, Oct. 5, 1949; P[S] to N. [U. Mayall], Dec. 19, 1949; SLO.

81. OS to ISB, Dec. 1, 1949, HHL; OS to WB, Dec. 15, 1949; OS to RMH, Dec. 15, 1949; RL.

82. [OS], "Some Details concerning Points Mentioned in My Letter to Mr. Hutchins of December 15, 1949" (memo to self), [Dec. 18, 1949], UCB; OS to PS, Dec. 29, 1949, YOA.

83. SC to RMH, Jan. 5, 1950, RL; OS to WB, Jan. 7, 1950, UCB; OS to [H]S, Jan. 22, 1950, HCO.

84. OS to A. R. Davis, Jan. 6, 1950; CDS to OS, Jan. 31, 1950; OS to CDS, Jan. 31, Feb. 21, 1950; SLO; OS to ISB, Jan. 22, 31, 1950; ISB to OS, Feb. 1, 1950; HHL; OS to A. R. Davis, Feb. 22, 1950; A. R. Davis to RGS, Mar. 21, 1950; F. A. Jenkins et al. to M. M. Davisson, Mar. 28, 1950; UCB.

85. OS to WB, Feb. 22, 1950, RL; OS to [GP]K, Feb. 3, 1950; SC to GPK, Mar. 28, 1950; ASC.

86. OS to [RM]H, Feb. 18, 1950; RL; OS to [GP]K, Feb. 3, 1950; SC to GPK, Mar. 28, 1950; ASC.

87. [S]C to GPK, Feb. 7, 1950; GPK to [S]C, Feb. 10, 1950; OS to [GP]K,

Feb. 21, 1950; [GPK] to [O]S, Feb. 25, 1950; ASC; WB to Members of the Department of Astronomy, Mar. 20, 1950, RL.

88. OS to HS, May 22, 1950, HCO.

89. SC to RWH, Apr. 21, 1950; OS to RWH, Apr. 24, 1950; RL.

90. Otto Struve, "W. F. Meyer's Work on Beta Canis Majoris," *ApJ* 112 (1950): 520.

91. OS to CDS, Feb. 21 (second letter of this date), Sept. 7, 1950, SLO; OS to RGS, July 1, 1950, UCB.

## Chapter Twelve    Epilogue: To the Centennial, 1950–1997

1. GPK to [S]C, Mar. 30, June 23, 1950; SC to [GP]K, Apr. 1, 1950; ASC; OS to ISB, Aug. 9, Sept. 1, Oct. 4, 1950, HHL; OS to CDS, Aug. 28, Sept. 24, 1950, SLO.

2. OS to "Dear Sir" (a change-of-address form letter), Mar. 1, 1950, SLO.

3. SC to WB, Mar. 7, 1950, RL; WB to OS, Mar. 23, 1950; OS to SC, Oct. 21, 1950; SC to OS, Oct. 27, 1950; UCB; OS to CDS, Sept. 7, 1950, SLO.

4. SC to RMH, Oct. 6, 1950, RL; J[HO] to GPK, Jan. 14, 1950; G[PK] to J[HO], Sept. 1, 1950, ASC.

5. OS to [S]C, Oct. 4, 1952, July 8, 1954, June 19, 1956, KK; OS to SC, Jan. 22, May 8, 1953; SC to OS, Apr. 15, May 15, 1953, UCB.

6. GPK to T. G. Cowling, Aug. 7, 1965, ASC.

7. Donald E. Osterbrock, "Su-shu Huang," *Astronomy Quarterly* 1 (1977): 261.

8. OS to HS, Oct. 12, 1950, May 31, 1951, HCO; OS to RGS, June 9, 1951; OS, "The Department" (text of a speech at a celebration of S. Einarsson's seventy-fifth birthday), May 28, 1955; UCB.

9. [J]AH to OS, Jan. 5, July 25, 1953; OS to JAH, Jan. 13, 23, July 29, 1953; OS to C. Kerr, Apr. 22, 1953; CDS to RGS, Apr. 29, 1953; OS to A. R. Davis, May 18, 1953; UCB.

10. James G. Hershberg, *James B. Conant: Harvard to Hiroshima and the Making of the Nuclear Age* (New York: Alfred A. Knopf, 1993).

11. OS to RGS, Oct. 31, 1950; OS to [A. R.] Davis, Nov. 8, 1953; OS to JLG, Nov. 11, 1953; C. Kerr, [untitled memoranda], Nov. 16, [Dec. 18], 1953; A. R. Davis to OS, Nov. 16, 1953; C. Kerr to [O]S, Nov. 18, 1953; OS to C. Kerr, Nov.19, 1953; UCB.

12. OS to PS, Feb. 16, 1953; P[S] to O[S], Nov. 6, 1953, Jan. 8, 19, 1954; OS to D. H. Menzel, Dec. 9, 16, 1953; OS to HS, Dec. 17, 1953; OS to Kerr, Dec. 23, 1953; OS to CPG, Dec. 23, 1953; UCB.

13. OS to PS, June 22, July 3, 1956; PS to OS, June 29, 1956; UCB.

14. L. V. Berkner to OS, Apr. 17, 1959; G. T. Seaborg to OS, Mar. 31, 1959; OS to I. I. Rabi, Oct. 31, 1961; UCB; Krisciunas (1992).

15. OS to BS, Oct. 17, 1961, UCB; Ira S. Bowen, "Annual Report of the Director, Mount Wilson and Palomar Observatories," *CIW Yearbook* 62 (1963): 1.

16. Otto Struve, B. Lynds, and H. Pillans, *Elementary Astronomy* (New York: Oxford University Press, 1959); Otto Struve and V. Zebergs, *Astronomy of the Twentieth Century* (New York: Macmillan, 1962); Otto Struve, *The Universe* (Cambridge: MIT Press, 1962).

17. L. A. Andrew to DEO, Nov. 21, 1994, Feb. 2, 1995.

18. "Yerkes and McDonald Observatories," *PASP* 63 (1951): 43; D. Hoffleit, "New Yerkes Director," *S&T* 10 (1951): 64; BS to CDS, Jan. 27, 1951, UCB.

19. OS to RMH, June 5, Oct. 4, 1953; RMH to OS, June 9, Aug. 21, 1953; UCB.

20. Dzuback (1991); McNeill (1991).

21. R. Kulsrud, "Bengt Strömgren," *American Philosophical Society Yearbook,* 1987, 216; M. Rudkjøbing, "Bengt Georg Daniel Strömgren (1908–1987)," *QJRAS* 29 (1988): 282.

22. Subrahmanyan Chandrasekhar and E. Fermi, "Magnetic Fields in Spiral Arms," *ApJ* 118 (1953): 113; Subrahmanyan Chandrasekhar and E. Fermi, "Problems of Gravitational Stability in the Presence of a Magnetic Field," *ApJ* 118 (1953): 116.

23. Osterbrock (1995a).

24. GPK to JHO, Sept. 8, 1950, ASC.

25. Cruikshank (1993).

26. William W. Morgan, "A Morphological Life," *ARA&A* 26 (1988): 1; Garrison (1995).

27. W. Albert Hiltner, "Polarization of Light from Distant Stars by Interstellar Medium," *Science* 109 (1949): 165; J. S. Hall, "Observations of the Polarized Light from Stars," *Science* 109 (1949): 166; HS to OS, Jan. 20, 1949; OS to HS, Jan. 25, Feb. 18, 1949; HCO.

28. Code (1992).

29. S. Arend, "In Memoriam Georges-Achille Van Biesbroeck," *Ciel et Terre* 90 (1974): 321; P. Mullerand P. Baize, "G. A. Van Biesbroeck (1880–1974)," *L'Astronomie* 88 (1974): 306.

30. Evans and Mulholland (1986).

31. F. K. Edmondson, "AURA and KPNO: The Evolution of an Idea," *JHA* 22 (1991): 68.

32. This is based in part on my own journal for the year 1966, when I had many conversations, discussions, and meetings with both of them and with others at Yerkes, concerned with joint telescope projects.

33. A. E. Whitford, "Joel Stebbins, 1878–1966," *BMNAS* 49 (1978): 293.

# BIBLIOGRAPHY

Adams, Walter S. 1938. "George Ellery Hale, 1868–1938." *ApJ* 87:369.
———. 1939. "George Ellery Hale, 1868–1938." *BMNAS* 21:181.
———. 1947. "Some Reminiscences of the Yerkes Observatory." *Science* 106:196.
———. 1954. "The Founding of the Mount Wilson Observatory." *PASP* 66:267.
Ashmore, Harry S. 1989. *Unreasonable Truths: The Life of Robert Maynard Hutchins*. Boston: Little, Brown.
Barnard, E. E. 1921. "Sherburne Wesley Burnham." *PA* 29:309.
Bok, Bart J. 1978. "Harlow Shapley." *BMNAS* 49:241.
Burnham, S. W. 1900. "A General Catalogue of Double Stars Discovered from 1871 to 1899: Introduction," *PYO* 1:vii.
———. 1906. *A General Catalogue of Double Stars within 121° of the North Pole*. Washington, D.C.: Carnegie Institution of Washington.
Christianson, Gale E. 1995. *Edwin Hubble: Mariner of the Nebulae*. New York: Farrar, Straus and Giroux.
Code, Arthur D. 1992. "William Albert Hiltner, 1914–1991." *BAAS* 24:1326.
[Compton], K[arl] T. 1944. "Henry Gordon Gale." *JOSA* 34:446.
Comstock, George C. 1922. "Benjamin Athrop Gould." *BMNAS* 17:153.
Crew, Henry. 1943. "Henry Gordon Gale, 1874–1942." *ApJ* 97:85.
Cruikshank, Dale P. 1993. "Gerard Peter Kuiper." *BMNAS* 62:259.
DeVorkin, David H. 1980. "The Maintenance of a Scientific Research Institution: Otto Struve, the Yerkes Observatory, and Its Optical Bureau during the Second World War." *Minerva* 62:259.
Douglas, A. Vibert. 1957. *The Life of Arthur Stanley Eddington*. London: Thomas Nelson.
Dzuback, Mary Ann. 1991. *Robert M. Hutchins: Portrait of an Educator*. Chicago: University of Chicago Press.
Evans, David S., and J. Derrall Mulholland. 1986. *Big and Bright: A History of McDonald Observatory*. Austin: University of Texas Press.
Fermi, Laura. 1968. *Illustrious Immigrants: The Intellectual Migration from Europe, 1930–41*. Chicago: University of Chicago Press.

Frost, Edwin B. 1921. "Sherburne Wesley Burnham, 1838–1921." *ApJ* 54:1.

———. 1924. "Edward Emerson Barnard." *MNAS* 21 (14th memoir): 1.

———. 1933. *An Astronomer's Life.* Boston: Houghton Mifflin.

Garrison, R. F. 1995. "William Wilson Morgan (1906–1994)." *PASP* 107:507.

Gingerich, Owen. 1990. "Through Rugged Ways to the Galaxies." *JHA* 21:77.

Goldberg, Leo. 1964. "Otto Struve." *QJRAS* 5:284.

Goodspeed, Thomas W. 1916. *A History of the University of Chicago: The First Quarter Century.* Chicago: University of Chicago Press.

Hale, George E. 1897a. "On the Comparative Value of Refracting and Reflecting Telescopes for Astrophysical Applications." *ApJ* 5:119.

———. 1897b. "The Aims of the Yerkes Observatory." *ApJ* 6:310.

Herbig, G. H., ed. 1970. *Spectroscopic Astrophysics: An Assessment of the Contributions of Otto Struve.* Berkeley: University of California Press.

Hughes, R. M. 1925. *A Study of the Graduate Schools of America.* Oxford, Ohio: Miami University Press.

Hynek, J. Allen, ed. 1951. *Astrophysics: A Topical Symposium Commemorating the Fiftieth Anniversary of the Yerkes Observatory and a Half Century of Progress in Astrophysics.* New York: McGraw-Hill.

Joy, Alfred H. 1958. "Walter Sydney Adams." *BMNAS* 31:1.

Kilminster, C. W. 1966. *Men of Physics: Sir Arthur Eddington.* Oxford: Pergamon Press.

Kuiper, Gerard P., ed. 1949. *The Atmospheres of the Earth and the Planets: Papers Presented at the Fiftieth Anniversary Symposium of the Yerkes Observatory, September 1947.* Chicago: University of Chicago Press.

Krisciunas, Kevin. 1992. "Otto Struve." *BMNAS* 61:351.

Ledoux, P. 1987. "Pol Swings et les débuts de l'astrophysique en Belgique." In *Quelques étapes de l'histoire de la géophysique en Belgique,* ed. A. Berger and A. Allard, 21–47. Actes du colloque du 14 mars 1986, Louvain-la-Neuve.

McCutcheon, Robert A. 1991. "The 1936–1937 Purge of Soviet Astronomers." *Slavic Studies* 50, no. 1:100.

McNeill, William H. 1991. *Hutchins' University: A Memoir of the University of Chicago, 1929–1950.* Chicago: University of Chicago Press.

Moore, E. H., H. Maschke, and H. S. White. 1896. *Mathematical Papers Read at the International Mathematical Congress Held in Connection with the World's Columbian Exposition, Chicago 1893.* New York: Macmillan.

Morgan, W. W. 1938. "Storrs B. Barrett." *PA* 46:125.

———. 1967. "Frank Elmore Ross." *BMNAS* 39:391.

Nicholson, Seth B. 1961. "Frank Elmore Ross, 1874–1960." *PASP* 73:181.

Osterbrock, Donald E. 1980. "America's First World Astronomy Meeting: Chicago 1893." *Chicago History* 9:176.

———. 1984a. "The Rise and Fall of Edward S. Holden." *JHA* 15:81, 151.

———. 1984b. *James E. Keeler, Pioneer American Astrophysicist: And the Early Development of American Astrophysics.* Cambridge: Cambridge University Press.

———. 1985a. "The Minus First Meeting of the American Astronomical Society." *Wisconsin Magazine of History* 68:108.

———. 1985b. "The Quest for More Photons: How Reflectors Replaced Refractors as the Monster Telescopes of the Future at the End of the Last Century." *Astronomy Quarterly* 5:87.

———. 1986. "Nicholas T. Bobrovnikoff and the Scientific Study of Comet Halley, 1910." *Mercury* 15:46.

———. 1993a. *Pauper and Prince: Ritchey, Hale and Big American Telescopes.* Tucson: University of Arizona Press.

———. 1993b. "A New Hampshire Yankee in Queen Urania's Court: S. W. Burnham, Double-Star Observer." *BAAS* 24:1236.

———. 1994. "Getting the Picture: Wide-Field Photography from Barnard to the Achromatic Schmidt." *JHA* 25:1.

———. 1995a. "Founded in 1895 by George E. Hale and James E. Keeler: The *Astrophysical Journal* Centennial." *ApJ* 438:1.

———. 1995b. "Fifty Years Ago: Astronomy; Yerkes Observatory; Morgan, Keenan and Kellman." In *The MK Process at Fifty Years,* ed. C. J. Corbally and R. F. Garrison, 199–214. San Francisco: Astronomical Society of the Pacific.

Osterbrock, Donald E., Ronald S. Brashear, and Joel A. Gwinn. 1990. "Self-Made Cosmologist: The Education of Edwin Hubble." In *Evolution of the Universe of Galaxies,* ed. Richard G. Kron, 1–18. San Francisco: Astronomical Society of the Pacific.

Osterbrock, Donald E., J. R. Gustafson, and W. J. S. Unruh. 1988. *Eye on the Sky: Lick Observatory's First Century.* Berkeley: University of California Press.

Payne-Gaposchkin, Cecilia. 1984. *An Autobiography and Other Recollections.* Ed. K. Haramundanis. Cambridge: Cambridge University Press.

Sheehan, William. 1995. *What Immortal Fire Within: The Life of Edward Emerson Barnard.* Cambridge: Cambridge University Press.

Storr, Richard J. 1966. *Harper's University: The Beginnings.* Chicago: University of Chicago Press.

Struve, Otto. 1938. "Edwin Brant Frost, 1866–1935." *BMNAS* 19:25.

———. 1940. "The Organization of the Observatory." *CMO,* no. 1:1.

———. 1947a. "The Story of an Observatory (the Fiftieth Anniversary of the Yerkes Observatory)." *PA* 55:227.

———. 1947b. "The Yerkes Observatory, 1897–1947." *PA* 55:413.

———. 1947c. "The Yerkes Observatory: Past, Present, and Future." *Science* 106:217.

———. 1950. *Stellar Evolution: An Exploration from the Observatory.* Princeton: Princeton University Press.

———. 1957. "About a Russian Astronomer." *S&T* 16:379.

———. 1958. "G. A. Shajn and Russian Astronomy." *S&T* 17:272.

———. 1962. "The Birth of McDonald Observatory." *S&T* 24:316.

Wali, Kameshwar C. 1991. *Chandra: A Biography of S. Chandrasekhar.* Chicago: University of Chicago Press.

Wolfmeyer, Ann, and Mary B. Gage. 1976. *Lake Geneva: Newport of the West, 1870–1920.* Lake Geneva, Wis.: Lake Geneva Historical Society.

Wright, Helen. 1966. *Explorer of the Universe: A Biography of George Ellery Hale.* New York: E. P. Dutton.

# INDEX

Prepared by Irene H. Osterbrock

*Page numbers in italics refer to photographs on those pages.*

A stars, 50, 98, 99
  peculiar, 102, 256
Abundance of elements in stars, 212,
  216, 221, 275, 282
Adams, Walter S., 250, 251, 262,
  275, 289, 296
  as adviser to McDonald Observatory plans, 127, 188, 191, 193
  as confidant of Gale, 60, 62, 70,
  116–17, 129
  consulted on Yerkes matters,
  90–91, 94–95, 116–17, 122,
  169, 288
  denies Struve's request for position,
  99–100
  education, 34–35, 40, 71
  moves to Mount Wilson, 44–45,
  48
  pro-American sentiments, 100,
  104, 117, 231
  receives honorary degree, 261–62
  research, 50–51, 61, 193, 222
  retirement, 267
Adel, Arthur, 232–33, 270–71
Adler Planetarium, 136, 195
Aitken, Robert G., 79, 86–87, 90,
  170
Allahabad University, 240, 249
Allegheny Observatory, 14, 48
Aller, Lawrence H., 290
ALSOS scientific mission, 259
Alvan Clark and Sons, 3, 9, 77.
  *See also* Clark, Alvan G.
American Association for the

Advancement of Science (AAAS),
  8, 38
  meeting at McDonald Observatory,
  203–4
  Moulton as secretary of, 167
American Astronomical Society (AAS)
  and *Astrophysical Journal*, 288,
  312
  founding of, 38–39
  meeting of 1922, 81; of 1925, 85;
  of 1926, 91–92; of 1927, 92; of
  1934, 154; of 1941, 242–44; of
  1947, 289–90
Anderson, John A., 57, 140
Andhra University, 276
Angell, James R., 108
Arcturus, 61, 137
Art Institute of Chicago, 12
Asteroids, 51, 140, 314
Astrometric program, 283, 284,
  315
*An Astronomer's Life* (Frost), 138,
  320–21
*Astronomical Journal*, 17
*Astronomy and Astro-Physics*,
  18–19
*Astrophysical Journal*
  editors of
  Chandrasekhar, 311–12
  Frost, 62
  Morgan, 288
  Struve, 236, 256, 273
  founding of, 18–19, 228
  turned over to AAS, 312

Astrophysical Research Consortium
(ARC), 315, 316
Astrophysics
beginnings of, 3–4, 9–10
Hale's emphasis on, 16–18, 316
Keeler's address on, 21
theoretical, 323–24
Yerkes research centered on, 316
*Astrophysics: A Topical Symposium*
(ed. Hynek), 290
*An Atlas of Stellar Spectra* (Morgan,
Keenan, Kellman), 255–56, 313
*Atlas of the Milky Way* (Barnard), 49
*Atlas of the Northern Milky Way*
(Ross), 57
*The Atmosphere of the Earth and
Planets* (Kuiper), 290, 312
Atomic bomb, 252, 260, 262, 268
Atomic Energy Commission (AEC),
293
ε Aurigae, 143, 216–17
Ayer, Edward E., 16, 63

B stars, 50
absolute magnitudes of, 88–89
calcium lines in, 85–86, 88, 96
distances of, 224, 281
radial velocities of, 84
Stark effect in, 98, 101
Baade, Walter, 100–101, 194, 230,
*230*
Babcock, Horace W., 207, 252, 267
Barnard, Edward E., 20, *22, 31, 37,
39, 52, 72,* 80
*Atlas of the Milky Way,* 49
death, 52
dedication to astronomical work,
31, 32, 316
early life, 30–31
hired on Yerkes staff, 32
Milky Way photography, 31, 49,
50, 51–52
Mount Wilson work, 44, 49
Barrett, Storrs B., 72
as aide to Frost, 68
endorses Crump, 70
radial-velocity observations of, 80,
84, 99
retirement, 68
as school board member, 75
Bartky, Walter, 183, 212, *239*
astronomy faculty member, on
campus in Chicago, 167–68,
226, 228

as associate dean of Physical Sciences, 260
as dean of Physical Sciences,
263–64, 274, 275, 288,
290–93, 299, 303–4
Struve's objections to, 238–40,
263–64, 272
Bates, David R., 229
Be stars, 143, 145, 150
Benedict, Harry Y.
advice from Frost, 123–24, 319
death, 204
and inception of McDonald Obser-
vatory, 123–26, 190
as president of University of Texas,
124
and See, 163
Benefactors of astronomy, 317,
322–23
Berkeley campus of University of
California. *See* University of
California, Berkeley
Betelgeuse, 61
Bethe, Hans, 227, 234
Beutler, Hans, 233, 251
Bidelman, William P., 261, 281
Binary stars
Burnham's observations of, 29–30,
32, 50
eclipsing, 235–36
Van Biesbroeck's observations of,
51
Birge, Raymond T., 296
Blaauw, Adriaan, 282
Blackstone, Isabel, 41–42
Blakslee, George G., 153
Blanco, Victor, *285*
Bobrovnikoff, Nicholas T., *83,* 290
as graduate student at Yerkes, 73
at Perkins Observatory, 144,
146–50, 233
as student at Kharkov University,
78
Bohr, Niels, 214, 215
Bok, Bart J., 306–7
advice on Yerkes and McDonald
faculty members, 170–72, 175,
196, 240–41
directorship possibility, 273
*The Distribution of Stars in Space,*
217
offered assistant directorship,
240–42, 249
Struve's relationship with, 240–41

as visiting lecturer at Yerkes, 216, 217
Bolometer, 25, 39
Bonney, Charles C., 12
Boss, Lewis, 41
Bowen, Ira S., 200
    as director of Mount Wilson Observatory, 267, 296–97, 299, 300
Brashear, John A., 5, 6, *13*, 14, 49
Breit, Gregory, 234
Brown, Harrison, 290
Bruce, Catherine W., 33–35, 49, 317
Bruce spectrograph, 35
Bruce telescope, 35, 49
Bundy, McGeorge, 307
Bunsen, Robert, 10
Burnham, Sherburne W., *13, 29*
    advice on site of Yerkes Observatory, 16
    as adviser to Hale, 3–4
    death, 51
    double-star observations, 3, 29, 30, 50, 51
    *General Catalogue of Double Stars,* 51
    honorary degree, 51
    at Lick Observatory, 6, 29–30
    retirement, 51
    on Yerkes staff, 28–30, 47
Burton, Ernest D., 58, 60, 107
Bush, Vannevar, 253, 289

Calcium clouds. *See* Interstellar calcium
Calhoun, John W., 204
California Institute of Technology, 59, 169, 263, 269, 284
Calvert, Mary R., 72, *83, 151,* 153, *215,* 283–84
Campbell, William Wallace
    and confirmation of theory of relativity, 65
    as director of Lick Observatory, 112
    radial velocity work, 34, 50–51
    as president of University of California, 85
β Canis Majoris, 301
α² Canum Venaticorum, 256
Carbon stars, 26–28, 50
Carleton College, 18, 71, 85
Carnegie, Andrew, 41, 317

Carnegie Institution of Washington, 253, 289
    grant for Yerkes staff, 35, 42
    support for Mount Wilson Observatory, 246
    support for sixty-inch telescope, 41, 43–45, 48
Case Institute of Technology, 152, 196
    gives honorary degree to Struve, 207
γ Cassiopeiae, 146
Center for Astrophysical Research in Antarctica (CARA), 316
Century of Progress Exposition. *See* Chicago World's Fair
β Cephei, 141
Cerro Tololo Interamerican Observatory, 248, 316
Cesco, Carlos, 261
δ Ceti, 141
Chamberlin, Rollin T., 290
Chamberlin, Thomas C., 7, 15–16
    planetesimal theory, 49, 164–65
Chandrasekhar, Lalitha Doraiswamy, 213
Chandrasekhar, Subrahmanyan, 159, *180, 215, 230,* 249, 273, 277, *285,* 292, 305
    accepts Yerkes offer, 184
    as acting chairman, 300–301, 303–4
    arrives at Yerkes, 212–14
    chooses to stay at Yerkes, 212–14
    death, 312
    disagreement with Eddington, 177, 230–31
    as editor of *Astrophysical Journal,* 311–12
    education, 177
    graduate students, 261, 280–81, 284
    honors bestowed on, 276, 279, 286
    at Institute for Advanced Study, 254–55
    *An Introduction to the Study of Stellar Structure,* 217–18, 221
    marriage, 213
    as member of Yerkes "council," 293–94, 299
    moves to Chicago, 311
    position offers, 249, 276–78
    promotions, 255

Chandrasekhar, Subrahmanyan
(*continued*)
racial prejudice against, 179–82,
212, 218–21
radiative transfer research, 181,
280
relationship with Struve, 276–78,
291, 303–4
and reorganization plan, 287–88
sought as Yerkes faculty member,
178–83, 324
stellar dynamics research, 212
stellar interiors research, 212
supports Kuiper, 291, 293
tension with Strömgren, 310
during World War II, 255
Chicago, housing segregation in,
180–81
Chicago, University of. *See* University
of Chicago
Chicago World's Fair, 137–38, 167,
218
Chrétien, Henri, 189
Clark, Alvan G., 9, *13*, 14, 19–20,
188
Cobb, Henry Ives, 16, 19
Code, Arthur D., 280, *285*
Columbian Exposition. *See* World's
Columbian Exposition
Colwell, Ernest C., 264–65, 275,
288, 301
Comets
Barnard's discovery of, 31
Bobrovnikoff's work on, 73
Herzberg's observations of, 271
Van Biesbroeck's work on, 51,
140, 314
Community involvement of Yerkes
personnel, 28, 74–75, 281
Compton, Arthur H., 70, 148, 212,
238, 271–72
as chancellor of Washington Uni-
versity, 263
as dean of Physical Sciences, 239,
242, 253, 271–72
at McDonald Observatory dedication, 205
as Nobel Prize winner, 148, 185
during World War II, 260, 268
Comstock, George C., 38, 85, 321
Conant, James B., 306
Conference on Novae, Supernovae,
and White Dwarfs (Paris),
229–30, *230*

Conference on Solar Research, 44
Conwell, Esther M., 281
Cooperative observatories
ARC group, 316
McDonald as example of, 247–48,
316
national observatories, 248, 316
Wisconsin-Indiana-Yale-National,
316
Copenhagen University, 175–76,
214, 276
Crane, Charles R., 79
Crew, Henry, 137
Crimean Astrophysical Observatory,
279
Crump, Clifford C., 72
career after Yerkes, 142
leaves Yerkes, 142
and McDonald Observatory inception, 122, 124–25
at Ohio Wesleyan, 68–69
relationship with Struve, 134,
140–42
at Yerkes, 69–70
Cuffey, James, 241–42
Curtis, Heber D., 150
Curtiss, Ralph H., 68
α Cygni, 202
P Cygni, 203

Davis, Dorothy N., 233
Dearborn Observatory, 4, 150, 284
Depression. *See* Great Depression
Dershem, Elmer, 269–70, 271
de Sitter, Willem, 117
Deslandres, Henri, 26
"Detached" lines of calcium. *See*
Interstellar calcium
Deutsch, Armin J., 281
de Vaucouleurs, Gerard, 314
*The Distribution of Stars in Space*
(Bok), 217
Double stars. *See also* Binary stars;
Spectroscopic binary stars
Burnham's observations of, 29–30,
32, 50
Kuiper's observations of, 211
Van Biesbroeck's observations of,
51
κ Draconis, 85–86
Dreiser, Theodore, 1
DuBridge, Lee A., 263
Duke, Douglas, 281
Dunham, Theodore, 101, 155, 226

Earlham College, 68, 142
Ebbinghausen, Edwin G., 261
Eclipse, solar
of 1900, 39
of 1918, 63, 69
of 1919, 65
of 1922, 66
of 1923, 63–64, 69
of 1932, 136
Eddington, Arthur S., 86, 98, *230*
and disagreement with Chandrasekhar, 177, 230–31
interstellar gas studies, 93–95
visit to Yerkes, 151–52
Edlén, Bengt, 208, 230
Edmonds, Frank N., 281, 314
Edmondson, Frank K., 248
Eighty-two-inch McDonald telescope
finished, 202
mirror for, 191–92, 200–202
plans for, 187–89
Struve's involvement in planning, 188, 190–91
Einarsson, Sturla, 295–97
Einstein, Albert, 65–66
Ellerman, Ferdinand, *37*
carbon star work, 50
community involvement, 28
as Hale's employee, 10–11
moves to Mount Wilson, 44–45
at solar eclipse of 1900, 39
at Yerkes, 25, 27, 40
Elvey, Christian T., *83*, *139*, 217, 251, 252, 296
as assistant director of McDonald, 246, 250
as graduate student at Yerkes, 79, 101–2, 113
leaves McDonald, 268–69
at McDonald, 129–30, 187, 192, 207
relations with Struve, 154, 257, 268–69
research, 192, 199, 221, 269
as staff member at Yerkes, 114, 140, 154
during World War II, 268–69
*Encyclopaedia Britannica*, 288
Enders, Gertrude, 152
Energy production in stars, 227, 234
European Southern Observatory, 316

Fairley, Arthur S., 135
Farnsworth, Alice, *83*

Fecker, James W., 144, 147, 149
Fermi, Enrico, 271
atomic bomb work, 251–52
collaboration with Chandrasekhar, 311
and Institute of Nuclear Studies, 265, 294
Filbey, Emery T., 184, 238–40, 247, 301
*The Financier* (Dreiser), 1
Flook, Lyman R., 142
Ford Foundation, 309
Förster, Wilhelm, 161
Forty-inch telescope, 27
first viewing through, 20
lens for, 8–9, 19–20
mounting exhibited at World Congress, 14
possibility of moving to McDonald, 246
stellar parallax work with, 35
suitability for astrometric program and spectroscopic work, 315
Fox, Philip, 50, 136–38
Franck, James, 238, 290
Fraunhofer, Josef, 9
Frost, Edwin B., *13*, *33*, *37*, 72, 82, *83*, 111, *115*, *139*, 318
as acting director of Yerkes, 47–48
aids Struve in coming to U.S., 78–79, 318
*An Astronomer's Life*, 138–39, 321
as *Astrophysical Journal* editor, 62
becomes director of Yerkes, 50
and Chicago World's Fair, 137–38
community involvement, 75
directorship qualities, 53, 58, 60, 67, 68, 152
education, 34
efforts to keep Struve, 87, 318
fund-raising efforts, 62–63
humanitarian qualities, 58, 318
illness and death, 75–76, 139, 321
joins Yerkes faculty, 33–34
lasting effects of directorship, 124, 318
loss of eyesight, 58
public outreach, 58, 74
as radial-velocity observer, 34, 35, 50, 84–85, 99, 143
respect of local residents for, 75, 139, 318, 320

Frost, Edwin B. (*continued*)
  retirement, 111–12, 133–34,
    319–20
  role in building of McDonald Ob-
    servatory, 122–24, 319
  role in choosing his replacement,
    114–15
  on solar eclipse expeditions, 39,
    63–64, 136–37
  at World Congress of Astronomy
    and Astrophysics, 14, 34
Frost, Mary H. (wife), 133, 136–37,
  243

Galactic structure, 281
Gale, Henry G., *135*, 226, 228, 318
  actions as dean, 87, 111, 135–36,
    150, 205, 226, 232
  becomes dean of Physical Sciences,
    62
  and friendship with Adams, 60, 62,
    116–17
  racial prejudice, 179–82, 212,
    218–21
  recommends Struve for director,
    177–21, 125
  and relations with Struve, 149, 153
  retirement, 238, 240
  scientific work, 60–61, 140, 193
Gallo, Joaquin, 205
Gamow, George, 226
Garner, Arch, 268, 270
Gates, Frederick T., 1
*General Catalogue of Double Stars*
  (Burnham), 51
Gerasimovich, Boris P., 96–97, 250
Gill, Jocelyn R., 235
Gingrich, Curvin H., 71
Goldberg, Leo, 243
Goodspeed, Thomas W., 28
Gothard, Eugen von, 14
Goudsmit, Samuel, 259
Gould, Benjamin A., 17
Graduate program, Yerkes Ob-
  servatory
  during Frost's tenure, 71–74, 79
  during Hale's tenure, 34
  during Struve's tenure, 213, 261
Graduate students, Yerkes Ob-
  servatory
  of Chandrasekhar, 261, 280–81,
    284
  first, 34–35, 71

  during Frost's directorship, 73–74,
    79
  of Herzberg, 284, 286
  of Kuiper, 284
  of Morgan, 261, 281
  of Strand, 283
  of Struve, 101–2, 149, 152–53,
    212, 232, 235, 261, 280,
    284–86
Great Depression, 153, 167
  effect on Mount Wilson, 100
  effect on University of Chicago, 74,
    110
  effect on Yerkes, 135–36
  job situation during, 141, 145,
    147
Greenstein, Jesse L., *215*, 244, *285*,
    286, 290
  and building of Caltech astronomy
    department, 284, 320
  and design of nebular spectro-
    graph, 197–98
  designs ultraviolet spectrograph,
    280
  joins Yerkes faculty, 234
  research, 221, 235–36
  during World War II, 252, 254
Guggenheim Fellowships, 94–95,
    234, 280, 306
Guthnick, Paul, 78

Hale, George Ellery, *4, 13, 22, 37, 43*
  builds Yerkes staff, 28–35
  and Carnegie Institution of Wash-
    ington, 44, 45
  childhood and education, 3, 4–5
  and conflict with Ritchey, 189
  and continuing interest in Yerkes,
    127, 140
  at dedication of Yerkes, 21–23
  on eclipse expedition of 1900, 39
  efforts to finance move of sixty-
    inch telescope, 40–42, 45, 48
  emphasis on astronomical excel-
    lence, 323
  emphasis on astrophysics, 3–4,
    16–17, 38–39
  at first viewing through forty-inch
    telescope, 20
  and formation of American Astro-
    nomical Society, 38–39
  and founding of *Astrophysical
    Journal,* 17–19, 38

and founding of Yerkes, 1, 9, 16–17
fund-raising skills of, 1, 3, 37, 44–45, 57, 64, 317
illness, 53
invents spectroheliograph, 5
joins University of Chicago faculty, 8
leaves Yerkes, 42–43
marriage, 5
organizes Conference on Solar Research, 44
organizes World Congress of Astronomy and Astrophysics, 11–12
resignation, 45, 50
secures funds for two-hundred-inch telescope, 57
solar research, 5, 6, 25–26
thesis, 5
visit to Lick Observatory, 5–6
Hale, William E. (father)
death of, 19
as financial supporter, 3, 4, 8, 19
Hale telescope, 297. See also Telescopes, Palomar two-hundred-inch
Hall, Asaph, 8
Hall, John S., 185, 313
Hall, R. Glenn, 283, 284
Hardie, Robert H., 286
Harnwell, Gaylord P., 263
Harper, William Rainey, 7
becomes president of University of Chicago, 6–7
and building of astronomy department, 8, 32, 159, 161, 163
death, 45, 50
dismisses Wadsworth, 18
early career, 6–7
efforts to retain Hale, 45, 47, 318
and founding of Yerkes, 1, 9, 20
illness, 48
and negotiations regarding sixty-inch telescope, 40–41, 45
and See, 159–63
at Yerkes dedication, 23
Harrell, William B., 142
Harris, Daniel E., 284
Harrison, Marjorie Hall, 280
Harrison, R. W., 290–91, 294, 301
Hartmann, Johannes, 85

Harvard College Observatory, 38, 182
assistant directorship offer to Struve, 118–22, 125–26
directorship offer to Struve, 306–7
Hale's volunteer work at, 5
Struve's visit to, 92–93
summer school at, 168
during World War II, 252–53
Hellweg, J. F., 189
Helmholtz, Hermann von, 12
Henrich, Louis R., 261
Henyey, Louis G., 215, 239, 285, 309
at Berkeley, 284, 295–96, 306–7, 309
and design and research with nebular spectrograph, 197–98, 221
as graduate student at Yerkes, 152–53
Guggenheim Fellowship, 234
joins Yerkes faculty, 212
as observer at Perkins Observatory, 153
thesis, 212
during World War II, 252, 254
Herbig, George H., 282–83
Hertzsprung, Ejnar, 170, 283
Herzberg, Gerhard, 233, 285, 287, 290
graduate students, 284–85
joins Yerkes staff, 171–72
Nobel Prize, 286
research, 251, 271–72
returns to Canada, 284
spectroscopic laboratory, 272, 279–80
High-proper-motion stars, 56, 139–40
Hiltner, W. Albert, 291, 305
as assistant director of Yerkes and McDonald, 274–75, 286
on "council" of Yerkes, 294, 299
death, 314
as director of University of Michigan Observatory, 313–14
as director of Yerkes, 313
discovers polarization by interstellar dust, 313
joins Yerkes faculty, 260–61
photoelectric photometry, 281, 313
role in Yerkes reorganization, 273–74

Holden, Edward S., 6, 29–31, 33
Horak, Henry G., 281
Hough, George W., 12
Hoyt, Frank C., 281
Huang, Su-shu, 280, *305, 306*
Hubble, Edwin P., 72, 116, 231
  galactic research, 206
  as graduate student, 73, 165
  nebular research, 199
  during World War II, 255
Huggins, William, 10
Hurlbert, Eri B., 160–61
Hussey, William J., 42, 69, 321
Hutchins, Francis (brother), 109
Hutchins, Maude (wife), 289, 309
Hutchins, Robert M., *110*, 196, 246,
  247, *310*
  accepts Struve's resignation,
  300–301
  and appointment of Bartky,
  263–64
  approves Yerkes reorganization
  plan, 287
  becomes chancellor of University of
  Chicago, 264, 273
  becomes president of University of
  Chicago, 107, 109–11
  convinces Chandrasekhar to remain, 276–78
  early career, 107–9
  education, 108–9
  and fiftieth anniversary of Yerkes,
  289
  and hiring of Yerkes faculty,
  171–73, 175, 177–79, 181, 184
  and McDonald Observatory agreement, 125–27, 192
  and McDonald Observatory dedication, 204–5
  personal problems, 288
  relationship with Struve, 132,
  134–35, 237–38, 240, 264,
  273, 279, 289, 300, 319
  resigns to take post at Ford Foundation, 309
  and Struve directorship offer,
  120–22, 131
  and Yerkes reorganization problems, 291, 293
Hutchins, Vesta (wife), *310*
Hutchins, Will (brother), 109
Hutchins, William J. (father), 109
Hutchinson, Charles L., 16, 62, 233

Huth, Carl F., 219
Hynek, J. Allen, 149, 150, 280
  *Astrophysics: A Topical Sympo-
  sium,* 290

IC 1318, 199
Indiana University, 248, 316
Individuals, significance in growth of
  astronomy, 317, 322–23
Infrared observations, 25, 39, 280
Ingersoll, Leonard, 235
Institute for Advanced Study,
  254–55, 263, 308, 311
Institute of Nuclear Studies, 265, 294
Insull, Samuel, 167
International Astronomical Union,
  45, 275
International Education Board,
  87–88
International Solar Union, 44–45
Interstellar calcium, 85–91, 93–94,
  96
Interstellar dust, 89, 94, 197, 206,
  224, 281–82
Interstellar gas, 94
Interstellar matter
  Barnard's observations of, 52
  Eddington's work on, 94
  Strömgren's research on, 221–22,
  282
Interstellar reddening, 94, 206, 224
*An Introduction to the Study of Stel-
  lar Structure* (Chandrasekhar),
  217–18, 221
Irwin, John B., 316
Isham, George S., 42, 49, 133

Janssen, Jules, 4
Jeans, Sir James, 166, 323
Jeffreys, Harold, 166
Jewish Charities of Chicago, 156,
  213
Johnson, Alice, *215*, 300, 306
Johnson, Harold L., 313
Johnson, Warren C., 277
Jose, Paul D., 271
Joy, Alfred H., 100, 256, 290
Judson, Harry P., 58, *59*, 60, 78,
  318
Jupiter, 97, 259

Kapteyn, Jacobus, 44, 323
Keeler, James E., *13*, 14, *22*

and *Astrophysical Journal,* 19
delivers Yerkes dedication address,
  21
at Lick Observatory, 6, 41
possibility of joining Yerkes faculty,
  33, 321
spectrograph designed by, 27
tests forty-inch lens, 19–20
Keenan, Philip C.
  *An Atlas of Stellar Spectra,*
    255–56
  education, 74
  on faculty at Yerkes, 142, 225, 234
  as graduate student at Yerkes, 73,
    140
  leaves Yerkes for Perkins Observa-
    tory, 184, 225, 267
  teaching assignment on campus,
    184
Kellman, Edith, *215, 255–56*
Kenwood Physical Observatory, 6, 8,
  *10,* 11
Keppel, Frederic, 248
Kerr, Clark, 307
Kiepenheuer, Karl-O., 291, *305*
Kimpton, Lawrence A., 277, 288,
  312
King, Arthur S., 62
Kirchoff, Gustav, 10
Kitt Peak Observatory, 248, 316
Klein, Felix, 12
Krogdahl, Margaret Kiess, 261
Krogdahl, Wasley S., 261
Kuiper, Gerard P., *171,* 177, *215,*
  *230,* 240, 244, 246, *285, 292,*
  *305*
  appointed director of Yerkes and
    McDonald Observatories, 288
  *The Atmospheres of the Earth and
    Planets,* 290, 312
  becomes director of Yerkes Obser-
    vatory, 312
  differences with Struve, 289–93,
    304
  double-star observing, 170,
    211–12
  as founding director of Lunar Plan-
    etary Laboratory, 313
  graduate students, 284
  intelligence work during World
    War II, 259
  joins Yerkes staff, 159, 169–70,
    171–72, 211

at Lick Observatory, 170
  marriage, 171, 211
  and planetary astronomy, 258–59,
    280
  as prospective director of Yerkes,
    273–74, 286–87
  recommends Chandrasekhar for
    Yerkes faculty, 179, 181–82,
    185
  resigns as director, 294
  tensions with Strömgren, 310
  as white dwarf expert, 170
  work on ε Aurigae, 216–17
  work on β Lyrae, 235–36
Kuiper, Sarah Fuller, 171, 211

Laing, Gordon J., 218
Lake Geneva, Wisconsin
  residents of, 15–16, 55, 62–63,
    233, 246–47
  as site for Yerkes Observatory,
    15–16
Langley, Samuel P., 10, 18, 41
Laplace, Pierre-Simon, 164–65
Laves, Kurt, 163–64
Ledoux, Paul, 249, 252
Lee, Florence E., *83,* 84
Lee, Oliver J., *72, 83*
  community service of, 75
  at Dearborn Observatory, 150
  as graduate student, 71
  on Yerkes staff, 50, 68, 71, 75, 82,
    150
Lee, Tsung-Dao, *305,* 311
17 Leporis, 202–3
Leuschner, Armin O., 53, 54
Levi, Edward H., 312
Lick, James, 5–6, 317
Lick Observatory, 170, 257, 282,
  322
  attempt to hire Struve, 86–87, 318
  Burnham and Barnard at, 29–31
  Hale's visit to, 5–6
  history of, *5–6*
  radial-velocity program, 34, 50
  Struve's Morrison Fellowship at,
    301, 303
Lilienthal, David E., 293–94
Lockyer, J. Norman, 10
Lowell, Percival, 162, 166
Lowell Observatory, 162, 270
Lunar and Planetary Laboratory,
  313

Lundin, C. A. Robert, 187–88, 200–202
Lundin, Carl A. R., 20, 47, 188
Lundmark, Knut, 230, *230*
Luyten, Willem J., 248
β Lyrae, 235–36

M 31 (Andromeda galaxy), 207
Mack, Julian E., 233
MacMillan, William D., 167–68
Mason, Max, 59–60, 107, 144
Massey, Harrie S. W., 229
Mayall, Nicholas U., 170, 251, 269, *305*
McCarthy, E. Lloyd, 201–2, *215*
McCuskey, Sidney D., 196
McDonald, William J., 122–23, 317
McDonald Observatory, *194,* 250.
    *See also* Eighty-two-inch Mc-
    Donald telescope; University of
    Texas
  cooperative plan for using, 247–48
  dedication of, 203–7
  end of joint operation, 314–15
  plans for building, 122–26
  site survey for, 130, 187
  staffing of, 192–93, 196–97, 206–7
McMath-Hulbert Solar Observatory, 276
McNamara, D. Harold, 303–4
Mehlin, Theodore G., 130, 187
Mellish, John, 250
Menzel, Donald H., 87, 129, 174, 218, 306–7
Merrill, Paul W., 87–88, 169, 244, *305*
Metallurgical Laboratory, 260–68
Michelson, Albert A., 19, 193
  Nobel Prize, 8, 60
  at Yerkes dedication, 23
Mikhailov, A. A., 275, 276
Milky Way, photographing of, 31, 49–53, 140
Millikan, Robert A., 60
Milne, Edward A., 206
Minkowski, Rudolph, 244
Minnaert, Marcel, 176–77, 259
  as visiting professor at Yerkes, 283
Miranda, 280
Moffitt, George W., 195–96
Molecular spectra, 271–72
  symposium on, 232–34
Moore, Carl R., 221

Moore, Joseph H., 67, 68, 90
Morgan, Helen Barrett, 73
Morgan, William W., *103,* 140, 154, 211, *215,* 218, *223,* 240, 244, 251, 273, 283, *285,* 286, 291, *305*
  as assistant director of Yerkes, 154, 172, 225, 227, 274
  *An Atlas of Stellar Spectra,* 255–56
  as director of Yerkes, 313
  collaboration with Blaauw, 282
  community service, 281
  on "council" of Yerkes, 294, 299
  death, 313
  develops UBV photometric system, 313
  as editor of *Astrophysical Journal,* 287–88
  education, 102
  as graduate student, 73–74, 102
  graduate students, 261, 281
  joins Yerkes staff, 74, 102
  marriage, 73
  observes Nova Herculis, 154
  peculiar A star work, 102
  promotions, 172, 256
  resigns directorship, 313
  retirement, 313
  spectral classification work, 73–74, 222–24, 233, 251, 255, 281, 310, 315
  thesis, 102
  at Tonantzintla dedication, 251
  during World War II, 254, 255–56
Morley, Edward W., 38
Moulton, Forest Ray, *165*
  as AAAS executive officer, 167
  and controversy with See, 166
  as graduate student, 164
  joins University of Chicago faculty, 164–65, 167
  leaves astronomy, 167
  origin of solar system theory of, 164–65, 167
  during World War II, 166
Mount Hamilton. *See* Lick Observatory
Mount Wilson, as site for Hale's observatory, 42–43
Mount Wilson Observatory
  Struve's trips to, 101, 296–97, 303
  during World War II, 250, 253
Mulliken, Robert S., 265

and hiring of Herzberg, 271–72
Nobel Prize, 238
organizes symposia on spectroscopy, 233, 251
as possible dean of Physical Sciences, 238, 260
Münch, Guido, 251, 280, 285, 286, 305

Nassau, Jason J., 207
National Academy of Sciences, 17, 140
Yerkes faculty as members of, 286
National Radio Astronomy Observatory (NRAO), 307–9
National Science Foundation, 248, 289
Naval Observatory, 162, 189–91, 313
Nebulae, photographing of, 36–37
Nebular research, 192–93, 198–200, 221
Nebular spectrograph, 197–99, 221
Negative hydrogen ion, 227, 229, 280
Nereid, 280
Ness, Lillian, 269
becomes secretary at Yerkes, 152
as confidante of Struve, 225, 287, 291, 310
leaves Yerkes, 300, 306
retirement, 309
Newcomb, Simon, 23, 38, 41, 54
Neyman, Jerzy, 296
Nicholson, Seth B., 53
Night sky research, 69, 192, 196, 269
Nobel laureates, 8, 60, 148, 238, 286, 311
North America Nebula, 198
Northwestern University, 136
bestows honorary degree on Burnham, 51
possible merger with University of Chicago, 150–51
Nova Aquilae, 92
Nova Herculis, 154

O stars, 50, 98, 221
spectral classification of, 224, 281, 282
O'Dell, C. Robert, 313
Office of Naval Research, 279, 292

Office of Scientific Research and Development, 253–54
Ohio State University, 257, 280, 306
Ohio Wesleyan University. See also Perkins Observatory
Crump at, 68–69, 141
telescope for, 68–69
Yerkes agreement with, 144–45, 148–49
O'Keefe, John A., 261
Oort, Jan H., 96, 138, 206, 259, 273, 276, 304, 306
as visiting professor at Yerkes, 283
Oppenheimer, J. Robert, 263, 306
Oppolzer, Egon von, 14
β Orionis, 146
δ Orionis, 85

Page, Leigh, 226
Page, Thornton, L., 244, 305
on Chicago campus, 225–26, 228, 274, 287
β Lyrae observations, 235–36
planetary nebulae paper, 260
during World War II, 252, 259–60
and Yerkes reorganization plan, 294, 299
Palomar Mountain, 297. See also Telescopes, Palomar two-hundred-inch
Parallaxes of stars, 35. See also Astrometric program
Parkhurst, John A., 37, 50, 72, 80, 82, 113
community service of, 75
Payne, Cecilia (later Payne-Gaposchkin), 230, 323
B star research, 92, 93
Stellar Atmospheres, 86, 99
thesis, 86
as visiting lecturer at Yerkes, 216
Payne, William W., 18
Pearce, Joseph A., 90
Pearson, Fred, 202, 254
Perkin, Richard S., 254
Perkin-Elmer Corporation, 196, 254
Perkins, Hiram M., 68
Perkins Observatory, 68–69, 306. See also Ohio Wesleyan University
operated with Ohio State, 150
spectrograph for, 145–47
Yerkes research at, 146–47
Phillips, John G., 284, 300, 306

Photoelectric cell, used in opening
    Chicago World's Fair, 137–38,
    *139*
Photoelectric photometry, 64, 281,
    310
    at McDonald Observatory, 156,
        192, 313
    Stebbins's work with, 112, 322
Pickering, Edward C., 5, 10, 19, 22,
    38, 41
Pickering, William H., 14
Pillans, Helen M., 235
Planck, Max, 259
Planetary astronomy, 57, 258–59,
    280. *See also individual planets*
Planetesimals, 164–66
Plaskett, Harry H., 90, 93, 118–19
Plaskett, John S., 94, *205*
    advice sought on McDonald tele-
        scope, 188, 191, 201–2
    interstellar calcium research, 85,
        90–91, 94
    at McDonald dedication, 205
Poincaré, Henri, 44
Polarization, interstellar, 313
Popper, Daniel M., 206–7, 254, 284
*Popular Astronomy,* 18, 66, 166, 289
Princeton University, 5, 34, 276–79
Pulkovo Observatory, 77, 96–97,
    250
Purinton, George D., 160, 161

Quadrangle Club, 220–21

Racial prejudice, 179–82, 212,
    218–21
Radial velocities of B stars, 84
Radial velocity programs
    at Lick Observatory, 50–51
    at Mount Wilson Observatory, 51
    at Yerkes Observatory, 34–35,
        50–51, 80–81, 318
Radiative transfer, 181, 280, 282
Radio astronomy, 236, 282, 307–8
Rainey, Homer P., 204, *205*
Raman, C. V., 221
Randers, Gunnar, 249, 252
Rao, K. Narahari, 284
Reber, Grote, 236
Refugee astronomers, 174, 249, 251,
    271, 276, 283. *See also* Rosen-
    berg, Hans
Ridell, Carl, 275

Ritchey, George Willis, 22, 37, 39, 47
    astrophotography, 50
    conflict with Hale, 189
    hired by Hale, 11
    moves to Mount Wilson, 44–45,
        48
    seeks to design McDonald tele-
        scope, 189–90
    at Yerkes, 20, 36–37, 40, 42
Ritchey-Chrétien telescope design,
    189–91
Roach, Franklin E.
    as graduate student, 131, 149
    at McDonald Observatory, 192,
        196, 203
    night-sky observations of, 192,
        196
    at Perkins Observatory, 145–46,
        149, 153
    during World War II, 269
Robertson, James, 189
Rockefeller, John D., 1, 247, 317
    finances University of Chicago,
        6–7
Rockefeller Foundation
    funds Perkins Observatory, 144,
        149
    funds two-hundred-inch telescope,
        57, 59, 246
    helps Rosenberg, 155–56
Rockets, World War II, 269, 280
Rojansky, Vladimir, 236
Rollefson, Ragnar, 233
Roman, Nancy G., 281
Romare, Oscar, 114
Rosenberg, Hans, 155–57, 213
Ross, Anna Olivia Lee (wife), 237
Ross, Elizabeth (wife), 211
Ross, Frank E., 56, 67, *83,* 118, 147,
    173
    appraises Yerkes situation, 70–71
    *Atlas of the Northern Milky Way,*
        57
    endorses hiring Kuiper, 169
    inventions of, 54
    joins Yerkes staff, 55–57
    Milky Way photography, 140
    planetary photography, 57
    retirement, 236–37
    search for high-proper-motion
        stars, 139
    at western observatories, 57, 140,
        169, 195, 211, 236

wide-field photographic program, 54, 56, 57
work on two-hundred-inch telescope, 57, 140
Rossby, Carl-Gustav, 290
Rosseland, Svein, 172–74, 276
Rotational velocities, 97, 143
Rowland, Henry, 5, 13, 14, 19
Royal Astronomical Society, 96, 297, 182
Rudnick, Paul, 197
Runge, Carl, 21, 22, 33, 321
Russell, Henry Norris, 205, 230, 251, 275, 276, 287
  advice on Harvard's offer to Struve, 119–22
  advice on McDonald Observatory, 127–28
  advice sought for new Yerkes staff, 168, 174, 175, 177, 178
  consulted by Frost, 53–54
  consulted by Gale, 117
  at McDonald dedication, 205
  praises Yerkes faculty, 217, 262
  urges Chandrasekhar's move to Princeton, 276, 278–79
  at Yerkes symposium, 243–44
Russia, 77–78, 249–50. See also Soviet astronomers
Ryerson, Martin, 16, 23, 49, 62
Ryerson Physical Laboratory, 23, 140, 225, 251

Sahade, Jorge, 261, 306
Sanford, Roscoe F., 100
Saturn, 258–59
Scheiner, Julius, 34
Schlesinger, Frank, 37, 155
  advice on McDonald agreement, 129
  as graduate student, 71
  joins Yerkes staff, 35, 42
  leaves Yerkes, 48
  role in choosing Frost's successor, 120, 122, 127
  stellar parallax work, 35, 41, 48
Schmidt, Bernhard, 194
Schmidt camera, 194–95
Schönberg, Mario, 261
Schwarzschild, Martin, 243, 255
Science, 289
τ Scorpii, 155
Seares, Frederick H., 53, 62, 91, 218

Secchi, Angelo, 10
Second Annual Conference of Astronomers and Astrophysicists, 38
See, Thomas Jefferson Jackson, 13
  advice to Benedict, 163
  career, 160–61, 162–63
  conflicts at Chicago, 162
  criticism of Moulton, 166
  as critic of Einstein's relativity theory, 66
  as first astronomy faculty member at Chicago, 159–60
  reputation at University of Missouri, 160–61
Segregation, housing, 180–81
Seyfert, Carl K., 196–97
Shajn, Grigory A., 97–98, 275, 276, 279
Shane, C. Donald, 295–97, 306
Shapley, Harlow, 155, 168, 205, 275, 276
  advice on new Yerkes faculty, 174, 196, 197
  advice on telescopes, 128, 188–89, 251
  appraisals of Yerkes, 70, 119–20
  and Chandrasekhar, 177–78, 182, 324
  as director of Harvard College Observatory, 92, 242, 248, 254, 279
  education, 92
  endorses Struve, 95
  offers Struve assistant directorship, 118–22, 126–29
  research, 81, 92
  retirement, 306
Shell stars, 202–3
Sherman, Frances, 232
Shortley, George H., 233
Sidereal Messenger, 18
Sixty-inch telescope
  mirror for, 36, 37
  move to Mount Wilson, 40–42, 44, 48, 317–18
  radial-velocity work with, 51
  site survey for, 42
Slettebak, Arne, 281, 285
Smith, Harlan J., 314–15
Snow, Helen, 42, 44, 49
Snow telescope, 42, 44, 48–49, 318
Solar prominences, 5

*The Solar System* (ed. Kuiper), 312
Solar system, origin of, 164–65,
    167
Soldner, Georg, 66
Soper, Edmund D., 145, 147–49
Soviet astronomers, 96–98, 275,
    279
Spectral classification, 222–25, 251,
    255–56, 315
  interference-filter system, 310
Spectrographs
  Bruce, 35, 73
  McDonald, 193, 195, 202–3,
    231–32, 275
  nebular, 197–99, 221
  Perkins, 145–47
  ultraviolet, 280
Spectroheliograph, 5
Spectroscopic binary stars, 50, 69
  Struve's investigations of, 81, 82,
    85, 97, 280, 303
Spitzer, Lyman, Jr., 279, 290, *305*
Sproul, Robert G., 296, 306
Stark, H. J. Lutcher, 190–91
Stark effect, 98–99, 101
*Stars and Stellar Systems: Com-
    pendium of Astronomy and As-
    trophysics* (ed. Kuiper), 312
Stebbins, Joel, 172, 234, 296
  at AAS meetings, 243, 290
  advice on staff for Yerkes, 53
  builds photometer, 114
  as candidate for Yerkes director-
    ship, 112–17, 321
  career, 112–13
  eclipse work, 64–65
  education, 112–13
  photoelectric work, 112–14, 206,
    224, 281
  scientific qualities and philosophy,
    64–65, 114, 116
*Stellar Atmospheres* (Payne), 86, 99
Stellar atmospheres, 86, 177, 181,
    221, 227, 275
Stellar dynamics, 217, 255
*Stellar Evolution* (Struve), 295
Stellar interiors, 175
  Chandrasekhar's research on, 177,
    212–13
Stellar rotation, 97–98, 251, 280
Stellar structure, Chandrasekhar's
    monograph on, 217–18, 221
Stetson, Harlan T., 145–49

Stillwell, Charles, 205
St. John, Charles, 62
Strand, Kaj A., 283, 284
Strömgren, Bengt, 159, *175, 215,
    230, 280, 298*
  abundance studies, 212, 216, 221,
    275, 282
  arrives in Chicago, 212, 216
  ε Aurigae work, 216–17
  becomes director of Yerkes,
    309–11
  declines directorship, 228–29
  early career, 175
  hosts Struves in Copenhagen,
    275–76
  interstellar matter research, 221,
    227, 229, 275
  moves to Yerkes, 216
  offer and acceptance of Yerkes po-
    sition, 175–76, 185, 324
  possibility as director, 293, 298,
    300–301
  resigns to go to Institute for Ad-
    vanced Study, 311
  returns to Denmark, 214, 226
  stellar atmospheres research, 221,
    227, 229, 275
Strömgren, Elis, 175–76, 185, 214,
    216, 226
Strömgren, Sigrid, 212, 214
Struve, Ludwig (father), 77–78
Struve, Mary Lanning (wife), 134,
    183, *188*
  illnesses, 93, 96, 125–26
  marriage, 84, 309–10
  move to Berkeley, 301
  travels abroad, 95, 276, 297–98
  travels to McDonald Observatory,
    130
  travels in United States, 88, 92–93,
    101, 296
Struve, Otto, 67, *83, 130, 139, 148,
    188, 205, 215, 253, 283, 285,
    292, 305, 308*
  accident while observing, 307
  army service (Russian), 77–78
  ε Aurigae observations, 216–17
  B star research, 84–86, 88, 93, 96,
    155
  Be star research, 143, 145, 150
  becomes "astrophysicist," 86, 95
  becomes U.S. citizen, 84
  becomes director of Yerkes, 131

and beginnings of McDonald Observatory, 125–28
books written by, 308–9
builds UC Berkeley astronomy department, 306–7
as chairman of observatory "council," 294
and Chandrasekhar, 181–83, 218–21, 276–78, 291, 303–4
collaboration with Gerasimovich, 96
collaboration with Swings, 104–5, 208, 209, 256–57
conditions for accepting directorship, 188
confides in Swings, 227, 229, 257, 295, 299
conflict with Crump, 134, 140–42
considered as director of Yerkes, 116–17
cooperative observatory plan, 247–48
differences with Kuiper, 289–91, 199
directorial style, 131–32, 134, 142, 153–54, 237–38, 286, 319
as director of NRAO, 307–8
disagreement with Plaskett, 90–91
κ Draconis research, 85–86
economies during Great Depression, 135–36, 184
as editor of *Astrophysical Journal*, 236, 256, 273
education, 77–78, 82
escape to America, 78–79
and fiftieth anniversary of Yerkes, 288–90
fiftieth birthday recognition, 289–92
graduate students, 101–2, 149, 152–53, 232, 261, 284–86
graduate work and thesis, 79–82
Guggenheim Fellowship, 94–95
and hiring astronomers on campus, 172–74, 176
historical articles by, 289
honors bestowed on, 207, 276, 279, 286, 295, 297, 299
hopes for continued ties to Yerkes and McDonald, 303–4
hopes for Mount Wilson appointment, 99–100
illness and death, 308, 321

interstellar calcium research, 85–86, 88–91, 93
joins Berkeley astronomy department, 295–97, 300
joins Yerkes staff, 79, 82
β Lyrae observations of, 235–36
marriage, 82, 84, 309–10
as Morrison Research Associate, 296–97, 301
Mount Wilson visits, 87–89, 101
nebular research, 199, 221
negotiates agreement with Ohio Wesleyan, 144–45
and new deanship appointment, 238–40
observes Nova Herculis, 154
offer from Harvard, 118–19, 122
offers from other observatories, 306–7
opposition to Bartky as dean, 263–64, 273
Perkins Observatory research, 145–49
plans for moving department, 246
plans for moving to campus, 228–29
as president of AAS, 290
promotions of, 87
radial-velocity work of, 67–68, 80–81, 84, 99, 143
relationship with Bok, 240–41
relations with Hutchins, 131–32, 134–35, 237–38, 319
reorganization plans for Yerkes, 273–74, 286–88
research method, 89–90, 193, 209
resigns as director of Yerkes, 292–93
resigns research associateship, 304
resigns from Yerkes, 299–301, 320
and Rosenberg, 155–57
scientific philosophy of, 191, 207, 213
shell star observations, 202–3
Stark effect research, 98–99
starts astrophysical monograph series, 217
*Stellar Evolution*, 295
stellar spectroscopy of, 193, 202–3, 280–97
and symposium on spectra, 251–52
and symposium at Yerkes, 243–44

Struve, Otto (*continued*)
  at Tonantzintla dedication, 251
  trip to IAU planning meeting,
    275–76
  visits Harvard, 92–93, 168
  World War II, 245, 249–50, 251,
    252, 265, 272
  work on rotational velocities of
    stars, 251, 280
Struve family history, 77
Sullivan, Frank R., 80, 136, *215*
Sun
  chromosphere, 26
  corona, 25–26, 39, 282
  ultraviolet spectrum, 280
Supergiant stars, 150, 202, 216–17,
    235
Swift, Harold H., 109–10, 112, 122,
    178
Swings, Pol, 230, *230*, 251, *285*, 286,
    290
  career in Europe, 102–3, 208, 320
  collaboration with Struve, 104–5,
    208–9, 256–57
  as confidant of Struve, 227, 229,
    257, 295, 299
  observing at McDonald, 257
  return to Belgium, 282
  stranded by World War II, 209,
    252, 257
  travels, 257
  as visiting professor at Yerkes,
    256–57
  visits to U.S., 104–5, 208–9
Symposium: The Atmospheres of the
    Stars and the Interpretation of
    Stellar Spectra, 242–44
Symposium on molecular spectra,
    232–34
Symposium on planetary atmo-
    spheres, 290
Symposium on spectroscopy,
    251–52, 268
Symposium on stellar atmospheres,
    290

Tacchini, Pietro, 14
Tate, John T., 233
Telescopes. *See also* Ritchey-Chrétien
    telescope design; Snow telescope
  advantages of reflectors over re-
    fractors, 14, 35–36, 40–41
  Lick one-hundred-twenty-inch,
    291; thirty-six inch, 322
  McDonald (*see* Eighty-two-inch
    telescope)
  Mount Wilson one-hundred-inch,
    101, 189, 201–2, 246, 248, 284
  Palomar two-hundred-inch (Hale),
    57, 99–100, 116, 246, 248, 284,
    291
  Sixty-inch (*see* Sixty-inch telescope)
  Yerkes Observatory (*see also* Forty-
    inch telescope; Twelve-inch tele-
    scope)
  Bruce, 35, 49
  horizontal, 25, 40
  twenty-four-inch reflector, 36, 267
Teller, Edward, 233, 235, 294
Terman, Frederick, 258
Thome, John, 14
*The Titan* (Dreiser), 1
Titan, 258
Titus, John, 252
Tonantzintla Observatory dedication,
    250–51
Truman, Harry S., 289
Trumpler, Robert J., 65, 206
Tuberg, Merle E., 280
Tave, Merle, 227
Twelve-inch telescope (Kenwood), 6,
    21, 318
  converted to twin instrument,
    66–67
  moved to Yerkes Observatory, 26
  shipped to McDonald Observatory,
    192, 193, 212–13

Ultraviolet observations, 280
Underhill, Anne B., 284, *285*
University of Arizona, 74, 313
University of California, Berkeley,
    282
  astrophysics department after
    Struve, 306, 320
  astrophysics department under
    Struve, 306, 320
  offers position to Struve, 295–96,
    300
University of Chicago, 306
  astronomy faculty on campus,
    159–62, 163, 167, 172–74,
    176, 184, 225–26, 238–39
  beginnings, 6–7

fiftieth anniversary, 242
financial difficulties, 110–11
graduate astronomy students on
    campus, 164, 167–68
reorganization into division, 111
during World War II, 251–53
University of Chicago Press, 313
    as publisher of *Astrophysical Jour-
    nal,* 19, 62, 311
    as publisher of astrophysical
    monographs, 217–18, 255
University of Liège, 103, 208
University of Michigan, 68–69, 248,
    270–71, 313–14
University of Minnesota, 69, 248,
    316
University of Missouri, 160–61
University of Southern California, 9,
    42
University of Texas, 248, 275, 288
    and building of McDonald Obser-
    vatory, 122–25
    negotiations with University of
    Chicago, 125–28
    termination of agreement with
    Chicago, 314–15
University of Virginia, 23, 248
University of Wisconsin, 15, 233. *See
    also* Washburn Observatory
Unsöld, Albrecht, 231–32, 307
Upper atmosphere research, 269
Uranus, 259
Urey, Harold C., 265
γ Ursae Minoris, 81

Van Arnam, Ralph, 136
Van Biesbroeck, George, 67, *80, 83,*
    117–18, 173, 177, *215,* 244,
    247, *285*
    assists with eighty-two-inch tele-
    scope, 201–2
    birthday telescope observing times,
    314
    boardinghouse, 267–68
    comet and asteroid observations,
    51, 140, 267, 314
    community service, 75
    death, 314
    double-star observing, 51, 140,
    267, 314
    eclipse travel, 136
    instrument designs, 199, 203

joins Yerkes faculty, 51
postretirement years, 267–68, 314
retirement, 267
    as teacher, 80
during World War II, 252, 254
Van Biesbroeck, Julia (wife), 172
    embarrassing letter, 154
    boardinghouse, 267–68
Van Biesbroeck, Marguerite (sister),
    215, 268
van de Hulst, Hendrik C., 282, 290
van de Kamp, Peter, 283
Visiting astronomers, postwar,
    282–83
Vogel, Hermann, 10, 34
von Neumann, John, 254
Vosatka, John, 275

Wadsworth, Frank L. O., 28, 36
Walker, George C., 15
Wares, Gordon W., 212, 261
Warner, Worcester R., 14
Warner and Swasey, 6
    as builder of eighty-two-inch tele-
    scope, 187–88, 191, 200–202,
    205
    as builder of forty-inch telescope,
    9, 20
Washburn Observatory, 29
    as site of AAS meeting, 92
    Stebbins's work at, 112, 321–22
Watson, Ernest C., 269
Weaver, Harold F., 307, 309
Weaver, Warren, 248
Weizsäcker, Carl von, 216
Whipple, Fred L., 290
White dwarfs, 170, 311
    Chandrasekhar-Eddington conflict
    about, 177, 230–31
Whitford, Albert E., 281
Wildt, Rupert, 243
    as possible Yerkes faculty member,
    174
    research of, 227, 233, 259, 280
Williams, Robley C., 243
Williams Bay, Wisconsin, 133, 214
    chosen as site for Yerkes Observa-
    tory, 15–16
    involvement of Yerkes staff in life
    of, 28, 74–75, 281
Williamson, Ralph S., 261
Wilson, Olin C., 169, 256

Wilson, Robert R., 268
Winans, John G., 233
Wolf, Max, 14
Women in astronomy, 323
Woodward, Frederic, *108*
   as acting president of University of
      Chicago, 107
   becomes vice president of Univer-
      sity of Chicago, 60
   relations with Yerkes, 67, 69–70,
      111–12, 184, 246
Wooten, Benjamin A., 73
World Congress of Astronomy and
      Astrophysics, 11–14, *13,* 14, 30,
      163
World's Columbian Exposition,
      11–12, 14
World War II
   America's entry into, 244, 249
   astronomers stranded by, 252
   beginnings of, 209, 227, 231–32
   end of, 262, 267
   Yerkes faculty during, 252–60,
      268–69
Wright, William H., 57, 199–200
   as director of Lick Observatory,
      170
   endorses Struve, 199–200
   as graduate student, 71
Wrigley, William, Jr., 63–64
Wrubel, Marshall H., 280
Wurm, Karl, 231–33
Wyse, Arthur B., 170, 244

Yale University, 6, 107–9, 248
Yerkes, Charles T., 2, 38
   asked to finance moving sixty-inch
      telescope, 41
   at dedication of Yerkes, 21, 23

and financing of Yerkes, 1–2, 9,
   44, 317
Yerkes Observatory, *37, 65, 72, 83,
   90, 215, 247, 285, 305*
   agreement to jointly operate Mc-
      Donald Observatory, 126–28
   astrometric program of, 283, 284
   astrophysical legacy, 10, 316, 324
   cooperative observatory idea, 316
   "council" formed to administer,
      294
   dedication, 21–23, *22*
   directors, 320. *See also* Frost;
      Hale; Hiltner; Kuiper; Morgan;
      Strömgren; Struve
   excellence of faculty, 217, 262,
      319
   exodus of Hale and others, 45, 47
   fiftieth anniversary, 288–89
   first staff, 37
   floor collapse, 20
   graduate program (*see* Graduate
      program, Yerkes Observatory)
   inception, 1, 9
   instrument shops, 17, 316
   optical bureau, 253–54
   photometric work at, 114
   present uses of, 315–16
   reorganization plan for administer-
      ing, 273–74
   selection of Frost's successor as di-
      rector, 112, 115–16
   shift of teaching to Chicago cam-
      pus, 315
   site selection for, 15
   work ethic of staff, 316, 324
   during World War II, 252–54
Young, Charles A., 5, 10, 18, 19, 34